PHYSICS, PHILOSOPHY, AND THE SCIENTIFIC COMMUNITY

BOSTON STUDIES IN THE PHILOSOPHY OF SCIENCE

VOLUME 163

BOSTON STUDIES IN THE PHILOSOPHY OF SCIENCE

Volume I *Physics, Philosophy, and the Scientific Community*
Essays in the philosophy and history of the natural sciences and mathematics

Volume II *Science, Politics and Social Practice*
Essays on Marxism and science, philosophy of culture and the social sciences

Volume III *Science, Mind and Art*
Essays on science and the humanistic understanding in art, epistemology, religion and ethics

ROBERT S. COHEN

PHYSICS, PHILOSOPHY, AND THE SCIENTIFIC COMMUNITY

Essays in the philosophy and history of the natural sciences and mathematics
In honor of Robert S. Cohen

Edited by

KOSTAS GAVROGLU
National Technical University, Athens

JOHN STACHEL
Boston University

MARX W. WARTOFSKY
Baruch College, The City University of New York

KLUWER ACADEMIC PUBLISHERS
DORDRECHT / BOSTON / LONDON

Library of Congress Cataloging-in-Publication Data

Physics, philosophy, and the scientific community : essays in the philosophy and history of the natural sciences and mathematics in honor of Robert S. Cohen / edited by Kōstas Gavroglu, John Stachel, Marx W. Wartofsky.
p. cm. -- (Boston studies in the philosophy of science ; v. 163)
Includes index.
ISBN 0-7923-2991-0 (set : alk. paper). -- ISBN 0-7923-2988-0 (alk. paper)
1. Science--History. 2. Physics--History. 3. Mathematics--History. 4. Natural history. I. Cohen, R. S. (Robert Sonné) II. Gavroglu, Kōstas. III. Stachel, John J., 1928- . IV. Wartofsky, Marx W. V. Series.
Q126.8.P48 1994
509--dc20 94-22250

ISBN 0-7923-2988-0
ISBN Set: 0-7923-2991-0

Published by Kluwer Academic Publishers,
P.O. Box 17, 3300 AA Dordrecht, The Netherlands.

Kluwer Academic Publishers incorporates
the publishing programmes of
D. Reidel, Martinus Nijhoff, Dr W. Junk and MTP Press.

Sold and distributed in the U.S.A. and Canada
by Kluwer Academic Publishers,
101 Philip Drive, Norwell, MA 02061, U.S.A.

In all other countries, sold and distributed
by Kluwer Academic Publishers Group,
P.O. Box 322, 3300 AH Dordrecht, The Netherlands.

Printed on acid-free paper

Printed in the Netherlands.

TABLE OF CONTENTS

C. LOGIC, LANGUAGE AND SCIENTIFIC RATIONALITY

D. HISTORY OF MATHEMATICS

EDITORIAL PREFACE

The essays in this *Festschrift* are celebrations of the human mind in its manifold expressions – philosophical, scientific, historical, aesthetic, political – and in its various modes – analytical, systematic, critical, imaginative, constructive. They are offered to Robert S. Cohen on the occasion of his 70th birthday, in acknowledgment of his own extra-ordinary participation in the life of the mind, and of his unfailing encouragement and facilitation of the participation of others. It is fitting that these volumes should appear in the *Boston Studies in the Philosophy of Science*, the series which he co-founded so many years ago, and of which he has been the principal editor for more than three decades. (These are perhaps the only volumes of that series which he has not edited or co-edited!)

The three volumes that constitute this *Festschrift* cover the range of Cohen's interests as a philosopher/scientist/humanist, as they also represent the spectrum of his professional and personal friendships. (Regretfully, the editors could not include contributions from more of them here.) The first volume centers around the philosophy and history of the natural sciences and mathematics; Volume Two collects essays related to Marxism and science, philosophy of culture and the social sciences; and the third volume focuses on science and the humanistic understanding in art, epistemology, religion and ethics.

The editors and the editorial committee express their thanks to Annie Kuipers, our editor, conscience and guide at Kluwer Academic Publishers, who has been guardian angel of the *Boston Studies* these many years, and a good friend of Robert's; to her able assistant, Evelien Bakker; to Carolyn Fawcett for apt translation and editorial assistance; and of course, to Robin Cohen for all around enthusiasm, timely revelations and steady support.

KOSTAS GAVROGLU *National Technical University, Athens*
JOHN STACHEL *Boston University*
MARX WARTOFSKY *Baruch College and the Graduate Center of The City University of New York*

KOSTAS GAVROGLU AND MARX W. WARTOFSKY

INTRODUCTION

Philosopher, physicist, historical sociologist of science, critical social thinker, master teacher, writer of clarity and wit, genial critic, prolific editor, organizer extraordinary, effective administrator, serious student of religion, of art, of history, vivid speaker, great listener, ubiquitous world-traveler and conferencier ("they seek him here, they seek him there, they seek him everywhere . . ."), inexhaustible conversationalist, teller of jokes, saxophonist and clarinetist, indefatigable reader – all of these, and more, and at once, is Robert S. Cohen, the butt of the three dedicatory volumes of which this is the first.

A *Festschrift* is an occasion of celebration, and here, many of Professor Cohen's friends, colleagues, admirers join to celebrate his 70th birthday with that form of homage most fitting for a scholar: their own scholarly contributions. These span the spectrum of Bob's intellectual interests, and mirror the richness and variety of his thinking over the years. This first volume collects essays in the philosophy and history of the natural sciences and mathematics; the second, essays on Marxism and science, the philosophy of culture and the social sciences; and the third, on epistemology, religion, ethics and art. That Professor Cohen is no simple dilettante in any of these fields, that his interests are persistent and his knowledge deep, that he has been a serious student, a critical commentator, and a contributor to thinking in these many areas is the mark of his broadly humanistic scholarship. Cohen is a scientist/humanist – that rarest of breeds. His humanism, like that of the Renaissance, is an effort at integration: of a scientific, analytic-empirical rationality with a normative-critical understanding of society and history, and with the work of the imagination in science and art.

By way of introduction, at this midpoint in his career, we would like to sketch, in broad strokes and with the cross-hatching of anecdote, something like an intellectual biography of Robert Cohen. To this end, one of us (Kostas Gavroglu) initially thought to do a taped interview with Bob, but realized quickly how naive it was to think that Bob could be constrained within the bounds of such a rational enterprise. The second move was to read much of Professor Cohen's published work. Despite

K. Gavroglu et al. (eds.), Physics, Philosophy and the Scientific Community, xi–xxvii.

the range of his themes, it was quickly obvious that Bob's work was not a collection of separate agendas. Rather, what Cohen has pursued so singlemindedly, and which in one way or another has touched so many of us in the community, (academic and beyond) is his unique and unified theoretical scientific and philosophical and educational and social and political agenda. What's more, it is obvious that Professor Cohen has had a great time pursuing it, enjoying every minute of it. He has been a happy man, moreover, a happy academic – a rather bizarre occurrence in this world of ours and a rather intriguing challenge to many of us.

EARLY YEARS

An *echt* New Yorker (a Manhattanite from Washington Heights yet!) Bob had an excellent high school education at DeWitt Clinton High School, where a not very insightful course in physics nevertheless inspired him to pursue the study of science. As he recounts it,

> Due to the instinctive genius of the New York City school system, I (from Manhattan) met Adolf Grünbaum (from the farthest reaches of Brooklyn) at the DeWitt Clinton High School (in the northernmost Bronx). . . . We had two-hour subway rides to school, more than ten-thousand fellow students (all male), a splendid four years of Latin, and world politics going down toward disaster.[1]

When he was sixteen, his teacher, who was a Wesleyan graduate, recommended Bob to Wesleyan University. He got a scholarship, and it was the start of a rewarding and tempestuous relationship with Wesleyan. After his first year, Bob was asked by the Dean of Admissions whether he knew any other students like himself to recommend for admission. He suggested Adolf Grünbaum, a German refugee who had been in the Arista (Honor) Society with him in high school, and Grünbaum was thereafter admitted. In 1939, he had been joined by another young physics student, Gerald Holton, whose family had just fled Austria also to escape the Nazis.

It is interesting to note that Bob's teaching of science began in 1939 during the Depression when he volunteered to teach high school physics and algebra in a Wesleyan project for a New Deal agency, the C.C.C. (Civilian Conservation Corps) in rural Connecticut. The students, unemployed and mostly school dropouts, were from very poor families, and Bob got first-hand knowledge of the barriers erected by poverty, both urban and rural, and by lack of education. It was Bob's first experience in teaching, with lectures, homework, grading, tutoring, blackboards and all the rest. He was encouraged by his experience; and indeed, this was a presentiment of his great success later on as a teacher of elementary physics to non-science students, and as the author of one of the great Introductory Physical Sciences texts for liberal arts students (about which more later).

The physicist V. E. Eaton and the philosopher Cornelius Krusé were very influential for Bob, during his undergraduate years at Wesleyan. He was a physics lab assistant for two years, and during his last two years he was a teaching assistant. During the summer of 1942 – his last at Wesleyan – Bob spent two months at Brown University's wartime Graduate School of Applied Mathematics and took courses in mathematics. Among his teachers there was Richard von Mises, and among his fellow students were Herbert Callen and J. M. Luttinger. He graduated in January of 1943 and went at once to Yale in an eight-month intensive Master of Science program in physics for war research, and at the same time was teaching physics full-time to students entering training in the Army and Navy air forces. But throughout this hectic period, philosophy continued to be a major interest. He studied Whitehead in a course at Yale Divinity School, and metaphysics as well as Kantian philosophy with Professor Ernst Cassirer, who had just arrived at Yale from his earlier refuge in Sweden. And then, he did some years of war work, mainly on magnetrons in radar countermeasures, as a sort of scientific secretary for an electronics and communications committee of the U.S. Joint Chiefs of Staff, headed by a physicist he admired, I. I. Rabi.

In 1946 Bob returned to Yale on a National Research Council fellowship and soon started working on his doctorate with Lyman Spitzer as his supervisor. It was in plasma physics, though the term had not been coined yet. Spitzer was an astrophysicist. One of his research interests was in interstellar clouds and the forces responsible for star-formation. Bob's thesis, "Electrical conductivity of completely ionized gases", involved the application of the "fascinating" Boltzmann equation. Though his work in plasma physics interested him, it did not intrigue him. In 1947, he received a post- doctoral fellowship from the American Council of Learned Societies to study philosophy for one year.

During that year he became friends with Paul Weiss (who had been a student of Whitehead, and whose course on Hegel was particularly influential on Bob), and also came to know Hempel, Blanshard, Margenau, Northrop, Breit. He took Tillich's course on existentialism, and one on Soviet legal theory with John Hazard at the Yale Law School. At the time, Yale had instituted a Program of Directed Studies for selected freshmen and sophomore humanities and social science majors who took a set of courses in literature, social science, history, physical and biological sciences, philosophy and art history. The philosopher's role was to integrate the various studies and guide the students in critical and comparative understanding of their courses. Bob was the philosopher-scientist whose responsibility was partly tutorial, partly seminar and lecture sessions. A memoir of Yale in those years by James C. Thomson (former Curator of the Nieman Foundation for Journalism at Harvard University) gives vivid testimony to the educational quality of this program:

> . . . despite frivolous rioting, noisiness about silence, and an aura of excessive political caution, Yale College continued to perform its magic. It was a marvelous teaching and caring institution in my day. We sixty-five Directed Studies freshmen were divided into two larger groups and then several smaller clusters. It is a sad commentary, not on Yale but on most education, that the best teaching I have ever experienced was freshman year at Yale from extraordinary men like the classicist Bernard Knox, the critic Maynard Mack, the philosopher and physicist, Robert Cohen.[2]

Bob stayed on for two years as an assistant professor in the philosophy department. But in 1949, the Yale Corporation refused to renew his appointment, and he was informed orally that the University could not continue to have a faculty member who was a communist. Bob confronted the Provost about this allegation, and the Provost told him that he was accused, among other things, of associating with Professor Haber of the Yale Law School (who later played a major role in the development of American civil liberties law) who was reported to be a communist. Bob calmly told him that he did not know the law professor, and that he himself was not a communist, that the information was false, but that whatever the outcome of his appeal against non-renewal of his appointment on these grounds, he would continue to defend in public the right of anyone to be one. The Philosophy Department and the Dean of the College were strongly supportive of Bob in his appeal, and eventually the University, despite its tradition of standing by personnel decisions once made, reversed its position.

A year after this crisis, another one erupted. During the Philosophy Department's discussion of new courses to be taught, Bob was invited to propose several of his own. Of these, his senior colleagues selected two: an advanced course in the philosophy of science (based on a study of the work of Émile Meyerson), and one on philosophy of Marxism. This was, to put it mildly, rather provocative in the political climate of that time. At the general faculty meeting which was to approve the new course proposals of the various departments, all hell broke loose. Many faculty thought this was a scandalous proposal. Others pointed out that similar material was already taught in the History and Economics departments, that Marxism was not a philosophy, but at best an economic theory. It was one thing to defend the decent young scholar and to resist the anti- intellectual climate that the Cold War had fostered among some University administrations; but it was quite another to give the impression of collective agreement by instituting such a course. The senior faculty of the Philosophy Department stood by Professor Cohen's proposal, and after repeated discussions, the course was approved by a slim margin. What made a difference at the end was a compromise by Cohen and his chairman, Brand Blanshard, to rename the course 'A Critical Study of Marxism'.[3] The course became a huge success, so much so that for some years, after Professor Cohen joined the faculty at Wesleyan University, he was called back every fall to Yale to initiate it.

Any intellectual biography which covers the 1950's in the United States has to cope with the traumatic effect of McCarthyism, and with the resistance to it which finally broke its back. With few exceptions, University administrations colluded in the witch hunt which had such a disastrous effect on American culture generally, and on education in particular. In Professor Cohen's case, a brilliant career as a young professor at Wesleyan, respected by his colleagues, loved by his students, was disrupted once again by the storm-winds of political repression that swept across the academic landscape. Mendacious charges, once again, of "communism", denial of tenure on blatantly political red-baiting grounds, weakness of will on the part of the University President (who years later apologized for his action) – all this added up to another instance of the destructiveness and academic immorality of McCarthyism on American campuses. Despite a principled fight by Bob and a number of faculty, the denial of tenure meant termination at Wesleyan.

In 1955, Bob left Wesleyan for a year on a research fellowship first in London and then at Cambridge University. In the event, there was a double irony: The time at Cambridge was an extraordinarily productive one, resulting in a range of important intellectual and personal friendships, and in a major essay on Carnap's philosophy of science. It was followed by Cohen's appointment to the Physics Department at Boston University, and to all that eventuated there. But there was a second irony to ensue. Years later, Cohen returned to Wesleyan, first to receive a citation as one of its most distinguished alumni, then to become a Trustee of the University for 15 years, and in 1986 to be awarded an honorary degree. In vulgar dialectical terms, one could construe this as a negation of the negation. But after all, ironies involve such reflexive doubling.

BOSTON

Bob came to Boston University in the fall of 1957, appointed by then President Harold Case, himself a civil libertarian with his roots in the Methodist social action tradition. The *curriculum vitae*-type information about Bob past this point can be summed up easily, even though it is extensive. It's what goes on between the lines that is interesting. At 35, Bob became chair of a physics department in flux, and built it into a very good department for a minimal budget, with imaginative curricula based on the classical hard-core physics courses, but going beyond, to draw in liberal arts students in a range of imaginative introductory physics courses, and to develop graduate study in the philosophical foundations of quantum mechanics and of relativity physics. Bob also did some excellent recruiting of top-notch research and teaching faculty, and pioneered in introducing (and teaching) a number of interdisciplinary courses and seminars on science, technology and society, science and religion, science and mysticism, among others. At the same time, Bob soon participated in the teaching of the Philosophy Department at Boston University. Thus, in addition to his 'routine' work – scare quotes because 'routine' doesn't seem to fit anything Bob does – teaching electrodynamics, thermodynamics, classical mechanics, etc., Bob also taught regularly in the Philosophy curriculum, offering courses and seminars on, e.g. Hume, Spinoza, Leibniz, Hegel, Mach, Russell, the development of the Vienna Circle, philosophy of Marxism, and on various themes in the philosophy and history of science, and on the social roots of science.

THE BOSTON COLLOQUIUM AND THE BOSTON STUDIES

In 1959–60, the Philosophy Department undertook to develop a modest M.A. program in the philosophy of science, to add a new dimension to the department's traditional focus on history of philosophy, ethics and philosophy of religion. Courses in the philosophy of science and in epistemology were being taught by Professors Marx Wartofsky and George Berry; work in symbolic logic, foundations of mathematics and philosophy of language, by George Berry; and cognate courses in foundations of physics by Professor Cohen. NDEA (National Defense Education Act) graduate fellowships in the philosophy of science had become available, in the aftermath of Sputnik. And the graduate program at Boston University, first at the Master's level, and then at the Ph.D. level, began to earn these fellowships in significant numbers. In order to familiarize the graduate students in the new program with current work in philosophy of science, and at Bob Cohen's suggestion, he and Marx Wartofsky submitted a proposal to the NSF to support a series of invited lectures by well-known philosophers of science, or scientists whose work was at the cutting edge of new conceptual developments; and also to support publication of the proceedings of such meetings in books. The idea was to form an inter-University group in the Boston area, which would, in a way, continue the activity represented earlier by the Unity of Science movement, itself derived from the Vienna Circle and inspired by Philipp Frank. After some discussions and organizing meetings with colleagues – e.g. Philipp Frank, Gerald Holton, Israel Scheffler, W. V. O. Quine, Hilary Putnam, Laszlo Tisza – and with the blessings of Charles Morris, the last president of the Institute for Unity of Science, it was decided to organize several initial meetings with invited speakers for the Spring Semester of 1960–61. Wartofsky suggested it be called a 'Colloquium' (a term not then in wide use in English, but which he had learned as a student at Columbia College, where upper level seminars were called Colloquia). Cohen and Wartofsky then proposed the title 'Boston Colloquium for the Philosophy of Science'; after initially contributing to the journal *Synthese*, and plagiarizing shamelessly from the title of the outstanding and pioneering series of volumes then coming from Herbert Feigl's Center for the Philosophy of Science at the University of Minnesota, they adopted the name, *Boston Studies in the Philosophy of Science* for the Proceedings volumes which were to eventuate from the Colloquium meetings. A venturesome and hearty independent Dutch publisher who focused on scientific books and journals, Anton Reidel, undertook to put out the first volume of the Proceedings of 1961–62.

This small snowball began to roll downhill, getting larger as it rolled. (One might say it's been downhill all the way since then, but that would give the wrong message because of the confusion of metaphors.) The Colloquium programs have grown and grown, and are now in their 35th year; and so too have the *Boston Studies in the Philosophy of Science*, now in their 169th volume. The organizing spirit behind all this activity was Cohen. And from the start, he and Wartofsky worked together by a kind of intuitive intellectual symbiosis, the kind that, in music, marks fine duos or quartets (a legitimate comparison, because it seemed, at times, that there were four of them!). The Colloquium has become a major feature of the academic scene, both locally, in Boston, and internationally. Some 600 Colloquia on a broad spectrum of issues in the philosophy and history of science, and on related interdisciplinary and transdisciplinary topics have been held since 1960. A wide range of scholars from the U.S. and abroad have found their lively, sophisticated and critical audiences in the Boston-Cambridge-Route 128 area, with its concentration of Universities, Colleges and scientific research institutions. Several generations of graduate students, and undergraduates as well, have been introduced not only to the substance of the philosophy of science, and to personal exposure to its leading practitioners, but also to the experience of critical philosophical discourse.

Inevitably, over the years, the fertile Boston Colloquium spawned its progeny. Locally, on the model of the philosophy of science colloquium, there were established the Boston Institute for the Philosophy of Religion, and the Boston Colloquium for Classical Philosophy. In the Philadelphia area, an inter-university philosophy colloquium developed on the inspiration of the Boston Colloquium; and in New York City, a small version – the Baruch Colloquium for Philosophy, Politics and the Social Sciences – was established when Prof. Wartofsky left Boston University to join the faculty at the City University of New York in 1983. In 1981–82 there was established in Israel the first 'overseas branch' of the BCPS, the Israel Colloquium for the History, Philosophy and Sociology of Science, under the leadership of Professors Yehuda Elkana, Michael Heyd, Asa Kasher and Edna Ullmann-Margalit.[4]

The volumes of the *Boston Studies in the Philosophy of Science* which began as proceedings of the Colloquium, later expanded, largely due to Cohen's wide acquaintance with outstanding scholars and with movements of scientific and philosophical thought in the U.S. and around the world, to include monographs by leading philosophers and historians of science, proceedings of significant conferences held elsewhere, and pioneering collections of essays in the philosophy of science from a variety of nations – e.g. Italy, Poland, Japan, Greece, China, the former Yugoslavia, Israel, as well as perhaps the widest publication of current research in these fields by scholars from the nations of the former Soviet Union and Eastern Europe. Also included were reprints or first translations of classical contributions to the philosophy and history of the sciences, by, e.g., Mach, Helmholtz, Meyerson, Leon Rosenfeld, Ludwik Fleck; Henry Mehlberg; *Festschriften* for Philipp Frank, Dirk Struik, Adolf Grünbaum, Lewis Feuer, John Watkins, Stillman Drake, Erwin Hiebert, Mario Bunge, Paul Feyerabend, Robert Butts, Marjorie Grene and Marx Wartofsky; and memorial volumes dedicated to Rudolf Carnap, Norwood Russell Hanson, Imre Lakatos, Benjamin Nelson, Yehoshua Bar-Hillel. Add to this the volumes inspired and edited by Cohen in the *Vienna Circle Collection* (also published by Reidel) – notably, the two volumes of Otto Neurath's essays (co-edited with Marie Neurath), those of Feigl, and of Hans Reichenbach (co-edited with Maria Reichenbach), and Eino Kaila, and more and more . . . and you have the most prodigious work of editorship, of intellectual inspiration and the dogged labor of publication, perhaps in the whole of our profession. Moreover, the philosophical prospect, as viewed from this editorial perspective, is generous, critical, original, sophisticated, undogmatic – perhaps in itself a paradigm of sweet scientific reason.

At the heart of the matter was a vision of the philosophy of science as a humanistic discipline. It was Cohen's ability, as a scientist-philosopher, to grasp this notion not as a rhetorical platitude nor as a curricular slogan, but in its hard, contentful sense, in the spirit in which Philipp Frank had conceived it a generation earlier. Bob's appreciation of Frank's views reveal not only his sympathies, but much of his own conception. At the memorial meeting held for Frank at Harvard (on October 25, 1966), Cohen expressed this vision eloquently:

> [Philipp Frank] fused sympathy, rigor, criticism and imagination throughout his career. . . . They show in his humanistic courses in physics for science students and in philosophy of science for students of humanities and the social sciences. They show in his

many discourses to theologians and metaphysicians. They show in his profound hope that a historical sociology of science might be cultivated in order to supplement logical analysis and thus to complete the philosophy of science.

He was of course devoted to the unity of the sciences, and to their positive part in private as well as public life. To Frank . . . unification properly suggested more than integration of separate specialties. . . . It meant also a synthesis of specialized learning with practical affairs and it meant also that joining of moral and political conceptions with those of the natural sciences and mathematics. To Frank, the scientific conception of the world was to be a guide to life, as pervasive and as humane as the classical philosophical tradition had ever sought. But it was a radical guide.[5]

With almost no editing of the rest of the account, this could very well have been written about Bob, and it would have been quite accurate. In a similar way, Cohen's editorial 'blurb' for the *Boston Studies in the Philosophy of Science* expresses this integrative intellectual vision vividly. It may be read on the back of the book jackets of many recent volumes of the Studies, but it is worth citing here. (Who reads back covers anyway?)

The series *Boston Studies in the Philosophy of Science* was conceived in the broadest framework of interdisciplinary and international concerns. Natural scientists, mathematicians, social scientists and philosophers have contributed to the series, as have historians and sociologists of science, linguists, psychologists physicians, and literary critics. . . . The editors believe that philosophy of science should itself be scientific, hypothetical as well as self-consciously critical, humane as well as rational, skeptic and undogmatic while also receptive to discussion of first principles.

In both of these cases, a rather typical displacement or objectification occurs: Cohen articulates his own thought in his account of the thought of others whom he takes as his intellectual, philosophical scientific heroes; and he gives external expression to his vision in the grand editorial project of the *Boston Studies* (as also in his contribution to the publication of the Vienna Circle series, and other publishing projects.) But apart from the vast editorial and organizational activity through which this broad conception of philosophy of science has been given life, and apart from the almost two hundred prefaces, introductions, commentaries in which Cohen has given us the critical reflection of his own thought in the mirror of other minds, how has Cohen's own scholarship – his research, lecturing, publication – articulated these ideas? What about what Cohen has to say for himself? Or (as our mutual friend and erstwhile colleague Agassi would say) what has Bob done for us lately?

PUBLICATION

'Publication', as some of us never tire of pointing out, means making one's ideas public. In this sense, Professor Cohen is one of the most widely 'published' minds in the philosophy of science, since he is intellectually ubiquitous. One of the most sought-after lecturers, one of the most lucid critics and commentators, one of the most widely-traveled and internationally known scholars in the field, and one of the most widely read and deeply cultured persons in the world, Bob is also impatient with the transformation of his own discourse into text. He'd rather talk, read, argue, criticize, question than write. This is strange, because in his many published writings, as well as in his editing, and in editorial discussion of ideas (e.g. in the more than one hundred prefaces to *Boston Studies* volumes), Bob's fluency as a writer is obvious. His prose is clear, his line is lean, his ability to illuminate difficult concepts, whether in physics or in philosophy, is extraordinary. Jeremy Bernstein, in his memorial tribute to his former teacher Philipp Frank, said 'He could explain so simply only because he understood so clearly.' This is true of Bob as well. Yet, he is diffident about writing, and some of us have become blue in the face trying to persuade Bob to, at least, transcribe a highly original lecture, or an illuminating talk or a perceptive commentary from his notes, or from the tapes, and to publish it in some more permanent and widely communicable form.

The corpus of published work – in the narrow sense of publication: in print and on paper – is rich enough despite this. It ranges from physics and the history of physics to philosophy and the history of philosophy; from philosophy of history and of the social sciences, to Marxism, and the Marxist philosophy of education; from philosophy of technology to the philosophy and history of the Vienna Circle and of the Frankfurt School as well. The remarkable thing is that Bob's essays in all of these varied fields are cutting-edge and original contributions to scholarship, at a level of high sophistication. One of us recalls an anecdote of many years ago, when Albrecht Wellmer, as a young representative of the Frankfurt School philosophy, and fresh from his graduate studies with Habermas, came to give his paper to the Boston Colloquium. As it turned out, Cohen was his commmentator, which left Wellmer a bit querulous, since all he knew at the time (from the Program announcement) was that Bob was then Chair of the physics department and a physicist. He was utterly confounded, as he said, when this physicist betrayed a close first-hand knowledge of the work of Horkheimer, Adorno, Marcuse, and others of the 'Old' Frankfurt school, and an even more astounding familiarity with the most current thinking of Habermas. What Wellmer was later to find out was that Bob was not only a close student of this movement, but a close acquaintance of Habermas, and a long-standing and close friend of Marcuse. Having studied the work of the Vienna Circle philosophers as well, and having personally befriended Hempel, Feigl, and Frank, over the years, Bob was able to produce a profound and thorough critical-comparative study of these otherwise separate movements which played such a major role in 20th century philosophy of science and in philosophy more generally, and to examine them against the background of Marxist philosophy.

In a major and masterful essay, 'Dialectical Materialism and Carnap's Logical Empiricism',[6] written in 1956 during his research fellowship year in England, Bob was among the first to present a philosophical appreciation of the relations among these two main currents of contemporary philosophy which were to have such a major intellectual influence in our century. At the same time, he noted Horkheimer's critique of positivist empiricism, and Marcuse's grounding of 'reason' in social life, from the side of the Frankfurt School's critical philosophy, and also Bachelard's historical conception of scientific knowledge. The footnotes to the essay presage themes and philosophies which were to rise to recognition on the American scene only two decades later, and the body of the argument introduces that classic issue between a phenomenalist conventionalism, a socially relativized constructivism, and a robust scientific realism which continues to preoccupy philosophers of science at present.

Cohen's published essays range over a broad philosophical landscape and we can do no more than note them, and group them here, and express the hope that a large number of them can be collected and reprinted in a volume, perhaps with Bob's contemporary appreciation (or critique) of his earlier work.

A. The essays most directly concerned with physics and the philosophy of physics, or with issues in contemporary philosophy and methodology of science. These include Bob's scientific paper based on his Ph.D. research,[7] as well as a series of early reviews of works in physics and the philosophy of science, e.g. reviews of Philipp Frank's *Modern Science and its Philosophy*,[8] Herbert Samuel's *Essay in Physics* and Erwin Schrödinger's *Science and Humanism*,[9] and the critical review essay, 'Epistemology and Cosmology: E. A. Milne's Theory of Relativity.'[10] Included here are also two exceptional introductions, each an independent essay, one on the tradition of explaining relativity without mathematics, from Einstein's own *Relativity – the Special and General Theory: a Popular Exposition* to Landau and Rumer's *What is Relativity*? for which Bob wrote this first introduction,[11] and the second, for the republication of Hertz's *Principles of Mechanics*.[12] In the latter essay Cohen reconstructs Hertz's sophisticated philosophy of science, his view of the nature of axiomatic systems, and of hypothetical entities or unobservables in physical theory, and of the importance of the analysis of concepts in physics. Cohen also offers an illuminating contrast of Hertz's theory of knowledge with that of Mach, and of Hertz's geometrization of physical reality with that of Descartes. As Cohen writes, "Hertz constructed, as an alternative, a system which permits purely nominal definitions of force and energy, and in which the sole primitive terms are mass, space and time. Just as d'Alembert reduced dynamics to statics, so Hertz reduced dynamics to kinematics."[13] Rounding out the series of essays on philosophy and history of physics are 'Ernst Mach: Physics, Perception and the Philosophy of Science,'[14] and 'Some Notes on Schrödinger and Mysticism,'[15] as well as a provocative reflection on the 'Needham problem', the question posed by Joseph Needham in his magisterial studies, *Science and Civilization in China* as to why modern science developed in the West and not in China. Bob has had a long-term interest in this question, and contributes to the methodological discussion of just what sort of question it is and what it would take to answer it. In Bob's hands, this becomes an analysis of the social contexts of scientific, philosophical and historical thought, and a scientific question in its own right.[16]

Among Cohen's essays on philosophy and history of science, there are also a number which deal with the relevance of social and historical contexts of scientific thinking. Far from proposing a relativisation of scientific truth, Cohen nevertheless points to constraints and ambiguities in science, the role of ideology in scientific concept-formation and the importance (conversely) of the philosophy of science for the history of science.[17, 18, 19, 20] In an oft-cited paper, 'Tacit, Social and Hopeful'[21] Cohen develops an important critique of Michael Polanyi's notion of 'tacit knowledge,' and its role in scientific thought. After a sharply critical yet appreciative discussion, in which Bob considers alternatives in Marxist epistemology, in Reichenbach and Carnap and in the phenomenologists, Cohen points approvingly to the feature of hope in Polanyi's view, contrasting it with Heideggerian or Sartrian despair. Polanyi discussed this essay at length with Bob, very positively, and it is indeed an illumination of the strengths and weaknesses in Polanyi's insight about scientific thinking. In a related essay, jointly authored with Joseph Agassi – 'Dinosaurs and Horses, or: Ways with Nature'[22] – Bob addresses the question of historically changing 'styles' in scientific thinking, or 'ways with nature' (a felicitous phrase with rich metaphorical resonances). Finally, in 'Social Implications of Recent Technological Innovations,'[23] Bob gives a wide-ranging survey of the globalisation of Western technology, in the context of a sketch of its historical development.

Discussion of Bob's writings on physics and philosophy of science would be incomplete without mention of a different genre of publication, yet one which embodies his commitment to physics education and to the wider understanding of the physical sciences and of scientific rationality in practice.[24] An early essay here is 'Individuality and Common Purpose,'[25] given as the Annual Address to the National Science Teachers Association in 1964. In that talk, Bob introduces the importance of the philosophy of science for the teaching of science. That genre is the Introductory Physics Text, and Bob's contribution to it is one of the most imaginative, accessible, culturally broad yet scientifically rigorous textbooks in the field. As this is written, the text, *Physical Science*[26] has just gone out of print, after eighteen years, with a final adoption last summer at the University of Iowa. Perhaps this is the appropriate place to note that Bob had been active for many years in the profession working towards the improvement of undergraduate teaching of physics, and with special interest in the teaching of physics to liberal arts students. The textbook, lucid, charming, sensitive to the social and historical contexts of the physical sciences, could only have been written by Bob Cohen.

B. The essays most directly concerned with Marxism , philosophy of the Social Sciences, and philosophy of Education. Bob is a leading student of Marx and of Marxism, and is known for this all over the world. The glib and often vicious assumption, especially during the period of intense red-baiting in the fifties, is that a strong or sympathetic interest in Marx and Marxism marks you as a communist, that is a member of the Communist Party. As we have seen, Bob was falsely accused of this on several occasions, since he was not a communist. But he was plainly and openly a sympathetic while critical student of the work of Marx and Engels, especially as it bore on the understanding of human society and history, and on the nature of human knowledge, particularly scientific knowledge, and he maintained colleagial relations with many other students of Marx and of Marxism all over the world. His interest in Marxist epistemology and its relation to scientific thought, as well as his concern with contemporary social movements in their relation to intellectual and educational life led him, as we have seen, to introduce some of the earliest teaching of Marxist philosophy in the American university. When the American Institute for Marxist Studies (AIMS) was founded (in 1964) Bob was asked to become chairman, and accepted. In the publications of that Institute, there appeared the papers of a symposium on Marxism and Democracy, to which Bob contributed a major paper (also entitled 'Marxism and Democracy').[27] He had already published a long classic study many years before (while still teaching at Wesleyan at the height of McCarthyism) entitled 'On the Marxist Philosophy of Education,'[28] in the Fifty-fourth Yearbook of the National Society for the Study of Education, and also at about that time, a hard critical review[29] of the then recent two-volume Princeton University Press collection of essays *Socialism and American Life* (Bob wrote: "What could have been a standard work on the American Left and its world setting fails in all but the inessential. These volumes are hopelessly inadequate. . . ." and so on). Having established his credentials very early, Bob participated in many conferences and colloquia, and gave innumerable talks, commentaries, on Marx and Marxism, and of course, continued to teach the course. But no publications emerged beyond those just mentioned, until 1978, when Bob was asked to contribute the biographical essays on Marx and on Engels to the encyclopedic *Dictionary of Scientific Biography*. This resulted in two splendid biographies[30] which give a sober, sophisticated and exceptionally well-informed account of these two thinkers. Bob's mastery of the material again pays off in the lucidity of his essays.

There is yet another essay, and this one is a sleeper. It comes out of left field, in a sense (not a political sense!), and it is entitled 'Causation in History.'[31] In it, Bob gives a personal account of how, as a physicist, he came to history and to the problematic of causation in history. It is an essay from the center of Bob's experience as a thinker, a scientist, a philosopher, a human being with strong social and moral commitments. But it is, for all the personal, intellectual history which it reveals, a rigorously analytical treatment of the deep question concerning the relation of 'external' factors to the scientific enterprise. It is a relatively short essay – a contributed chapter in a book – but it exhibits the breadth and suppleness of Bob's mind, his easy combination of the analytic, the empirical, the normative, the anecdotal . . . all in the clutch of a problem. It is a dialectical essay – an inquiry in the best Socratic sense. (There is, in fact, a discussion which follows the essay, which includes Bondi, Yourgrau, Tennessen and Törnebohm, among others.) And it gives us, in short compass, the vivid image of the happy philosopher-scientist, now past his three-score and ten and on his way to more and yet livelier engagements of the mind and heart.

Athens and New York, 1994

NOTES

1 R. S. Cohen, 'Adolf Grünbaum, a Memoir', in *Physics, Philosophy and Psychoanalysis, Essays in Honor of Adolf Grünbaum* (*Boston Studies* **76**), 1983.

2 James C. Thomson Jr., 'Neither Here nor There', in Diana Dubois, ed., *My Harvard, My Yale*, New York: Random House, 1982, p. 228.

3 With the exception of Sidney Hook's course at New York University on 'Modern Materialism', and Edgar Sheffield Brightman's year-long graduate seminar on Marx's *Capital* in the philosophy department at Boston University, taught many years earlier, Cohen's seems to be among the first such courses taught at an American University. Courses on Marxism were of course taught at a number of Catholic Universities, with a well-defined ideological aim.

4 For a fuller account of the Israeli Colloquium, see Edna Margalit, ed., *The Kaleidoscope of Science* (Proceedngs of the Israel Colloquium . . .), *Boston Studies* **94** (1986).

5 R. S. Cohen, 'In Memory of Philipp Frank', in *Philipp Frank, 1884–1966*, October 25, 1966, Cambridge, Mass., 1966.

6 'Dialectical Materialism and Carnap's Logical Empiricism', in Paul A. Schilpp, ed., *The Philosophy of Rudolf Carnap* (The Library of Living Philosophers, LaSalle Ill.: Open Court, 1963, pp. 93–158.

7 Robert S. Cohen, Lyman Spitzer, Jr. and Paul McR.Routly, 'The Electrical Conductivity of a completely Ionized Gas,' in *The Physical Review* **80**(2), 230–238 (1950).

8 Robert S. Cohen, 'A Philosophy of Science,' in *The Yale Review* **XXXIX**(4), 738–741 (Summer 1950).

9 Robert S,. Cohen, 'Reflections on Physics,' in *The Yale Review* **XLII**(1), 133–135 (Autumn 1952).

10 Robert S. Cohen, in *The Review of Metaphysics*, March 1950, pp. 385–405.

11 L. D. Landau and G. B. Rumer, *What is Relativity?* tr. by N. Kemmer, Introduction by Robert S. Cohen (Greenwich CT: Fawcett Publications), 1966, pp. 13–31.

12 Heinrich Hertz, *The Principles of Mechanics*, with a new Introduction by Robert S. Cohen (New York: Dover Publications, 1956), 20 pp.

[13] R. S. Cohen, *Ibid.*
[14] Robert S. Cohen, 'Ernst Mach: Physics, Perception and the Philosophy of Science', in R. S. Cohen and R. J. Seeger, eds., *Ernst Mach: Physicist and Philosopher* (*Boston Studies* **6**, 1970), pp. 126–164.
[15] R. S. Cohen, 'Some Notes on Schrödinger and Mysticism,' in M. Bitbol and O. Darrigol, eds., *Erwin Schrödinger, Philosophy and the Birth of Quantum Mechanics* (Gif-sur-Yvette: Editions Frontières, 1992) pp. 95–100.
[16] 'Robert S. Cohen, 'The problem of 19(k)', *Journal of Chinese Philosophy* **1**, 103–117 (1992).
[17] R. S. Cohen, 'Science: Open Problems and Uncertain Answers', *Boston University Journal of Education* **145**(3), 43–50 (Feb. 1963).
[18] R. S. Cohen, 'Constraints on Science,' in R. S. Cohen, P. Feyerabend and M. Wartofsky, eds., *Essays in Memory of Imre Lakatos* (*Boston Studies* **39**, 1976), pp. 79–86.
[19] R. S. Cohen, 'Reflections on the Ambiguity of Science', in Leroy S. Rouner, ed., *Foundations of Ethics* (Notre Dame: Notre Dame University Press, 1983), pp. 223–234.
[20] R. S. Cohen, 'Is Philosophy of Science Germane to the History of Science? The Work of Meyerson and Needham', in *Ithaca* (10th International Congress History of Science, 1962, Paris: Hermann), pp. 213–223.
[21] R. S. Cohen, 'Tacit, Social and Hopeful', in M. Grene, ed., *Interpretations of Life and Mind* (London, 1971).
[22] Robert S. Cohen and Joseph Agassi, 'Dinosaurs and Horses, or: Ways with Nature', *Synthese* **32**, 233–247 (1975).
[23] Robert S. Cohen, 'Social Implications of Recent Technological Innovations', in P. T. Durbin and F. Rapp, eds., *Philosophy and Technology* (*Boston Studies* **80**, 1983), pp. 35–47.
[24] R. S. Cohen, *Physical Science* (New York: Holt, Rinehart and Winston, 1976).
[25] R. S. Cohen, 'Physical Ideas, their Content, Logic and Social Context', *American Journal of Physics* **18**(9), 570–575 (December 1950).
[26] R. S. Cohen, 'Individuality and Common Purpose: the Philosophy of Science', Annual Address to the National Science Teachers Association, March 21, 1964, *The Science Teacher* **31** (1964). To be reprinted in the Annual 'Golden Oldies' series, *Science and Education* **3** (1994).
[27] Robert S. Cohen, 'Marxism and Democracy,' in Herbert Aptheker, ed., *Marxism and Democracy – A Symposium* (New York: Humanities Press, 1965), pp. 1–17.
[28] Robert S. Cohen, 'On the Marxist Philosophy of Education', in *Modern Philosophies and Education*, The Fifty-Fourth Yearbook of the National Society for the Study of Education, ed. N. Henry (Chicago, Ill.: The University of Chicago Press, 1955), pp. 175–211.
[29] Robert S. Cohen, 'Socialism and American Life – A Review', *Monthly Review*, 289–304 (1954).
[30] Robert S. Cohen, 'Karl Marx' and 'Friedrich Engels', in *Dictionary of Scientific Biography* **XV** (Supplement 1), (New York: Charles Scribner's Sons, 1978), pp. 403–417 and pp. 131–147, respectively.
[31] Robert S. Cohen, 'Causation in History,' in W. Yourgrau, ed., *Physics, Logic and History* (New York: Plenum Press, 1970), pp. 231–245.

JOAN LISA BROMBERG

EXPERIMENT VIS-A-VIS THEORY IN SUPERCONDUCTIVITY RESEARCH. THE CASE OF BERND MATTHIAS[1]

Historians of physics have recently been taking a searching look at experimental practice. We should expect, however, that previously unexamined areas of science will yield previously unnoted aspects of the "dialogue between theory and experiment."[2] I contend in this paper that the events in superconductivity research after the second world war do just this. More specifically, I will try to show that the work of U.S. experimentalist Bernd T. Matthias confronts us with questions such as whether experimentalists always accept theoreticians' theories as useful; whether they share with theorists common criteria for what makes a theory useful or successful; whether, indeed, theoretical practice and experimental practice actually lead to disparate or competing theories.

These are issues that we usually think of as separating scientists from engineers, rather than dividing the experimentalists and theorists who cohabit a single discipline. Yet disciplinary boundaries change. And a central fact of twentieth century science is the growth in and around physics of the materials sciences, a congeries of disciplines including physics, chemistry, metallurgy, and engineering. Pushed vigorously by industry and, in the United States at any rate, by the military, it has been passing from an empirical craft to a subject accessible to theory. Now, new materials can be useful, even when they are not well understood. That means that the empirical side of materials sciences has been funded by industry and the military even as the scientific side has been promoted. Thus we find that during materials sciences' transition to maturity, some physicists functioned more as empiricists or craftsmen than as workers within a mature science. These crypto-materials scientists, however, were still lodged within the disciplinary structure of physics. They battled the theorists as fellow-physicists and their fights constitute another, proper, part of the same "dialogue between theory and experiment."

The first stance Matthias took with respect to superconductor theory was to try to provide the evidentiary base upon which it could be constructed. This story begins in 1949, when Matthias and John K. Hulm

K. Gavroglu et al. (eds.), Physics, Philosophy and the Scientific Community, 1–10.

met at a dinner party at the University of Chicago.[3] Hulm was a Briton who had just completed a Ph.D. at Cambridge University and had come to Chicago's Institute for the Study of Metals as a post-doctoral fellow. Matthias was a German-born and Swiss-trained physicist who had arrived in the United States two years earlier, and was at Chicago as an assistant professor of physics, on leave from Bell Telephone Laboratories.[4] The two decided to begin a collaboration on ferroelectrics, an area in which each had worked as a graduate student. They soon switched, however to superconductivity. As the one phenomenon in solid state physics that had not yet received a theoretical explanation, superconductivity appeared to them both more challenging and more rewarding.

Most of the other scientists in superconductivity were working on metallic elements at this point. And most were concerned with the physical properties of their samples . . . their specific heats, thermal conduc-tivities, electrical conductivities. Matthias and Hulm instead took up the superconducting compounds and alloys of the transition elements. (Transition metals are distinguished from nontransition elements in their electron configurations. They have unfilled shells of electrons with high angular momenta.) Rather than studying their chemical parameters, they looked at factors like: the type of chemical bonding involved; the crystal structure; the relation between superconductivity and the electronic configuration of the elements that entered into the compounds or alloys.

In this work, they made use of a method that Matthias had begun to use as a research associate at the *Eidgenossische Technische Hochschule* in Zurich in the mid-1940s. Matthias had written his thesis on the ferroelectric crystal potassium sodium tartrate [Rochelle salt], and the frequency-dependence of its piezoelectric response to external fields.[5] Shortly after, he and a collaborator, W. Merz, published an experiment in which they introduced a sequence of different ions into Rochelle salt, of similar size, valence, and coordination number but different dielectric polarizability, and measured the differences in piezoelectric response.[6] This technique of "substitutional chemistry",[7] whereby some of the atoms of a solid lattice are replaced by others that differ in stipulated properties, had continued to function as part of Matthias' armamentarium as he worked on ferroelectricity. When they turned to superconductivity, Hulm, who had done some superconductivity research at Cambridge, was providing the low-temperature techniques, and Matthias was providing the chemical method and, I conjecture, the program.

That program was to use both the known compound superconductors and the new ones that they were concocting "to throw new light on the chemical and structural conditions governing the occurrence of superconductivity."[8] The two scientists appear to have hoped to amass a dense base of information from which empirical regularities could be deduced. Such rules or regularities would then provide the materials from which theorists could erect a theory. Colleagues of Matthias knew him as an admirer of Kepler, whose laws had ultimately found their theoretical basis in the work of Newton. Quite possibly, Matthias hoped to do for superconductivity what Kepler had done for astronomy.

Matthias returned to Bell Laboratories in 1951 and continued his superconductivity research program, with a gradually expanding group of assistants and collaborators.[9] He continued to use the technique of substitutional chemistry, developing it into an ever more subtle and powerful tool. The method entailed, of course, a thorough knowledge of the properties of the elements. Matthias became notorious among his physicist-colleagues for his chemist's familiarity with every detail of the periodic table. He also became habituated to scouring the literature for papers on new materials. This is to be ranked as the second of his characteristic methods. Matthias' immersion in this data reminds us again of Kepler. As Kepler's profound knowledge of the data of Tycho Brahe led him past the shoals of his own errors to his laws of planetary motion,[10] Matthias' intimate knowledge of the properties of materials produced an intuitive sense of where to look for superconductivity.

And his achievements were impressive. During the middle 1950s, he synthesized hundreds of new superconductors. The most notable was Nb_3Sn. Building on work of G. Hardy and Hulm, Matthias and his Bell Laboratories collaborators discovered this compound in 1954. It had the highest transition temperature, 18K, of any superconductor discovered to that date, and it was to become, in the 1960s, the foundation of a new technology of high-field superconducting magnets.[11] Matthias also succeeded in his goal of formulating rules for the superconductors. The most famous of these illustrates the kind of result he was achieving. It relates the transition temperature of a superconductor to the average number of valence electrons of its constituent atoms. Matthias stipulated that the highest transition temperature obtained for compounds with an average number of valence electrons a little below 5 and about 7.[12] The Baconian program he and Hulm had conceived for superconductivity, of amassing facts and discerning regularities as a precondition for a theory, was on track.

In the event, however, the program was derailed. That is, the Newton of superconductivity, in His manifestation in the three persons of John Bardeen, Leon N. Cooper, and J. Robert Schrieffer, published a theory of superconductivity in 1957 that failed to build on the regularities Matthias was inducing from his data. Among experimental observations the three instead gave pride of place to the isotope effect, discovered independently by scientists at Rutgers University and at the Naval Research Laboratory in 1950. These investigators had found that the transition temperature for isotopes of mercury varies with mass according to the equation $T_c = M^{-1/2}$.[13] This effect suggested that the massive ions of the crystal lattice were involved, as well as the conduction electrons. Bardeen, Cooper, and Schrieffer posited that the electrons were attracted to each other by a secondary effect, mediated by the lattice, and linked up in pairs, which then became coordinated in a collective state that gave rise to the typical zero resistance and perfect diamagnetism that characterizes superconductors. That is, superconductivity was due to electron-phonon interactions, where the phonons are quanta associated with the lattice and its vibrations.

This is not to say that Bardeen, Cooper and Schrieffer were ignorant of Matthias' work, which was widely recognized. Rather my claim is that Bardeen, Cooper and Schrieffer did not make use of Matthias' results to reach their theory. Indeed, they would have been hard put to do so. The transition elements and compounds which Matthias, Hulm, and the others of this group favored, though interesting both for their technological applications and the new phenomena that they presented, were "dirty." They contained chemical impurities and physical imperfections. Nor did they all have properties that were uniform throughout their bulk; some were a mixture of different phases. Matthias and his associates were doing "dirt physics."

With the BCS theory, Matthias now saw the results he had achieved within the field of superconductivity demoted in importance and, more generally, he saw the role of the experimentalist being overshadowed by that of the theorist. Reputation was important to him; he was not above campaigning for that supreme sign of recognition in the scientific community, the Nobel prize. He was, moreover, a contentious man. On the one hand, he was by nature quick to take offense; on the other, he enjoyed the stimulation of controversy and revelled in the arena it provided for his native theatricality. His response was to attempt to demonstrate experimentally that the BCS theory was wrong.

He attacked along several lines. In one, he undertook a series of experiments to show that it was not the electron-phonon interaction alone that underlay superconductivity. He did this by making solid solutions in which iron, an element with a magnetic moment, was dissolved in a transition metal, like titanium. When the transition temperature of this solid solution was measured, it was found that it rose as more of the magnetic material was introduced. This suggested to Matthias that the superconductivity was not impeded by magnetic ions, a result that conflicted with the BCS theory.

In a second line of attack, he and his colleagues set out to show that the isotope effect that had been found for mercury, and subsequently for lead, thallium, and tin, was not universal. In 1961, Theodore Geballe, Matthias, George Hull and Ernest Corenzwit published results that showed that ruthenium had an isotope effect that was below 10% of that of mercury;[14] in 1962, the group got similar results for osmium. Since tin, mercury, thallium and lead are nontransition elements and osmium and ruthenium are transition elements, Matthias could now partition the field of superconductors. The nontransition elements formed a jurisdiction he ceded to the BCS theory and the electron-phonon interaction; the transition elements and their combinations were a territory he proposed to reign over himself.

The theorists rose to the challenge. Already in 1958, David Pines published an article, which he worked up even as Bardeen, Cooper, and Schrieffer were putting the finishing touches on their paper, that tried to account for Matthias' rules on the ground of BCS theory. In 1959 Matthias' neighbor at Bell Laboratories, the theorist Philip W. Anderson, published a BCS-based "Theory of Dirty Superconductors." Later, in collaboration with the French physicist Pierre Morel, he began calculations to yield the ruthenium data. This work was perfected by James Garland in 1963.[15] As for Matthias' other line of attack, metallurgical work by his collaborator, the German scientist Ernst Raub, demonstrated that the iron dissolved in titanium had not been homogeneously dispersed throughout the lattice, but had precipitated out in superconducting threads.

Meanwhile, in 1961, Matthias moved from Bell Laboratories to the new University of California campus that Roger Revelle was organizing on the outskirts of San Diego. He continued his relation with Bell

Laboratories as consultant[16] but he now needed to find financial support for his University of California work.

It is significant in this regard that Department of Defense funding agencies were then creating new institutions and new funding for materials sciences. The Department of Defense's Advanced Research Projects Agency (ARPA), itself recently established as part of the response to Sputnik, was organizing Interdisciplinary Laboratories for the synthesis and study of materials. Its aim – and that of the Department of Defense in its materials program – was to force the pace of materials technology in those areas that were of interest to the military, to join academic science as tightly as possible to applied engineering, and to encourage the creation of "a new generation of engineers and scientists in materials technology [that would] acquire and assure respect for applied problems".[17]

Matthias and the newly forming UCSD Physics Department applied to ARPA. In the event, it was the Air Force Office of Scientific Research that generously funded Matthias' La Jolla projects. This funding added a new member to Matthias' group of clients. Previously his audience had been, on the one hand, the physics community. Secondly, it had been the management of the Bell Telephone Laboratories, which believed that superconductivity, with its promise of resistance-free low cost electric transmission and electronic devices with vastly lowered heat dissipation demands, was an obvious technology for a company with its market interests, and which gave its scientists a free hand to pursue whatever phase of superconductivity struck their interest. Now Matthias had an additional client: an agency that was working consciously to forge a new joining of scientists and technologists in the new interdisciplinary area of materials science.

His new position within the emerging complex of materials science may have played a part in the stance Matthias took in an unusual and blunt exchange with Philip Anderson that the journal *Science* published in 1964.[18] The ostensible topic was whether the BCS theory was adequate to the facts of superconductivity. The quarrel turned, in actuality, on the correct criteria for evaluating a theory.

Matthias maintained that the most important data to be explained were, first, which substances were capable of exhibiting superconductivity, and second, the temperatures at which the transition from normal to superconducting state occurs. In particular, the theory should explain the circumstance that experiment was gradually revealing that almost all metals could become superconducting if they were sufficiently cooled. The BCS theory had not been able to give any prognostications at all in these areas. "[T]he present 'theories' are unable to state the rules for the occurrence of superconductivity, which, in my opinion, are essential in any explanation of the phenomenon." (p. 379). These "rules" were precisely what Matthias aimed at in his own research program, and what he was finding in regularities like that which linked the number of valence electrons per atom to the transition temperature. Moreover, the identification of materials that are superconducting and of their transition temperatures are vital results for any research aimed at technological applications.

Anderson, on the other hand, found most decisive the fact that the BCS theory brings the electrons' quantum field to the fore. It is a field theory of electrons that parallels the field theory of electromagnetism and it also throws light on a wide range of phenomena, from superfluidity to the recently discovered Josephson effect, and flux quantization.

It is interesting that Anderson felt the necessity of dealing with a Baconianism that he took to be universally accepted within the physics community. "The scientific method advances – we all learned – by induction from experimental facts." (p. 373). One might guess that he had encountered precisely this Baconian faith in his Bell Laboratories colleague, Matthias. Anderson's defense against it was to maintain that the first requirement for theory is that it be consistent with older established principles, like quantum mechanics or relativity theory, principles that are themselves the embodiment of prior experimental results. That is, true Baconianism demands that "[t]he experiments against which a theory must be tested are not merely those under direct consideration but the ones carried out over the past 50 to 100 years." (*ibid.*) Consistency with these older principles makes a theory plausible and brings it to the first stage along its trajectory. The second historical state is qualitative agreement with experiment: just those results for phenomena like those cited above that the BCS theory has successfully illuminated. Only the third stage requires quantitative agreement with experiment. Here is where Matthias' demand for prediction of transition temperatures would be met. Anderson expected that BCS theory would soon be at this point.

Matthias, on the other hand, concluded that the BCS theory was at best a "model", valid for a limited area of superconductors. In particular, he predicted that it would not hold for the transition metals. Here the facts "would suggest to an unprejudiced observer that there is a drastic difference in the mechanisms causing superconductivity in the two groups of metals." (p. 380). Thus Anderson stressed the structure of the theory and its consistency with other theories. "[A theory] which is consistent with [accepted theories] but disagrees with experiment is often not wrong, for we often find that experimental results change, and then the results fit the theory." (p. 373). Matthias stressed the multiplicity of phenomena manifested through experiment. If theories could not explain them, so much the worse for the theorist. The experimentalist could still codify them in a manner useful to the technologist.

Matthias, who had begun his superconductivity work with the aim of assembling facts that could underlie theory, now began increasingly to withdraw from a dialogue with theory. At this time, a number of theorists were beginning to use BCS theory to explore whether novel materials might be concocted that had higher transition temperatures than those then available. Matthias seized on this work to define his own program more sharply. He began to label his "the empirical approach" and to oppose it to a "theoretically motivated" approach.[19] With scorn he noted that only the first approach had been fruitful. It had attained the highest yet known transition temperature and it had uncovered a never ending procession of new superconducting materials and new physical phenomena. As for the theoretical approach, "alas, the experimental fallout from all these hundreds of learned and imaginative treatises has been totally, and without exception, nil. Through them, not a single existing transition temperature was ever raised, not to mention the absence of any discovery of new superconductors, high or low."[20] The best prescription for further progress, he kept pointing out to his Air Force sponsors, was to continue his own "enlightened" empirical approach.

We thus find within solid state physics in the 1950s through the 1980s at least two experimental traditions. One interacts with theory and works with materials amenable to theory. The other, represented by Matthias, is either anterior to theory (as he and his collaborators first conceived it) or parallel to and distinct from it, as he later saw it. Matthias' motives for placing himself within such a tradition were several. Certainly egoism was one – a desire to occupy a central place within his areas of solid state physics. Another was a fascination with new phenomena and new materials, whether or not they were currently explicable. A third was the willingness of his funders at Bell Telephone

Laboratories and the Air Force Office of Scientific Research to support a type of research that was not leading directly to new physics, but did lead directly to new technology.

We are left with a number of questions. How strong or, alternately, how idiosyncratic a tradition did Matthias represent within physics? And what were the sources of this tradition in chemistry, metallurgy, and in physics itself? Does military support for Matthias' "empirical program" qualify as one more example of Paul Forman's thesis that Department of Defense money and efforts skewed American physics toward the applied, the pragmatic, and the fragmented in the 1950s and 1960s?[21] Most interesting of all, how did the "dirty physics" of Matthias and his colleagues eventually become clean physics? That is to say, how did the materials revolution play itself out, and what role, if any, is Matthias to be assigned in this drama?

About five or six years after Matthias's death, a radically new discovery appeared: the oxide superconductors, with transition temperatures that by now have reached well over 100° Absolute. Matthias certainly had not anticipated these. But these discoveries caused great consternation among the theorists, who agreed on only one thing: the lattice vibrations underlying BCS theory could not possibly account for these new superconductors. Exotic new theories sprang up like desert flowers after the rainy season. Not one of them has found universal acceptance. So Matthias' scepticism on the predictive value of superconductivity theory has been vindicated. His colleagues are sure that had he lived, he would have said "I told you so."

NOTES

[1] This work, a preliminary and tentative aperçu over the material, was supported by the National Science Foundation. I am grateful to several colleagues of Professor Matthias who talked to me about him. I alone bear responsibility for the interpretations and errors.

[2] Timothy Lenoir and Yehuda Elkana, eds., "Practice, Context, and the Dialogue Between Theory and Experiment," special issue, *Science in Context* **2**, 1 (1988).

[3] John K. Hulm, J. Eugene Kunzler, and Bernd T. Matthias, "The Road to Superconducting Materials," *Physics Today*, January 1981, 34–43.

[4] "Obituaries: Bernd T. Matthias", by Albert Clogston, Theodore H. Geballe, and John K. Hulm, *Physics Today*, January 1981, p. 84.

[5] *Helvetica Physica Acta* **16** (1943), 99–135. Ferroelectrics are electric analogues of ferromagnetics. They preserve a dielectric polarization in the absence of an external electric field and show the phenomena of hysteresis.

[6] P. Bärtschi, B. T. Matthias, W. Merz, and P. Scherrer, "Verschiebung des Sprungpunkts der NH_4 Rotationsumwandlung," *Helv. Phys. Act.* **18**, 238–240.
[7] See the obituary, cited above in note 4.
[8] B. T. Matthias and J. K. Hulm, "A Search for New Superconducting Compounds," *Physical Review* **87** (1952), 799.
[9] He also worked at the same time in ferroelectricity and, later, in ferromagnetism, although these investigations are not discussed here.
[10] Arthur Koestler, *The Sleepwalkers* (NY: Macmillan, 1959).
[11] B. T. Matthias, T. H. Geballe, S. Geller, and E. Corenzwit, "Superconductivity of Nb_3Sn," *Phys. Rev.* **95**, p. 1435.
[12] B. T. Matthias, "Empirical Relation between Superconductivity and the Number of Valence Electrons per Atom," *Phys. Rev.* **97** (1955), pp. 74–76.
[13] J. Bardeen, L. N. Cooper, and J. R. Schrieffer, "Theory of Conductivity," *Phys. Rev.* **108** (Dec. 1957), 1175 ff. C. A. Reynolds, B. Serin, W. H. Wright, and L. B. Nesbitt, "Superconductivity of Isotopes of Mercury", and Emanuel Maxwell, "Isotope Effect in the Superconductivity of Mercury," *Phys. Rev.* **78** (1950), pp. 487 and 477.
[14] T. H. Geballe, B. T. Matthias, G. W. Hull, Jr., and E. Corenzwit, "Absence of an Isotope Effect on Superconducting Ruthenium," *Phys. Rev. Letters* **6** (1961), 275–277.
[15] D. Pines, "Superconductivity in the Periodic System", *Phys. Rev.* **109** (1958), 280–287. Philip W. Anderson, "Theory of Dirty Superconductors", *Jour. of Phys. Chem. Solids* **11** (1959), 26–30; James W. Garland, Jr., "Mechanisms for Superconductivity in the Transition Metals" and "Isotope Effect in Superconductivity", *Phys. Rev. Letters* **11** (1963), 111–119. P. Morel and P. W. Anderson, *Phys. Rev.* **125** (1962), 1263.
[16] Where, in fact, he was directing research, rather than consulting on that of others. He also continued a consultancy at Los Alamos National Laboratory, which he had recently begun. In effect, the "Matthias school" of superconductivity now was spread over three institutions.
[17] Enclosure to Memorandum from E. Creutz to G. Arrhenius *et al.*, Dec. 17, 1964, Box 2, folder 9, and Walter Kohn to Charles Yost, *ibid.*, June 8, 1962, Box 25, folder 9. Matthias papers, Mandeville Department of Special Collections, University of California, San Diego.
[18] "Superconductivity," *Science* **144** (1964), 373–381.
[19] Undated, untitled manuscript, Matthias papers, box 40, folder 1.
[20] B. Matthias, "High Temperature Superconductivity?", *Comments on Solid State Physics* **3** (Oct.–Nov. 1970), 93–96, p. 94. He concludes, "I have tried to point out that the present theoretical attempts to raise the superconducting transition temperature are the opium in the real world of superconductivity where the highest TC at present and at best, is 21 K. Unless we accept this fact and submit to a dose of reality, honest and not so honest speculations will persist until all that is left in this field will be these scientific opium addicts, dreaming and reading one another's absurdities in a blue haze." We find here another aspect of a scientist's role within his/her discipline: the scientist as entertainer.
[21] Paul Forman, "Beyond Quantum Mechanics: National Security as a Basis for Physical Research in the United States, 1940–1969," *Historical Studies in the Physical and Biological Sciences* **18**(1), 149–229.

CATHERINE CHEVALLEY

PHILOSOPHY AND THE BIRTH OF QUANTUM THEORY

In a little-known paper published one month before his major article on complementarity and the interpretation of quantum mechanics, Niels Bohr made some comments about the influence of philosophy on the genesis of quantum theory. The main line of his argument went as follows. The connection between philosophy and the natural sciences was not only met, he said, at the earliest stages of science, but it was also a permanent one: "Cross-fertilization has taken place during the entire development of science up to recent times, where the vast accumulation of empirical data in all fields and the high standards of methods required for the acquisition and analysis of the data have necessitated extensive specialization in science". Such a cross-fertilization occurred each time physics was faced with problems "which call for a revision of *the concepts underlying our understanding of observations*". Then, clearly pointing to quantum mechanics, Bohr added that it had been of "inestimable significance that scientists were able to find *support and points of departure* for new advances in the endeavors of philosophers to make clear the *foundation and limits of human intellectual activity*", and that these endeavors were passed on to scientists "through the formation of words which embodies the thought of these philosophers and *had now become part of the common language*".[1]

If we look at these remarks a little closely, we soon realize that there is something about them which is both astonishing and enigmatic from our contemporary point of view. First, they run counter to most received views concerning the part that philosophy played in the emergence of quantum theory. Philosophy has been said either to have had no influence at all, or to have acted *via* a somewhat rigid or affective commitment of quantum physicists to logical empiricism or to the widely spread epidemic of *Lebensphilosophie*.[2] But what sense can we make of Bohr's statement that the problematic of the limits of the understanding gave support and points of departure for the revision of such concepts as those which underlie our notion of what observation is? On the other hand, the way in which Bohr describes philosophy's influence is also very unusual. When we try to conceive of an influence of philo-

K. Gavroglu et al. (eds.), Physics, Philosophy and the Scientific Community, 11–37.

sophical conceptions on the construction of a physical theory, there seems to be a limited spectrum of possibilities, each of which follows from our personal and sometimes unconscious view of what philosophy is. We can think, for instance, of the foundational framework of a doctrine, or restrict philosophy to the heuristics of a general orientation (be it 'metaphysics' or positivism), or try to grasp the insidious effect of a flux of ideas. But what could be the underlying presupposition of the idea that philosophy acts from the inside through ordinary language, that it invests science through an action-by-(linguistic)-contact mode of propagation? Clearly the reason why we are puzzled by Bohr's remarks is that they bear on two controversial issues. One is that of the philosophical context in which quantum mechanics was originally formulated, and this touches directly upon the difficulties of our present understanding of the so-called 'Copenhagen interpretation', or 'orthodox' view, which was derived from Bohr's ideas and accepted by most physicists. The other one bears on one of the most thorny questions about the relationship between science and philosophy, that of how does a new physical theory achieve philosophical consistency if, as seems plausible, it cannot rely on a former philosophical scheme.

In the limited scope of this paper, I shall mainly attempt to provide an interpretation of Bohr's comments, and to give some evidence to support them. My intention will not be to make them look like unquestionable gospel truth, but I want to use them as a guide, to explore the insights which they might possibly give, and to spell out some of their implications. In order to do so, I shall draw first on the conceptual history of quantum theory,[3] second on the history of philosophy. During the genesis of quantum theory, epistemological thinking was essential because it allowed to analyze which concealed presuppositions of classical theories had to be renounced; Bohr's special strategy was a mixture of conceptual analysis and of close attention to experiments, and I shall first describe the main stages of the evolution which led him to the complementarity idea. This will not, however, be enough to understand why Bohr thought that philosophy acted as a language, through words that were hidden philosophical concepts. To gain such understanding, it will be necessary to read again the same story from the point of view of the history of philosophy; this will reveal a kind of inner consistency in Bohr's successive moves, thanks to which he was led to criticize from the inside the former philosophical language that he inherited, to bend it, so to speak, against itself and to emphasize the inconsistencies that would result from reifying a mode of thinking that had become inadequate. This strange experience of keeping the words without keeping the meaning of the words, as Heisenberg once put it graphically, finally suggests two conjectures, which I shall state in the end, one about what happened to the Copenhagen interpretation when it was translated into another philosophical language, and the other one about the part that general philosophy might play in contemporary philosophy of science.

'*A revision of the concepts underlying our understanding of observation*'.

What exactly did Bohr have in mind when he talked about the 'revision of the concepts underlying our understanding of observations'? To grasp the meaning of this expression, one should not rely too much on recurrent history. Hindsight comments on quantum theory usually fail to encapsulate the core of the conceptual work that was necessary to break with classical theories because they tend to compare two maps: one is the map of quantum theory as we know it today, and the other one is a map of classical physics as we understand it today. We can now move in one or the other of these two worlds, with some anxiety maybe about their relationship, but without contradiction. To the physicists who created quantum mechanics, the situation was different since they had only one map, which was obviously wrong. New experiments contradicted the only available formal schemes, but these new experiments were described in the language of classical theories, they had emerged from within classical physics – not like anything purely 'given', just ready to be made intelligible through a bottom-up process of conceptualization – and no other set of concepts was at hands to interpret them. The situation was thus one in which people experienced the foundering of a whole set of representations without being able yet to grasp where and why they foundered, and in each phase of the building of quantum theory internal criticism of classical physics put new epistemological demands. In what follows, I shall begin by sketching the story of Bohr's struggle to meet some of these demands, by looking first at the situation which was created by his own 1913 theory of the constitution of molecules, second at the two-stages strategy which he adopted during the years 1914–26, and third at the complementarity argument and subsequent paradoxes which he introduced in 1928.[4]

Conceptual change amounts to a break-down of the unavoidable 'hermeneutic circle'[5] in the definition of what is observation, and it entails a destructuration of the theoretical language. As Bohr put it as a leitmotiv, "the task with which Planck's discovery confronted physicists was nothing less than, by means of a thorough *analysis of the presuppositions* on which our most *elementary concepts* are based, to provide room for the quantum of action in a rational generalization of the classical physical description".[6] This program to analyze the applicability of classical concepts was started by Bohr in 1913 for reasons which are easy to understand. The 1913 theory had shed crude light on the fact that no coherent set of concepts was available to account for the motion of electrons in the atom. Since it created a well-localized contradiction, Bohr's hydrogen atom model was the real starting point of quantum theory. Numerous examples of unintelligible empirical data had occurred previously, such as the numerical harmonies of the Balmer–Ritz–Rydberg formulas, the break in the experimental curve which described the spectrum of black-body radiation, the way electrons recognized only colors (frequencies) or the astonishing discovery that alpha particles collided with something hard inside the atoms.[7] However, when Bohr took over J. J. Thomson's program of understanding how the distribution of a number of identical 'elements' (electrons) could explain the physical and chemical properties of the atoms, he came up with conclusions which clearly contradicted basic principles of classical physics in at least four ways: (a) in 'stationary states', the electron in accelerated motion was forbidden to radiate; (b) emission and absorption of radiation by an atom occurred through a discontinuous transition of an electron from one stationary state to another, a process which seemed to have no cause and was described by an equation ($\Delta E = h\nu$) which connected continuity (variation of energy) with discontinuity; (c) the classical concept of frequency was split into two concepts: one of mechanical frequency and one of optical frequency; (d) two distinct kinds of laws appeared to rule the atom: classical laws as long as there was no interaction with radiation (and therefore no observation), and quantum laws otherwise. This looked incoherent, but met experimental success. During the ten years that followed, the difficulty was therefore to find a way to put together classical concepts and quantum discontinuity, in order to elaborate a theory of the mechanism of transitions.[8]

Bohr's strategy of conceptual analysis had several components: to lower the number of new hypotheses, to keep working as long as possible with classical concepts (such as the electro-magnetic field, or position, momentum and energy), to state the range where they could be defined consistently, to isolate the hypotheses which did not require recourse to classical laws (hence the 'postulates' formulated in 1917–18), to multiply points of view, and always to look for a 'rational generalisation' of classical theories (not merely, as Sommerfeld did, for a set of formal rules).[9] In the realization of this program up to 1928, two stages seem to have been of special importance.

The first one is well characterized by the notion of *formal analogy*, which was the core of what Bohr called the principle of correspondence from 1920 on. As introduced in the years 1918–23, formal analogies embodied Bohr's will to look for a relationship as close as possible between classical electrodynamics and quantum theory. However, analogies did not mean vague similarities, nor did they refer to visualizable models. Analogies were 'formal', that is they amounted to the conjecture that some parts of the classical formalism could be transferred into the atomic realm. Bohr's idea was that there was a 'far-reaching correspondence' between the kinds of transitions which were possible in the atom an the harmonic components of the motion of the electron,[10] and the principle of correspondence was a principle of analogy inasmuch as it allowed to carry over formal relationships from one theory to another under given conditions. This strategy of analogies ruled most of the numerous episodes of the history of 'old quantum theory', such as the whole problem of selection rules, Bohr's and Kramer's helium model, Bohr's second atomic theory, Bohr's and Pauli's work on the anomalous Zeeman effect, etc. A famous letter from Bohr to H. Høffding describes analogy as the most useful method for the construction of a new theory, but it is essential to underline that analogies were meant to test the transfer of a *formal* structure.

By 1924–25, full recognition of the failure of 'ordinary mechanics' was knowledged by Bohr. This started a second stage in conceptual analysis, during which Bohr renounced two 'basic presuppositions' of classical theories, which were also, as will be clear further, basic presuppositions of the theory of knowledge.

First, in view of the difficulties which had accumulated during the years 1923–24, the suspicion which had been cast from the beginning on the relevance of classical space-time description and the concept of trajectory for the electron ended up in definite rejection of '*spatio-temporal representations or pictures* in favor of *symbolic analogies*. The main event which cristallyzed this rejection was the break-down of the Bohr–Kramers–Slater theory, which had been Bohr's ultimate attempt to save continuity (at the cost of strict conservation of energy and momentum in individual processes). *Rather than reinforcing the discrete image of the light-quanta*, Bothe's and Geiger's experiments were interpreted by Bohr – and Pauli – as a general failure of 'the ordinary space-time/intuitive (*anschaulicher*) pictures, by means of which the description of natural phenomena has hitherto been attempted'[11] – a statement which was immediately followed by the prediction that generalization of classical electrodynamics would require 'a fundamental revolution in the concepts upon which the description of nature has been based until now'.[12] At this point precisely, Bohr began to use the expression '*symbolic analogies*', referring thus to the only possible approach of processes which resisted space-time description.[13] Symbolic analogies provided freedom from 'intuitive representations' (especially orbits), and they prepared the road to Heisenberg's *Umdeutung* paper and to Heisenberg's, Born's and Jordan's radical move of eliminating the geometry from the interior of the atom and of representing physical quantities by matrices. The conceptual condition of possibility for the 'non-intuitive' formalism was thus the rejection of the constraint of spatio-temporal representations.

A year later, in 1926, Bohr was led to a second renunciation, or rather limitation, of a theoretical and epistemological fundamental concept of classical physics, that of *continuity*. Continuity was not only a feature of the formalism of classical theories, *via* the assumption that physical quantities could be represented by continuous functions with causal evolution; it also enjoyed a very high power of conviction through the successes of optics. However, it had been shown untenable in all cases where quantum discontinuity was the only choice, and the continuity-discontinuity debate developed throughout twenty-five years. But it came to a peak in 1926, when Bohr and Schrödinger argued about the situation created by the mathematical equivalence of wave and matrix mechanics. The epistemological impact of this episode, which is not yet as well known as it should be, was enormous.[14] Schrödinger's work in 1926, which took support from de Broglie's 1924 thesis, was embedded in the claim that it was possible to give a 'continuous and intuitive (*anschaulich*)' picture of atomic phenomena and that all discontinuous conceptions (such as stationary states, quantum jumps plus, as Heisenberg complained, the photo-electric effect, Franck collisions or the Stern–Gerlach experiments) had to be eliminated from physics. This claim was explicitly based by Schrödinger on arguments taken from the 'theory of knowledge', which he thought introduced an asymmetry between his own *anschaulich*-continuous theory and the *unanschaulich*-discontinuous matrix mechanics. In face of this, Bohr differed from Heisenberg and Born, who decided to stress the discrete approach even more .[15] Indeed Bohr accepted the wave-packet representation with enthusiasm as reinstating a (restricted) possibility for the application of space-time pictures in quantum theory.[16] However, he did not accept Schrödinger's interpretation, which tended to a wave ontology. Rather, he suggested that wave mechanics and matrix mechanics were both 'correspondence theories', that is theories based on formal analogies, and that the crucial obstacle was that of the conditions for unambiguous definition of concepts. As he put it in a letter to Schrödinger, "the definition of *every concept or rather every word* presupposes the *continuity* of the phenomena and hence becomes ambiguous as soon as this presupposition cannot be upheld".[17] A year later, when announcing to Einstein the results of Heisenberg's work on indeterminacy, Bohr was to use exactly the same phrasing: "it has of course long been recognized how intimately the difficulties of quantum theory are connected with *the concepts, or rather with the words* that are used in the customary description of nature, and which all have their origin in classical theories".[18] To Bohr, the essential insight which was provided by the discussion with Schrödinger was that univocal description of concepts (such as wave and particle, the superposition principle and the 'atomicity' of light-effects) was not attainable in atomic physics.

Up to 1926, Bohr's epistemological response to the internal breakdown of the 'harmony' between mathematics, physics and philosophical presuppositions which characterized classical theories thus went through successive commitment to *formal analogies, symbolic analogies* and acceptance of the unavoidable *ambiguïty* of classical concepts in the atomic realm. What was the next step? in 1927–28, Bohr presented a general conception of the limitation of classical concepts which he thought allowed one to "harmonize the different views, apparently so divergent, concerning this subject",[19] and he called this conception *complementarity*. Complementarity was clearly an answer to the battle about matrix and wave mechanics which had taken place during the year 1926: "superposition Schrödinger Heisenberg"![20] Posterior 'translations' of the Complementarity idea in quantum mechanics textbooks tend naturally to present it in the language of observables and operators in Hilbert spaces; complementarity only expresses the fact that observables can be associated to operators the product of which is not always commutative; in 1927, however, Complementarity was formulated in the language of space-time description versus causality. Manuscripts give evidence of Bohr's reasoning: all classical concepts are defined through space-time pictures; quantum theory has a 'feature of discontinuity' which is in sharp contrast with the principles underlying the ordinary description of phenomena; hence, since definition of concepts require continuity, "*every concept* used in accounting for the experimental evidence will have a *restricted validity* when dealing with atomic phenomena".[21] Once granted, this restricted validity of all classical concepts allowed for a new understanding of the relation between space-time coordination on the one hand and applicability of the conservation of energy-momentum on the other hand. In Bohr's phrasing, in quantum theory "we must regard the space-time coordination and the claim of causality, *the union of which* characterizes the classical theories, as *complementary but exclusive* features of the description, symbolizing the idealization of observation and definition respectively".[22] Shifting from classical to quantum theory was to shift from the conjunction of two modes of description to their exclusive disjunction. From AND to OR. Each mode was an idealization or abstraction: space-time coordination idealized observation while causality (fixation of the total energy of a closed system) idealized definition. But whereas in classical physics, they provided together a complete description (in Bohr's terms, a 'unique image') of a system (e.g. described by generalized coordinates), in quantum theory, these two modes became incompatible or rather reciprocal: both were necessary for a complete description ('complementary'), but 'exclusive'. Heisenberg's indetermination principle was the formal transcription of a similar line of reasoning. But it is important to understand that Bohr was also suggesting here a new way to understand the intellectual process by which we go backwards from our experience to some knowledge of the objects under investigation. In our experience, we make use of space-time description and causality; classical theories had defined a non-ambiguous use of such concepts for the field of our direct everyday experience; quantum processes were known to us through their irreversible amplification effects on the measurement apparatus, and hence required that such concepts be maintained; however, they could not any more be understood as referring to properties or 'absolute attributes' of the objects, and knowledge had to move backwards in a statistical fashion. Complementarity described the conditions which ruled a non-ambiguous use of ordinary concepts in such a situation, which Bohr thought had no precedent in the history of physics.

The genesis of the Complementarity idea, when seen as rooted in Bohr's realization that classical physics extrapolated ordinary language to intrinsic properties (words into ontology), explains why he was led to qualify 'quantum paradoxes' as language paradoxes. As early as 1927–28, Bohr identified the main paradox 'from the point of view of the theory of knowledge' as linked to the implicit influence of our linguistic practice, to the pregnancy of what we would call today 'entrenched predicates' – a move which has been long neglected in Bohrian exegesis but which touches upon a recurrent worry in the recent literature. If space-time description, spatio-temporal separability, common-cause models and talk of objects and properties fail to provide an adequate understanding of quantum processes, then one of the many tasks of philosophical interpretation remains to construe a coherent criticism of our persistent naturalistic ontologies. This worry is often expressed today by saying that quantum theory challenges our 'most basic metaphysical assumptions' or that it is the endgame for understanding if understanding has to require pictorial physical mechanisms and if it turns out to be impossible to 'purge mankind of this need for a causal (. . .) explanation of physical phenomena'.[23] Though this need for laws and necessity may well be nothing but historical conditioning,[24] the worry is challenging. How was it expressed in Bohr's interpretation of the epistemological paradoxes raised by the new quantum mechanics? In 1928, Bohr noted at the beginning of his complementarity paper that, because our interpretation of the experimental material rests essentially upon the classical concepts, there were "difficulties involved in the *formulation* of the quantum theory".[25] These difficulties were disclosed at the end of the paper: they implied our "*adapting our forms of intuition*", a task especially difficult since "every word in our language is bound to our forms of intuition".[26] This, Bohr added, was in analogy with the "general difficulty in the formation of human ideas, inherent in the distinction between subject and object".[27] In his next paper, Bohr wrote that "in consequence of this state of affairs, even words like to be and to know lose their unambiguous meaning".[28] Such assertions gave testimony of how important to Bohr had been the realization that, as Schrödinger put it, "the words and concepts used until now no longer suffice",[29] but they also provided the guiding line of Bohr's thinking afterwards – for instance when it came to answer the EPR argument: Bohr argued that the very fact of talking of "disturbing the phenomena by observation" (as in EPR criterion of reality) amounted to maintaining a tacit presupposition of ordinary language which needed to be criticized (hence the redefinition of the word 'phenomenon').[30] A fundamental feature of Bohr's philosophy was thus to attract attention to how much of our mode of thinking is linked to our mode of expression: if the latter failed to account for a new field of experience, this could not but affect the former. However, granted that quantum theory shook "the very foundations underlying the *building up* of concepts, on which not only the classical description of physics rests, but also all our ordinary mode of thinking",[31] how did Bohr conceive of the task of adapting our forms of intuition? It is difficult even now to give a concise assessment of his solution to the language paradox. But I shall suggest that this solution relied heavily on the assumption that ordinary language developed along two different lines, which Bohr called *analysis* and *synthesis*. In analysis, one is engaged in the process of localizing things in space and time and describing their causal evolution, in the frame of a sharp distinction between subject and object; this process originates in our daily experience and developed historically into a non-ambiguous language through classical physics. In synthesis, on the contrary, a different kind of knowledge is obtained (in art, conversation, emotions, states of mind) *via* a juxtaposition of words which would be incompatible in analysis; this kind of knowledge allows a description of objects which are not immediately representable in our intuition (space and time). As a 'rational synthesis' of incompatible predicates, complementarity reinstated physics as 'what we can say about nature', while it liberated philosophy from the impossible task of providing an ontological description of the objects 'behind' the operators.

We can now turn back to Bohr's 1928 remarks about philosophy and science; there he wrote that cross-fertilization occurred each time physics was faced with a revision of the concepts underlying our understanding of observation. At this point there is a possible exegesis of his remarks. By 1928, it was indeed clear that the break with classical theories had involved criticism of many tacit presuppositions of the formal theoretical language which dealt with the visualization of the basic entities, the representation of characteristic physical quantities, the nature of determinism as a deductive connection between the initial state of a system and its state at any time past or future, the negligible effect of measurement or the replicability of the 'same' phenomena. Such criticism had required thorough epistemological analysis, or, in Bohr's terms, 'logical criticism of concepts'. From formal analogies to language paradoxes and rational synthesis, Bohr's response to this requirement had adapted to each advance that was made in the development of quantum theory, and in many cases it had made such advances possible. If we understand philosophy as a general 'anti-natural attitude',[32] then philosophy did "give support and points of departure" for the construction of quantum theory.

But obviously this would be only a *weak* version of a satisfactory interpretation of Bohr's remarks. The weakness here does not comes from viewing philosophy as an attitude rather than as a position,[33] but from viewing philosophy only as an attitude, while it is also experienced in the history of philosophy as a language: such a weak version is definitely not enough to account for Bohr's enigmatic reference to words of ordinary language and to hidden concepts. Is it then possible to construe a strong version of his remarks?

Philosophy as giving support and points of departure through ordinary language.

The question is: how can we make sense of Bohr's assertion that the endeavors of philosophers to clarify the foundation and limits of human intellectual activity gave inestimable support and points of departure to quantum theory through the formulation of terms which became part of ordinary language? This question can be split into three: which words? which support? which philosophers? I shall not attempt here to give anything like a systematic answer to these questions. However, I shall try to uncover some unexpected connections with philosophy that appear in the very story which was briefly sketched above. This requires a change in the point of view, looking at the same story from another standpoint. My conjecture will be that Bohr's response to the crisis of classical theories did not obey only one internal logic, but *two*. Its first logic was to take experiments seriously, to make as few hypotheses as possible, and not to constrain the mathematics by former epistemological requirements, while its second logic was rooted in a consistent use and reinterpretation of the *philosophical language* which he inherited, which led him to empty this language of its former meaning.

Can such a philosophical language be easily identified? To this preliminary question, the answer is clearly: No. One reason for this is that Bohr said very little. He wrote no laudatory autobiography. He resisted all attempts to label his ideas with a systematic – *ism*, and especially disliked such endeavors, as L. Rosenfeld, W. Heisenberg and O. Klein noted many times. He thought that the epistemological situation was a new one, and therefore could not be expressed properly in a former conceptual scheme. Probably, like many of us, he managed as well as he could with one basic problem and several different forms of expression ('forms of life'), gathered at random from his education, readings and friends. Another reason why it is not easy to identify Bohr's philosophical heritage is that, while Bohr said very little, subsequent commentators said very much. There have been many attempts to reconstruct Bohr's ideas as a general position, which ended up many times by reshaping Bohr as a positivist, or a realist, or a pragmatist, as philosophy of science itself moved on from positivism to realism or pragmatism. A similar predicament affected the interpretation of Einstein's work; hence D. Howard's and A. Fine's recent pleas for epistemological neutrality.[34] No really neutral interpretation of any author can ever be achieved; however, there must be an intermediary human standpoint between God-like absolute historical truth and Mephistophelian distortion of an existing thought.

To look for such an intermediary standpoint it is necessary, I believe, to look first into Bohr's *vocabulary* (especially if one wants to understand his 1928 remarks), that is, to look into how Bohr formulated the difficulties which he met in his work: why did he use such words in connection with such problems? Much clarification has already been achieved recently by doing this. N. Wise has studied Bohr's early use in his physics of words such as 'irrationality' or 'irrational whole or totality', 'individuality', 'causality' with respect to continuous conservation of energy, 'symbolic truth', 'harmony', 'continuity-discontinuity', or a general anthropomorphic terminology, and he has shown how closely Bohr followed Høffding's philosophy in such cases.[35] J. Faye[36] has much reinforced the suggestion that Høffding's search for a theory of knowledge and a psychology different from both Kant's and Mill's views had an unmistakable influence on Bohr. Like Høffding, Bohr made a distinction between 'principle' and 'ideal' of causality and incorporated analogy in scientific work; like Høffding, he thought that the *a priori* forms of intuition were not *a priori* but had empirical genesis, and he criticized the correspondence or adequation conception of truth, treating 'reality' as a predicate which expressed a maximization of legal connections and an achievement of the greatest possible harmony between multiple points of view; like Høffding again, he renounced any conception of the subject-spectator to the profit of a shifting line (depending on the context) between subject and object, etc . . . Undoubtedly, Bohr knew modern philosophy, Kant, W. James or S. Kierkegaard, at least partly through Høffding, and he also shared in the main Høffding's conception of epistemology.

However, when confronted with Bohr's vocabulary, one should maybe take one step more, and identify the philosophical language in which quantum theory was born and formulated as the general *erkenntnistheoretisch* language which pervaded philosophy and physics during the XIXth Century, especially in the German tradition. This language disappeared: one feature of post-war translations of Bohr's ideas in the literature was that words and concepts tended to be disconnected from their original widely accepted meaning. To make this plausible, one can give a few simple examples of 'words of ordinary language' that are used to hide specific philosophical concepts in Bohr's (or Pauli's, Heisenberg's, Born's and Schrödinger's) papers. 'Intuition' (*Anschauung*), for instance, always referred to Kant's concept of the mode through which thought immediately refers to its object,[37] while *Anschauungsformen* referred to space and time as the general forms of sensibility through which we grasp all phenomena. 'Presentation' (*Darstellung*) referred to the operation through which a concept is presented in pure intuition (as in mathematics) or in empirical intuition (as in physics), and therefore was different from 'representation' (*Vorstellung*), which was the generic term for all the different kinds of knowledge.[38] 'Analogy' (*Analogie*) referred to either of two Kantian concepts. It could refer first to the 'analogies of experience' which served to describe Kant's epistemology of physics in the 'Analytic of Principles' of the *Critique of Pure Reason*. But when 'analogy' and 'symbol' were used together, they referred to the very different concept of analogy which Kant qualified in the *Critique of Judgment* as a second kind of act that made a concept sensible (see further). 'Phenomenon-object' (*Erscheinung-Phänomenon/ Gegenstand-Object*) referred, sometimes very precisely, to the different stages of the process of objectivation as described by Kant. Behind the word *Erklärung* and its contrast (many times reassessed by Bohr) with *Beschreibung*, there was a discussion which had been crucial to the development of the *Geisteswissenschaften* during the second half of the XIXth Century. Words like 'non ambiguïty' (*Eindeutigkeit*), 'coordination' (*Zuordnung*), 'causality' (*Kausalität-Geseztlichkeit-Ursache und Wirkung*), 'symbol' (*Symbol*), 'image' (*Bild*), 'acceptability' (*Zulässigkeit*), 'exactness' (*Richtigkeit*), 'appropriateness' (*Zweckmäßigkeit*), were all widely discussed in the epistemological works of H. von Helmholtz, H. Hertz, E. Mach, J. Petzoldt, and many others.[39] The physicists who created quantum mechanics thus inherited from the widespread influence of Kant's work[40] a general language the conceptual sharpness of which had been already tested and modified by their predecessors. In this sense ordinary language did provide philosophical concepts, and we have an explanation of Bohr's strange emphasis on innocent-looking words.

But this only answers, provisionally, the first question: which words? The second question, the most difficult to deal with, is: *which support*? How did these words help in any way to understand the breakdown of classical theories? This is where I would like to suggest that, as I said above, the evolution of Bohr's epistemological response had a logic of its own which developed consistently through the philosophical terms which he used. What was the inner consistency of Bohr's thinking when he moved from formal analogies, symbolic analogies and rejection of space-time intuitions, to the idea of a reciprocal applicability of space-time description and causality, and finally to the analysis and synthesis solution of the language paradox? If we change our standpoint and look back at the genesis of quantum theory as if it belonged to the history of philosophy, we see that the destructuration of the theoretical language of classical theories had as its counterpart a destructuration of the philosophical language of Kant's theory of knowledge. This internal breakdown took support from a tradition that is little visible now but can be identified, and it left almost nothing intact except Kant's fundamental stance that we are finite beings and hence cannot ever compare our representations to essences.

The first thing to consider here is the initial state of the language which finally lost its effectiveness in the 1920's. In doing so, I shall only give a rough sketch of Kant's epistemology, in which my main purpose will be to emphasize that it had two sides, or two faces.

The *first side* was the theory of scientific knowledge which Kant provided in his *Critique of Pure Reason*. Kant wanted to get rid of dogmatic metaphysics[41] and to provide a proper foundation for natural sciences,[42] he wanted to kill the two spectres of dogmatism and scepticism and to put an end to a situation where philosophers committed sceptical suicide while physicists despised philosophy (though practicing it unconsciously). Therefore he sought to eliminate 'pompous ontology' and replace it by a 'simple analytics of the Understanding'.[43] The way to do this, he thought, was to draw a map of the sources and mechanisms of knowledge: a transcendental topology.[44] Hence Kant's abstract divisions and compartimentalizations. Sensibility, the ability to receive representations, and Understanding, the ability to think objects, were the only two sources for (scientific) knowledge: the first provided the intuitions, the second provided the concepts.[45] Reason in itself, the power of principles, was not a source of objective knowledge, since it was not restricted by the field of possible experience; however, it provided architectonic unity. Objective knowledge, which required that intuitions could be subsumed under pure concepts, was made possible through the crucial process of Schematism: in Schematism, a phenomenon, redefined by Kant as the 'indeterminate object of an empirical intuition',[46] could become an object – this was the task of transcendental imagination and of determining judgment. Physical knowledge was basically objectivation of the phenomena, which meant that to the empirical intuition provided by presentation in pure intuition (space and time) plus a sensation, the Understanding applied its *a priori* concepts, the *Denkformen* of quantity, quality, relation (substance-accident, causality-dependence, community), and modality (possibility, existence, necessity). Thus Nature could be defined as the connection of phenomena through necessary rules, e.g., laws.[47] Finally, one overall presupposition of this description was Continuity. Continuity was a fundamental feature of space and time, hence of all phenomena,[48] and there also existed a *lex continui in Natura*, which conditioned the unity of natural sciences.[49] There could be absolutely no room in physics for discontinuity, an assumption which provided the ground for Kant's rejection of atomism and his plea in favor of a continuous 'dynamist' physics based on the concept of ether.

There was, however, *another side* of Kant's conception of knowledge which he developed in the *Critique of Judgement*. Here again, I shall only mention what is relevant to the perspective of this paper. In the third *Critique*, Kant considered non-theoretical or non-objective knowledge such as was provided by the use of the concept of 'formal finality'. While 'determining judgement' acted in theoretical knowledge by applying universal laws to particulars, 'reflective judgement' started with the particulars to seek the universal: for instance, in aesthetics and teleology (living organisms).[50] There, Kant introduced subjectivity, a form of holism and the notion of finality as the 'lawfulness of the contingent'.[51] He also gave central importance to a counterpart of Schematism, which he called Analogy or Symbolism. In § 59, Kant wrote that human intellect made use of *two* kinds of presentations of a concept in the intuition, that is, made a concept sensible in two different ways. The first was the one described above for objective knowledge; the second prevailed in cases where one dealt with concepts which only Reason could think (such as the concept of God, or any abstract concept like that of a despotic state). Here, the judgment provided an indirect presentation, through which it first applied the concept to sensible intuition and then applied the *mere rule* by which it had reflected on the intuition to another entirely different object, of which the former thus became the symbol. While 'schematic' presentation provided a way to apply the Categories to pure or empirical intuitions (and therefore held for scientific knowledge), 'symbolic analogies' provided an indirect presentation of the concept in the intuition (and held in all cases, such as theology, art or language itself, where no direct intuition was available).[52] Analogy and symbolism were a way to *transfer a relationship* without keeping the poles of the relationship. As defined already in the *Prolegomena*, Analogy was "not an imperfect similarity of two things, but a perfect similarity of relations between two quite dissimilar things".[53] It had the full dignity of a process of knowledge, but not of scientific knowledge, since it held in cases where *no direct intuition* of the abstract concept was available. It especially applied to language, as Kant noted briefly. Finally, it helped to establish a sharp distinction between, on the one hand, objective knowledge and on the other hand, art, religion, language, teleology.

One reason for Kant's astonishing influence over the development of philosophy, mathematics, physics, physiology, philology or aesthetics during the XIXth Century and , as Max Born once put it, 'up to our time', was that he created a language. What happened to this language in the genesis of quantum theory? My conjecture, as stated above, is that Bohr made a consistent use of it which however ended up in an internal break-down of Kant's description of objectivation. How was this possible?

If we recall the story that was sketched in the first part of this paper, we can now read it anew from a philosophical point of view. During the period of the correspondence principle, Bohr's problem was to build quantum theory by attempting to transfer formal relationships: this, he called *formal analogies*. In 1924–25, Bohr finally acknowledged the failure of 'all spatio-temporal pictures' and renounced *Anschaulichkeit*: this he called 'having recourse to *symbolic analogies*'. More generally, he always used the word 'symbolic' to refer to a situation where the object was not directly given in the intuition (hence, the idea that wave and matrix mechanics were both symbolic representations).[54] Thus, he merely applied the lexicon of the *Critique of Judgment*, although *in a heretic way* since he applied it to physical science. Then, in 1926 Bohr realized that continuity, which was a necessary condition for our representation of phenomena, was not a necessary condition for the formalism which would describe the behaviour of quantum processes; the language had to split into two parts. On one side, it would be necessary to keep working with classical concepts; on the other side, what we would say about nature did not have to be restricted to what made our own field of experience representable. Thus Bohr renounced the overall presupposition of Kant's theory of objective knowledge, asserting that to maintain it was to make physics 'ambiguous'. In 1927–28, he suggested a reciprocal applicability of space-time description and causality; hence, he denied that intuitions and concepts could be made homogeneous, which disproved schematism. Finally, Bohr emphasized that language on its analysis side created paradoxes if extended to atomic physics; all classical physics and Kant's theory of knowledge were ruled by classical logic, but to Bohr, paradoxes disappeared once one realized that objectivation provided by language was a more complex process which involved synthesis. This was a very special way of understanding what J. von Neumann and G. Birkhoff were to describe later as a conflict between classical logic and the quantum formalism;[55] but it amounted to a blow of similar impact on what had been a centuries-old pre-sup-position. What can we conclude? In the sequence of the various epistemological stages of the formation of quantum theory, there was a definite philosophical consistency, but this consistency was not one any Kantian would expect. Bohr took support and points of departure in concepts of the philosophical language that he inherited, but he applied them contrary to their *mode d'emploi*. It was like twisting the language to use it against itself. Bohr interpreted the problem-situations which he met, or created, in physics in the terms that he thought were adapted, however strange these situations could seem: words – hidden philosophical concepts – worked as models in ordinary language for these situations. Hence it is not surprising that Bohr referred to them, and never to a philosophical framework; there was no existing systematic philosophy in which objectivation of phenomena could make sense under such conditions. Bohr did not adjust Kant's description of objectivation to a new situation, rather he emptied it from the inside. His way of thinking was like building an entirely new house with the same building blocks. This went on after 1928; though Bohr never formulated in an articulated philosophy the wholly different (probabilistic) epistemology for which he had made room, he tried to stabilize a new epistemological lexicon adapted to the situation, notably through his redefinition of 'classical philosophical' concepts such as phenomenon, objectivity, or nature.[56] But the most striking feature of this story is certainly the slow dislocation of Kant's original views on knowledge that occurred during the genesis of quantum theory and which, at least in Bohr's opinion, contributed to make it possible.

How much did Bohr borrow from the 'endeavors of philosophers to understand the limits of human intellectual activity' in the course of his own thinking? That would be the third question: *which philosophers*? I shall not answer this question in any detail here, but only point to some conjectures. From the internal logic of Bohr's line of thought as I have just suggested to reconstruct it, it follows that the eccentric interpolation of Kant's *Critique of Judgment* into the problem of the objectivation of phenomena was crucial for that logic. I have argued elsewhere that there existed a tradition in which such an interpolation was a more or less explicit move: this was the tradition started by Goethe's emphasis on 'symbolical anthropomorphism' in science – and Bohr's extensive knowledge of Goethe's works is well evidenced. Other obvious connections ought to be mentioned. Rejection of *Anschaulichkeit* and recourse to the *Symbol* concept was explicit in Helmholtz's papers on non-Euclidean geometries and philosophy, and influential afterwards, for instance, on Hertz and Husserl.[57] Bohr's conception of language exhibits striking similarities to the work of W. von Humboldt, who notably described language as a process of analysis and synthesis, formation of concepts as taking place *via* images, symbols or analogies, and language in general as being essential in objectivation.[58] Indeed, we might say that in order to undermine the very basis of the metaphysics (including Kantian metaphysics) associated with classical theories, Bohr took support on the line of thought which goes from Humboldt to Wittgenstein and attempts to criticize ontological prejudices.[59] That authors such as Goethe, Humboldt, Helmholtz and Hertz appeared in the 1920's as belonging to the same line of thought is clearly evidenced by Cassirer's books. Høffding's own philosophy owed much to this tradition, so that his original contribution to Bohr's feeling of philosophical concepts seems to have been essentially his criticism of the continuity presupposition. To my knowledge, the complementarity idea has no exact equivalent; but what it required, Bohr borrowed from these 'endeavors of philosophers'. Bohr's debt should thus be appreciated with respect to a whole line of thought which had inherent philosophical consistency. It had not developed into a new epistemology, but it enjoyed that kind of empirical reality with which philosophical language is endowed when it is alive, transmitted from one to another and thus constantly changing our perception of ordinary language.

At this point, what interpretation can we suggest for Bohr's 1928 remarks? In the midst of his major break-through in the interpretation of quantum theory, Bohr wrote that philosophy had given inestimable support and points of departure for the revision of fundamental concepts, and had acted through words. This, I believe, meant much more than just a tribute paid to epistemological thoughtfulness. I have tried to construe a strong version of these remarks, evidenced by the internal development of both physics and philosophy, which leads to three conclusions. *First*, quantum mechanics was built and interpreted by Bohr (and most of the members of the Copenhagen group) in a philosophical language that came from the tradition of *Erkenntnistheorie*, that is from a tradition that focused on how much of Kant's legacy could be maintained or had to be renounced in view of the developments that had occurred in sciences and in philosophy. *Second*, this philosophical language, which was common knowledge at this time, was used consistently by Bohr with respect to the problems that he met in physics. But this led him to contradict openly and indeed to dislocate some Kant's most fundamental stances about the conditions for objective knowledge, the aprioricity of space, time and causality, or the sharp distinction between science *versus* art, language and religion. These were starting points for further developments in Bohr's thinking, which I have not attempted to explicate here, such as his general elaboration of something close to a 'linguistic turn' mixed with a very unusual empiricist stand, or his contention that physics ought to be considered at the same level as all other human activities, or his efforts to formulate in terms of the history of 'modern philosophy' the meaning of the break with the 'conceptual frame'[60] of classical physics. Hence, *third*, Bohr's feeling that philosophy had been acting through words, that is, *not* as a ready-made doctrine or a general inspiration or a vague trend of ideas. If Bohr 'thought as philosophers do',[61] it was because he was acutely aware that philosophy can be experienced as a language in which we live – a precise language, though submitted to change. In ordinary life, some words can suddenly recall all sorts of reminiscences of happiness or misery, of understanding or opacity – as soon as we talk, we project on the world sedimented meanings. This *Illiers-phenomenon* has its counterpart in philosophy. In Bohr's use of philosophical concepts, there were precise reminiscences of Kant, Goethe, Helmholtz or Høffding which acted as sedimented ideas. Each concept or word was thus like a point of view that made the landscape look different; and Bohr's overall strategy is well-known to have been a systematic multiplication of points of views and languages.

This is, after all, an extremely original conception of the influence of philosophy in science. How original it is can be seen from a brief comparison with received stances on this matter. For logical positivism, there could be no question of an influence of philosophy in such a sense, since philosophy was seen as being either 'metaphysics' or the empiricist theory of meaning. The latter, being defined as the very method of science, was by *petitio principii* effective in science; but the former, even in Philip Frank's tolerant definition of metaphysics as a semantical interpretation of the cognitive content of science,[62] could not provide anything but some psychological help, some sort of a crutch. Being already too old, the words of philosophy were seen as impotent with respect to new advances. A. Koyré suggested an antagonistic view of the influence of philosophy, but it would be just as hard to find there an analysis of the part played by philosophical concepts embodied in ordinary language. To Koyré the influence of philosophy in science was basically the influence of a fundamental orientation, that of 'platonism' or 'mathematical realism', which was the substructure of science; all scientific revolutions were always 'determined' by philosophical conceptions, more specifically by the rediscovery of the essential value of mathematical realism, a schema which he extrapolated from Galileo to Einstein.[63] Philosophy's influence being perennial left no room in such a conception for the potential affectivity of words and of conflictual traditions. Bohr's remarks, on the contrary, imply that philosophy does not act as a fundamental orientation nor as a doctrine, but as a set of problems that have historical existence in language.[64] To make sense of this, we must look at philosophical concepts as *cluster-concepts*,[65] namely as the association of a well-defined primary meaning (here, Kant's lexicon) and of a cluster of questions that arise out of the discussion of this primary meaning. But then an interesting fact emerges. If philosophy acts in science *via* cluster-concepts, it can happen that its special power comes from the *critical* import of such concepts. That is, it can happen that words provide the means to criticize the primary context of the concepts to which they were originally linked;[66] or, to put it differently, 'old words' if seen as sets of open questions may have a critical import which turns out to be crucial in the understanding of new developments in science. Bohr's outdated remarks thus may contain the seeds for a reinterpretation of the part played by philosophical thinking in the construction of a physical theory.

My intention here was mainly to clarify Bohr's 1928 comments about the influence of philosophy on the formation of quantum theory. But as I said at the beginning, these remarks suggest two conjectures.

The first one concerns our present unsatisfactory understanding of the Copenhagen interpretation of quantum mechanics. The fact that Bohr's ideas have been 'either overlooked or distorted beyond recognition', caricatured into a short-sighted positivism, and are unheard of in most textbooks,[67] is by now well acknowledged. But why does the invocation of Copenhagen orthodoxy in the foundational literature, *pro* or *contra*, look so different from Bohr's original stance? Besides the difficulties introduced by melting into one pot Bohr's and Heisenberg's positions, which were occasionally nearly incompatible, one main reason which I have tried to suggest is that the initial philosophical subtlety and complexity of Bohr's (or others') approach was lost in the course of its translation into a language, that of post-war philosophy of science, that was very different from the one in which it was originally formulated. This is to say that the roots of unintelligibility are in the difference between two philosophical traditions.

However, such unintelligibility would be innocuous if we could assume that the philosophical context in which quantum mechanics was born was neutral with respect to its interpretation. If Bohr's comments are to be taken seriously (the more since his style of understatement and constant earnestness is legendary) and if they have evidence to support them – namely if physics and philosophy intermingled from the beginning – this is not the case. But then the way in which philosophy acted becomes important: what did philosophical consistency mean for quantum theory, in Bohr's conception of it? I have tried to show that Bohr took support from a number of 'cluster-concepts' (words of ordinary language that embodied sedimented meanings and open questions) in order to analyze some of the most basic assumptions of classical theories, and that this also led him to question the coherence of the former epistemology by dissolving the glue which held the concepts together. Put otherwise, to Bohr philosophical consistency required two things: it required first that we criticize the language, ordinary and philosophical, from within, so as to point to some vestigial ways in which we use it; and second it required us to construct a new kind of epistemology. Bohr pursued these goals by borrowing from a tradition that had already moved a long way in the direction of the criticism of both Kantian foundationalism and naïve empiricism, but which has almost disappeared from contemporary philosophy of science. What is the situation today? On the one hand, it is by no means clear that we have learned even now to be free, in the philosophy of science, from some tacit presuppositions of language that are entrenched in very old ways of thinking; on the other hand, general philosophy has made much progress in the understanding of how these presuppositions are linked to some main lines of thought in modern philosophy, while it has more or less renounced the task of the interpretation of science itself. The second conjecture that I would like to extrapolate from Bohr's 1928 remarks is thus a little more than the first: besides the fact that the full import of Bohr's views on quantum theory remain difficult to understand if one is not aware of its philosophical context, one could also assume that philosophy of science and general philosophy would much benefit from each other, including their most sophisticated queries.

NOTES

1 N. Bohr, 'Harald Høffdings 85 Aars-Dag', *Berlingske Tidende*, March 10th 1928, quoted by J. Faye, *Niels Bohr. His Heritage and Legacy*, Kluwer, 1991, 53 (emphasis mine). I want to thank J. Faye for sending me the original text, and C. Nielsen for helping me with the Danish.

2 That quantum theory, or rather the Copenhagen interpretation, was influenced by logical positivism has been asserted by many philosophers of science. There is no historical or epistemological support for such an assertion, however, as Bohr and Heisenberg made clear afterwards and as shown, more evidently, by internal history of quantum theory.

The opposite assertion that quantum theory had been influenced by a *Lebensphilosophie* embodied in O. Spengler's book, *The Decline of the West*, has been advocated by P. Forman in his extremely well-documented paper, 'Weimar culture, causality and quantum theory, 1919–21', *Historical Studies in the Physical Sciences* **3** (1971), 1–116. For criticism, see J. Hendry, 'Weimar culture and quantum causality', *History of Science* **18** (1980), 155–180.

[3] Up to 1928, since Bohr's remarks are dated 1928. They would also apply afterwards, as would be much more easy to show.

[4] My description will be much too elliptic. For further detail, see N. Bohr, *Physique atomique et connaissance humaine*, Paris, Gallimard, 1991, Introduction and Glossary (quoted hereafter as N. Bohr [1991]). I agree with most of M. Beller's analysis, especially about Bohr's interpretation of the Bothe–Geiger 1924–25 experiments, the prevalent importance of stationary states, quantum jumps and collision problems over the wave-particle dilemma or the indeterminism issue, and the influence of the 1926 debate with Schrödinger in the genesis of the complementarity argument. See M. Beller, 'Matrix theory before Schrödinger', *Isis* **74** (1983), 469–491 and 'The birth of Bohr's Complementarity. The context and the dialogues', *Studies in the History and Philosophy of Science*, Vol. 23, No. 1 (1992), 147–180. Epistemological history of QM, which started with M. Jammer's work, is crucial to contemporary philosophy of science through its illustration of problems such as reduction of theories, undetermination by evidence, or the fruitfulness of a very specific brand of anti-realism.

[5] Science itself designates certain areas in the picture of the world as observable. I borrow this expression not from H. G. Gadamer, but from B. van Fraassen, *The Scientific Image*, Clarendon Press, 1980, 57.

[6] N. Bohr, 'Atoms and human knowledge', in *Atomic Physics and Human Knowledge*, New York, Wiley, 1958 (quoted hereafter as *APHK*), 85.

[7] These four anomalies served as the premises of Bohr's Trilogy in 1913, which assembled Ritz's combination principle, the Plank–Einstein hypothesis about the discontinuity of radiation and the Rutherford's 1911 experiments.

[8] At this stage Bohr was essentially preoccupied with the problem of transitions, and with the problem of a quantum perturbation theory, more than with the problem of the nature of light.

[9] O. Darrigol has given a most thorough and illuminating description of this strategy in his book, *From c-Numbers to q-Numbers*, University of California Press, 1992.

[10] Bohr's hypothesis mixed a generalization of *statistical* asymptotic agreement between classical and quantum theory in the limit of high quantum numbers with a strong emphasis on the *contrast* between the two theories.

[11] N. Bohr, Addendum to 'On the behaviour of atoms in collisions', in *N. Bohr Collected Works*, North-Holland Pub. Co. (quoted hereafter as *BCW*), vol. 5, 204. Bohr used a similar phrasing as early as 1920.

[12] *Ibid.*, 205.

[13] See, e.g., his letter to M. Born, May 1st 1925, in *BCW* 5, 310.

[14] See, e.g., W. Heisenberg, 'The development of the interpretation of quantum theory' in *N. Bohr and the Development of Physics*, W. Pauli *et al.* eds., Pergamon Press, 1955, 14. On the various levels of the 1926 discussion, see e.g. C. Chevalley, 'Continu et discontinu dans la construction de la théorie quantique. Un exemple', in

H. Sinaceur and J. M. Salanskis, eds., *Le Continu Mathématique*, Berlin, Springer, 1992.

15 However, Heisenberg's position is sometimes very close to Bohr's; see, e.g., his 'Quantenmenchanik', *Die Naturwissenschaften* **14** (1926), 989–994, which attempts to disarticulate Schrödinger's views together with the classical concept of 'reality'.

16 Provided wave-packets, which dispersed, were not seen as representing individual processes, and provided also the idea of separate stationary states could be maintained.

17 N. Bohr to E. Schrödinger, December 2nd, 1926, in *BCW* **6**, 460 and 46. See O. Darrigol's comments on Bohr's analogy with Riemann surfaces in his [1992], 134 sq.

18 Letter from N. Bohr to Einstein of April 13th, 1927, in *BCW* **6**, 418 and 21.

19 N. Bohr, 'The quantum postulate and the recent development of atomic theory', *Nature* (April 14th 1928), 580.

20 N. Bohr, Ms, in *BCW* **6**, 80. Complementarity does not 'translate' Heisenberg's principle since it was achieved independently by Bohr, in close connection with the debate with Schrödinger; see C. Chevalley, in N. Bohr [1991], 64–73.

21 N. Bohr, Ms, 'Philosophical foundations of quantum theory', in *BCW* **6**, 69 (emphasis mine). The reconstruction of Bohr's reasoning requires extensive analysis of all manuscripts and published versions of the complementarity paper. I only want to stress here in which way the complementarity idea was originally formulated.

22 N. Bohr, *loc., cit.* note 19, 580 (emphasis mine). In the Ms quoted above, Bohr wrote that the complementary nature of the space-time description and the claims of causality was the *symbolical* expression of "a general reciprocal relation between the maximum sharpness of definition of the space-time and energy-momentum vectors associated with individuals" (*BCW* **6**, 69).

23 R. I. G. Hughes, *The Structure and Interpretation of Quantum Mechanics*, Harvard University Press, 1989, xii (also 1–8), and J. Cushing, 'Quantum theory and explanatory discourse', *Philosophy of Science*, vol. 58–3 (1991), 337–358, here 353. See also P. Teller, 'Relational holism and quantum mechanics', *British Journal for the Philosophy of Science* **37** (1986), 71–81.

24 See B. van Fraassen, *Laws and Symmetry*, Oxford U. P., 1989, Introduction.

25 N. Bohr, *loc. cit.* note 19, 580 (emphasis mine).

26 *Ibid.*, 590. The German says *Anschauungsformen.*

27 *Ibid.*, 590.

28 N. Bohr, *Atomic Theory and the Description of Nature*, Cambridge University Press, 1934, 19.

29 Letter from E. Schrödinger to N. Bohr, November 23rd 1926, in *BCW* **6**, 459.

30 N. Bohr, [1939]. On this see E. Scheibe, *The Logical Analysis of Quantum Mechanics*, Pergamon Press, 1973; C. Chevalley, in N. Bohr [1991], 85–88; 517–520.

31 N. Bohr, *ATDN, loc. cit.* note 28, 101.

32 Such as defined by E. Husserl, e.g., in *Logical Researches*, II, Chap. 1, § 2–3 and 7: philosophy makes the very act of thinking an object, and is suspicious about all presuppositions coming from natural sciences or the history of metaphysics.

33 For an elaboration of the idea that philosophy is more an attitude than a position, see B. van Fraassen, 'Against Transcendental Empiricism', in T. J. Stapledon, ed., *The Question of Hermeneutics*, Dordrecht, Kluwer, sched. 1994.

34 See, e.g., D. Howard, 'Was Einstein really a Realist? in A. Pagnini ed., *Realismo,*

empirismo, construttivismo, Florence, 1994 (forthcoming); A. Fine, *The Shaky Game*, Chicago University Press, 1986. Interpretation has been made dependent on decisions about who and what is philosophically 'right'. In some respect, criticism of this way of thinking is similar to D. Bloor's and H. Collins's criticism of normative history of science.

35 See N. Wise, 'How do sums count? On the cultural origins of statistical causality', in L. Krüger *et al.*, eds., *The Probabilistic Revolution*, The MIT Press, Cambridge Ma., 1987, I, 395–425. I have adopted a similar principle in the Glossary of N. Bohr [1991], 350–567.

36 J. Faye, *Niels Bohr. His Heritage and Legacy*, Kluwer, 1991.

37 E. Kant, *Kritik der reinen Vernunft*, A 20.

38 On the Kantian division of representation in general, see *KrV*, A 319-B 376. For comments on Helmholtz's discussion of it, see C. Chevalley, *Glossary* to N. Bohr [1991], 428 sq.

39 On the history of *Eindeutigkeit*, see D. Howard, 'Einstein and *Eindeutigkeit.* A neglected theme in the background of general relativity', in J. Eisenstaedt and A. Kox, eds., *Einstein and the History of General Relativity*, Boston, Birkhauser, 1991. On Helmholtz's and Hertz's epistemologies as influential in Bohr's writings, see C. Chevalley, in N. Bohr [1991], *Glossary*. On the meaning of the concepts of *Zuordnung* or *Symbol* around the turn of the XXth Century, see T. Ryckman, 'Conditio sine qua non?', *Synthese* 88 (1991), 57–95; C. Chevalley, 'Physique quantique et Philosophie', *Le Débat*, Paris, Gallimard (November 1992).

40 To my knowledge, there is no systematic study of Kant's influence among physicists during the XIXth Century.

41 Dogmatism is defined by Kant as the assumption that one can make progress with the sole help of a 'pure knowledge by concepts'; see *KrV* B 35.

42 This is the 'received' view on Kant, based on the *Prolegomena* and the second edition of the *Critique*. It is relevant here; however, Heidegger elaborated a totally different interpretation of Kant's project, which ought to be discussed in any complete account of the interpretation of Kant.

43 E. Kant, *KrV* A 247, B 303.

44 E. Kant, *KrV* A 268, B 324.

45 E. Kant, *KrV* A 50, B 74.

46 E. Kant, *KrV* A 20, B 34.

47 E. Kant, *KrV* B 263.

48 E. Kant, *KrV* A 170, B 212.

49 E. Kant, *KrV* A 658, B 686.

50 Kant had some trouble reordering his map: see *Kritik der Urteilskraft (KU)*, first and second introduction.

51 E. Kant, *KU*, first introduction, VI.

52 See E. Kant, *KU* § 59.

53 E. Kant, *Prolegomena*, § 58.

54 For further detail on Bohr's use of the concept of 'symbol', see C. Chevalley, 'N. Bohr's Words', in J. Faye and H. Folse, eds., *Niels Bohr and Contemporary Philosophy*, Dordrecht, Kluwer, 1993.

55 See the discussion with J. von Neumann in *Réunion sur les Nouvelles Théories de*

la Physique. Varsovie 1938, Paris, Institut de Coopération Intellectuelle, 1939. From the point of view of the history of philosophy, the question can be formulated as: Why should logic be transcendental? For a remarkable elaboration of this problem, see C. Imbert, *Phénoménologie et langages formulaires*, Paris, Presses Universitaires de France, 1992.

[56] Each case would need a special inquiry, with special attention to the extreme thoughtfulness of Bohr's style of writing and to the discontinuous processes by which his words are given new meanings, and set in phrases which rarely change afterwards. As Heisenberg noted many times, Bohr struggled to 'create a new language'. How conscious Bohr was of the necessity to change the philosophical language can easily be appreciated in his almost obsessional emphasis on 'questions of terminology' and reassessment of the idea that "every analysis of the conditions of human knowledge must rest on considerations of the character and scope or our means of communication "(*APHK*, 88).

[57] When they redefined *Anschaulichkeit* in their 1927 Solvay paper, W. Heisenberg and M. Born merely applied to physics Helmholtz's former suggestion for mathematics.

[58] See C. Chevalley, in N. Bohr [1991], Glossary, 480–502.

[59] For a recent assessment of this line of thought, see J. Habermas, *Nachmetaphysisches Denken. Philosophische Aufsätze*, Suhrkamp Verlag, Frankfurt, 1988.

[60] The notions of conceptual frame and *Weltanschauung* were widely discussed in German philosophy in the 1920–30's. A conceptual frame, in Bohr's lexicon, is an 'unambiguous logical representation of relations between experiences' (*APHK*, 85). It is a language in a broad sense, including mathematical relationships, physical interpretation, and epistemology; in a conceptual frame, there is no sharp distinction between 'formal logic', 'semantics' and 'philological syntax' (*Ibid.*, 68). As a language, a conceptual frame is created by men, and therefore free from the illusion of absolute objectivity: 'we are suspended in language'. On the other hand, it is adapted to a given field of experience, which entails that it is also submitted to change: 'the main point to realize is that all knowledge presents itself *within a conceptual framework* adapted to account for *previous experience* and that any such frame may prove too narrow to comprehend new experiences' (*Ibid.*, 67). The fact that all knowledge be given within a conceptual frame does not, I believe, give testimony of Bohr's 'transcendental' thinking, as J. Honner has suggested. Conceptual frames have empirical genesis, and Bohr should rather be seen as someone who tried to elaborate a new 'post-Kantian' brand of empiricism, wholly different however from logical empiricism. In any case, Bohr thought that conceptual change, such as was required by quantum mechanics, asked for the creation of a new harmony between formalism, physics and philosophy, and that conversely such a new harmony could only be born out of a criticism of the totality of the former framework. What he called the 'mechanical conception of Nature' included words of ordinary language that provided visualization of the basic entities – particles or waves – a mathematical formalism (the Hamilton–Jacobi representation for the equations of mechanics), an epistemological 'ideal for scientific explanation', philosophical presuppositions such as the exclusion of the 'subject' from the description, philosophical implications, like the schism between science, religion, values and art, and a complex history, starting in the Renaissance. Therefore, each of these parts of the language of the conceptual frame had to be subjected to close analysis, if one was to gain full understanding of what was at stake in quantum theory – namely, the 'departure from customary principles of natural philosophy' *ibid.*, 32) and 'our entire mode of thinking'. It would probably be illuminating to engage in a systematic comparison between Bohr's notion of conceptual frames and

various competing notions such as *Weltanschauungen*, paradigms/disciplinary matrices, networks or spider's webs.

[61] As Heidegger once said. See *Die Frage nach dem Ding*, Tübingen, M. Niemeyer, 1962, 51.

[62] See P. Frank, 'The place of logic and metaphysics in the advancement of modern science', *Philosophy of Science* **15** (1948), 279.

[63] I refer to A. Koyré, 'De l'influence des conceptions philosophiques sur l'évolution des théories scientifiques', *Etudes d' histoire de la pensée philosophique*, Paris, Gallimard, 1971, 253–269.

[64] To my knowledge, nothing similar to this can be found in the works of N. R. Hanson, S. Toulmin or T. S. Kuhn, despite their interest in the language-shifts which occur in scientific change.

[65] One could transfer to philosophical language H. Putnam's attempt to interpret major concepts in physics (like kinetic energy) as 'cluster-concepts'. See H. Putnam, 'The Analytic and the Synthetic' (1962), in *Philosophical Papers* II, 33-70. See also G. G. Granger's remarks on what are philosophical concepts in *Pour la connaissance philosophique*, Paris, O. Jacob, 1988, Chap. 6.

[66] History of science would easily illustrate this phenomenon. For instance, obvious examples of cluster-concepts in Kepler's work would be the concept of *Species*, by which he does not refer to Plato's *eidos*, but to XIIIth Century Oxford optics (R. Grosseteste and R. Bacon), or the concept of *Emanatio*, which he takes over from the Neo-Platonic tradition in order to criticize the Aristotelian conception of light. See D. Lindberg, 'The Genesis of Kepler's Theory of Light: Light Metaphysics from Plotinus to Kepler', *Osiris* **2** (1986), 5–42; C. Chevalley, *Les Fondements de l'Optique Moderne*, Paris, Vrin, 1980.

[67] See C. A. Hooker, 'The nature of quantum-mechanical reality', in R. G. Colodny, *Paradigms and Paradoxes*, Pittsburgh, 1972; J. Honner, *Niels Bohr and the philosophy of quantum physics*, Oxford University Press, 1987; D. Murdoch, *Niels Bohr's philosophy of physics*, Cambridge University Press, 1987; H. Folse, *The philosophy of Niels Bohr*, North-Holland Pub. Co., 1985; J. Faye, *loc. cit.*, note 36; or E. Scheibe, *The Logical Analysis of Quantum Mechanics*, Pergamon Press, 1973. N. R. Hanson and P. K. Feyreabend made similar statements.

MARIA L. DALLA CHIARA AND G. TORALDO DI FRANCIA

IDENTITY QUESTIONS FROM QUANTUM THEORY

To what extent does quantum mechanics give rise to violations of Leibniz' theory of identity? This is a highly controversial problem. With different arguments, the question has been positively answered by Margenau, van Fraassen, Mittelstaedt, da Costa, Krause and other scholars. A negative answer, instead, has been defended by Aerts, Piron, Costantini, Garibaldi.

In our opinion, a formal analysis of the identity problem in quantum theory can be adequately developed in the general framework of an *intensional semantics*, that admits of applications in different fields, from the theory of propositional attitudes to linguistics. We will sketch here only the main intuitive ideas of this approach, which has been presented in [9], [10], [11].

1. AN INTENSIONAL SEMANTICS

Intensions are dealt with as autonomous abstract entities, whose behaviour is supposed to be governed by an *intensional set theory*: a convenient generalization of a standard set theory, like Zermelo–Fraenkel or von Neumann–Bernays-Gödel.[1] Accordingly, there will be no attempt to reduce intensions to more or less complicated systems of extensional objects, as happens in the usual possible world approaches. At the same time, *extensions* are described as limit-cases of intensions.

An intensional object may have an extension in a given situation; and is called *extensional* when it coincides with its extension. Following the tradition, we assume that the extension of an individual concept is either an individual or a class; the extension of a concept is a class; the extension of a proposition is a truth-value.

The basic semantic concept is the notion of *intensional structure*:

DEFINITION 1. An *intensional structure* is a system

$$\mathbf{M} = \langle I, \vec{R}, \vec{o}, D, E, ref, count, Prop, val, Conc\rangle$$

where:

K. Gavroglu et al. (eds.), Physics, Philosophy and the Scientific Community, 39–46.

(1) I is a non empty set of contexts, that on different occasions may play the role of *pieces of information, possible worlds, situations, states of physical systems*, and so on.

(2) $\vec{R}$ and $\vec{o}$ are respectively a possibly empty sequence of context-relations and of context-operations.

(3) D is the domain-function, that assigns to any context i the non empty set D_i of the individual concepts at i.

(4) E is the extensional domain-function, that assigns to any context i the possibly empty set E_i of the extensional individuals at i. For any context $i : D_i \supseteq E_i$.

(5) *ref*, the reference-function, assigns to any context i a partial function ref_i. If defined, the reference at i of an individual concept d is an extensional individual of i: $ref_i(d) \in E_i$. The reference of an extensional individual is the individual itself (for any d of E_i, $ref_i(d) = d$).

(6) *count*, the counterpart-function, assigns to any pair of contexts $\langle i, j \rangle$ a partial function $count^{i,j}$. If defined for an individual concept d of i, $count^{i,j}(d)$ represents the counterpart of d at the context j. From an intuitive point of view, the counterpart of an individual concept in a different context represents a kind of "alter ego" in an alternative situation.

(7) *Prop*, the proposition-function, assigns to any context i the set of the propositions $Prop_i$ at i. Any $Prop_i$ is supposed to contain at least the truth-values *truth* and *falsity* ($\{\mathbf{1}, \mathbf{0}\} \subseteq Prop_i$). further, any $Prop_i$ is closed under a sequence of propositional operators (negation, conjunction, etc.).

(8) *val*, the valuation-function, assigns to any context i a partial function val_i. If defined, the valuation at i of a proposition X ($val_i(X)$) is a truth-value. Let $True_i$ represent the set of the *true* propositions X of i (such that $val_i(X) = \mathbf{1}$). Membership conditions to the set $True_i$ may depend on other contexts j and on the context-relations and operations (as happens in the usual possible world semantics).

(9) *Conc* is the concept-function, that assigns to any context i the set of the concepts at i. Any concept has an n-arity ($1 \leq n$). If defined for a sequence of individual concepts $d_1, \ldots, d_n$ of i, the application of an n-ary concept C to $d_1, \ldots, d_n$ is a proposition.

As expected, different conditions required for the elements of particular classes of intensional structures give rise to different semantic situations and permit to characterize different logics.

In this framework, identity questions may be dealt with either in a metatheoretical or in a theoretical way. Metatheoretically, two individual concepts of a context i will be called *indiscernible* (or *indistinguishable*) when they share all the same properties at i:

DEFINITION 2. For any $d, e \in D_i$: *Ind(d, e)* at i (d and e are *indiscernible* at i) iff for any monadic concept C of i, $C(d) \in True_i$ iff $C(e) \in True_i$.

At the same time, d and e will be called *identical* at i when they have the same reference at i:

DEFINITION 3. For any $d, e \in D_i$: *Id(d,e)* at i (d and e are *identical* at i) iff $ref_i(\mathrm{d}) = ref_i(\mathrm{e})$.

Theoretically, the indiscernibility and the identity relations can be dealt with as particular binary concepts of any context i.

Now, a *Leibnizian semantic situation* will be naturally characterized by the simultaneous validity of the following principles:

THE PRINCIPLE OF INDISCERNIBLES
Indiscernibility implies identity in any context
(for any context i and for any pair of individual concepts d,e: if d and e are indiscernible at i, then d and e are identical at i).

THE SUBSTITUTIVITY PRINCIPLE
Identity implies indiscernibility in any context
(for any context i and for any pair of individual concepts d,e: if d and e are identical at i, then d and e are indiscernible at i).

Both principles admit of counterexamples in the framework of this semantics. Violations of the substitutivity principle were well known to Leibniz himself and can be easily constructed by referring to oblique contexts (let us think of the classical example of "the morning and the evening star"). Violations of the principle of indiscernibles may arise, whenever, for different reasons, the class of the admitted properties is not sufficiently rich (as, for instance, in Henkin's semantics for second order classical logic).

2. IDENTITY AND INDISCERNIBILITY IN QUANTUM THEORY

Suppose we want to describe in our semantics a compound physical system S, which evolves in time and may assume different states. For simplicity, let us refer to a two-particle system $S = S_1 + S_2$ (for instance, the electron and the nucleus of a given hydrogen atom). Here the contexts i may be naturally identified with possible information of the observer about the system under investigation. In the happiest situation, i may correspond to *maximal* information: *a pure state*, that according to the mathematical formalism of quantum theory (QT) is represented by a normalized vector ϕ in the Hilbert space H_S associated to our system. Since S consists of the two systems S_1 and S_2, the Hilbert space H_S will be the tensor product of the two Hilbert spaces H_{S_1} and H_{S_2}, associated respectively to S_1 and to S_2. In a case where the information is not maximal, a context i will be represented by a *mixed state* (a density operator in H_S), or more generally by a property corresponding to a partial description of the system.

As to the mathematical representation of the physical properties of our systems (S, S_1 and S_2), one may refer to two canonical choices. According to von Neumann's classical proposal, a quantum property is represented by a projector P in the Hilbert space associated to the system. More generally – following the *operational approach* to QT, first proposed by Ludwig – properties may be put in correspondence with the so called *effects*, which are particular generalizations of projectors. Namely, according to the "Born rule" of orthodox QT, the probability that a physical system in a given (pure or mixed) state W satisfies a physical property represented by a projector P is determined by the number $Tr(WP)$, where Tr represents the *trace functional*. However, projectors are not the only operators for which a Born probability can be defined. Any linear bounded operator E of H such that $Tr(WE) \in [0, 1]$ is called an *effect*. It turns out that the class of the effects properly includes the class of the projectors. In a sense, effects represent the "maximal" possible notion of physical property, in agreement with the probabilistic rules of QT; from an intuitive point of view, they can be regarded as produced by yes-no measurement devices, testing questions about our physical system. An important difference between projectors and proper effects is the following: projectors can be associated to *sharp* questions having the form "the value of the physical quantity Q lies in the Borel set Δ", whereas effects may represent also *fuzzy* questions like "the value of the quantity Q lies in the fuzzy Borel set Δ". Projectors and effects turn out to have different algebraic structures and consequently give rise to different logics.[2]

Each context i, corresponding either to a state or more generally to a physical property (a projector or an effect), provides information about the whole system. Information about the component systems can be derived from the information about the whole system, in accordance with the mathematical formalism of QT. As a consequence, one can say that each i determines, in our intensional semantics, two individual concepts d_1 and d_2, belonging to the set of the individual concepts D_i. As to the reference function, one may conventionally assume that ref_i is defined for an individual concept d belonging to D_i only if the context i determines for the component system corresponding to d a pure state that provides a space localization within a given precision. On this basis, also the counterpart function will receive a natural physical interpretation.

Now, a typical quantum mechanical situation that may occur is the following: a context i provides maximal information about the whole system. However, the information determined by i about the component systems is not maximal: d_1 and d_2 do not correspond to pure states. This is exactly what happens in the case of a compound physical system consisting of indistinguishable particles of the same kind. For instance, let us consider a system consisting of two helium 4 nuclei, which obey the Bose-Einstein statistics. Since we are dealing with a situation where the two component systems are *bosons*, any pure state of the whole system will have the *entangled* form

$$1/\sqrt{2}(\phi \otimes \psi + \psi \otimes \phi).$$

At the same time a factorized state $\phi \otimes \psi$ (which would permit to distinguish one component from the other one) is impossible. As a consequence, the states of the two component systems will not be pure and will be represented by two equal mixtures (non maximal information).

On this basis one easily obtains a violation of the principle of indiscernibles: d_1 and d_2 will be indiscernible at i, since they are represented by the same mixed state. However, they will not be identical, because the reference at i of both d_1 and d_2 is strongly undefined. In spite of this, the pure state of the total system informs us that one is dealing with two individual particles, although characterized by equal descriptions. It is worthwhile to notice that the monadic concept that describes the universe of i (the total concept of i) seems to be hardly tractable as a set in a standard set-theoretical sense, since its "elements" are in principle not distinguishable from one another and cannot be ordered. Collections of this kind, that we have termed *quasets*,[3] turn out to be particular examples of monadic concepts that are essentially intensional.

Is it possible to avoid anti-Leibnizian conclusions in QT? One may refer to at least two canonical ways out. According to a first attempt, identity is to be reduced, in this framework, to equiintensionality. As a consequence, the individual concepts d_1 and d_2 of our example turn out to be identical, since they are represented by the same mixed state. Hence, the principle of indiscernibles trivially holds. However, in our opinion, such a solution seems to lead to a somewhat pathological use of the term "identity": after all we are dealing with a "collection" of distinct objects, whose cardinality is two. And this is decided by the pure state associated to the whole system.

A second attempt stresses the fact that d_1 and d_2 take into account only the "statistical properties" of the component systems. In principle, one could guess that "other" properties should permit to distinguish the two subsystems. In fact, even in macroscopic situations one may build up appropriate statistics, where different individuals turn out to be statistically indiscernible. The attitude underlying such a solution seems to be very close to a kind of "hidden variable spirit". However, one cannot forget that, differently from similar macroscopic cases, here a context i may correspond to *maximal* information about the whole system, that in principle cannot be extended to a richer one. If we trust orthodox QT, it has no sense to maintain that there must be a property which *in mente Dei* holds for one particle and does not hold for the other one. If QT is valid, even an omniscient mind should be bound to the limitations imposed by the theory. One cannot conclude: God is Leibnizian, even if the quantum physicist is compelled to violate the principle of indiscernibles.

Università di Firenze

NOTES

[1] For instance the intensional set theory presented in [9], that has some features in common with other theories proposed by Aczel, Feferman, Cantini, Parsons.

[2] Projectors determine natural models for orthodox quantum logic (first proposed by

Birkhoff and von Neumann). Effects instead have suggested non-standard versions of quantum logic, where the excluded middle and the non contradiction laws are possibly violated.

[3] See [12]. A theory of *quasisets* has been independently proposed by da Costa and Krause ([8], [9]). In spite of many similarities, the notions of quaset and of quasiset are not equivalent.

REFERENCES

1. Aczel, P., "Algebraic semantics for intensional logics", in *Property Theories, Type Theories, and Semantics* (ed. by G. Chierchia, B. Partee, and R. Turner), Dordrecht, Reidel, 1988.
2. Aerts, D. and Piron, C., "Physical justification for using the antisymmetrical tensor product", preprint.
3. Bealer, G. and Moennich, U., "Property theories", in *Handbook of Philosophical Logic IV* (ed. by D. Gabbay and F. Guenthner), Dordrecht, Reidel, 1989.
4. Beltrametti, E. and Cassinelli, G., *The Logic of Quantum Mechanics*, London, Addison-Wesley, 1981.
5. Cantini, A., *Proprietà e operazioni*, Napoli, Bibliopolis, 1983 (English version to appear, North Holland).
6. Castellani, E. and Mittelstaedt, P., "Das Principium identitatis indiscernibilium in der Physik", to appear.
7. Costantini, D. and Garibaldi, U., "The non-frequency approach to elementary particle statistics", in *Statistics in Science* (ed. by R. Cooke and D. Costantini), Dordrecht, Kluwer, 1990.
8. Da Costa, N. and Krause, D., "Schrödinger logics", to appear in *Studia Logica*.
9. Dalla Chiara, M. L., "An approach to intensional semantics", *Synthese* **73** (1987), pp. 479–496.
10. Dalla Chiara, M. L., "Possible worlds, counterfactuals and epistemic operators", in *Knowledge, Belief and Strategic Interaction* (ed. by C. Bicchieri and M. L. Dalla Chiara), Cambridge University Press, 1992.
11. Dalla Chiara, M. L., "Epistemic logic without logical omniscience", in *Advances in Scientific Philosophy* (ed. by G. S. Schurz and G. Dorn), Amsterdam, Rodopi, 1991.
12. Dalla Chiara, M. L. and Toraldo di Francia, G., "Individuals, kinds and names in physics", *Versus* **40** (1985), pp. 29–50. Reprinted in *Bridging the Gap: Philosophy, Mathematics and Physics* (ed. by G. Corsi, M. L. Dalla Chiara, and G. Ghirardi), Dordrecht, Kluwer, 1992.
13. De Martini, F., "The stocastic interferometry and the foundations of the Bose-Einstein statistics", in *Quantum Measurement in Optics* (ed. by P. Tombesi and D. F. Wallis), New York, Pleumn Press, 1992.
14. Feferman, S., "Towards useful typefree theories", *The Journal of Symbolic Logic* **49** (1985), pp. 75–111.
15. Feferman, S., "Intensionality in mathematics", *Journal of Philosophical Logic* **14** (1985), pp. 41–55.

16. French, S. and Krause, D., "A formal framework for quantum non-individuality", to appear in *Synthese*.
17. French, S. and Redhead, M. L. G., "Quantum physics and the identity of indiscernibles", *British Journal for the Philosophy of Science* **39** (1988), pp. 233–246.
18. Jauch, J. M., *Foundations of Quantum Mechanics*, London, Addison-Wesley, 1968.
19. Krause, D., "On a quasi-set theory", *Notre Dame Journal of Formal Logic* **33** (1927), pp. 402–411.
20. Margenau, H., *The Nature of Physical Reality*, New York, McGraw-Hill, 1950.
21. Mittelstaedt, P., "Constituting, naming and identity in quantum logic", in *Recent Developments in Quantum Logic* (ed. by P. Mittelstaedt and E. W. Stachow), Mannheim, Bibliographisches Instiut, 1985.
22. Parsons, C., "Intensional logic in extensional language", *The Journal of Symbolic Logic* **47** (1982), pp. 289–328.
23. Stoecker, M., "Individualität, Identität, Unentscheidbarkeit", *Conceptus* **12** (1988), pp. 5–29.
24. Toraldo di Francia, G., *Le cose e i nomi*, Bari, Laterza, 1986.
25. Van Fraassen, B., "Statistical behaviour of indistinguishable particles: problems of interpretation", in *Recent Developments of Quantum Logic* (ed. by P. Mittelstaedt and E. W. Stachow), Mannheim, Bibliographisches Institut, 1985.
26. Van Fraassen, B., *Quantum Mechanics: an Empiricist View*, Oxford, Oxford University Press, 1991.
27. Weingartner, P., "A new theory of intension", in *Scientific Philosophy Today* (ed. by J. Agassi and R. Cohen), Dordrecht, Reidel, 1981.

DEAN S. EDMONDS

SOME REMINISCENCES OF ROBERT COHEN'S PHYSICS DEPARTMENT

I. MEETING BOB COHEN – MOLECULAR BEAMS

In 1961, three years after obtaining my doctorate at MIT, I was looking for the tenure-track faculty position I had always wanted. My first job had been a post-doctoral appointment at the Cambridge Electron Accelerator, a position made particularly attractive by the fact that the director, Prof. M. Stanley Livingston, had been my registration officer during my graduate years. But now the machine was finished, I was neither a machine operator nor a high-energy physicist, and in any case I wanted the combination of research and teaching that a faculty position represents. What did one do in such a case? Well, one went to the New York Meeting of the *American Physical Society*, so called because it was then held in New York's Manhattan Center and the neighboring Hotel New Yorker, the sixth floor of which was generally referred to as the flesh market because that's where all the people who wanted to hire physicists had their suites and thus where all the unemployed foregathered. I well remember looking down the list of prospective employers and noting that with one exception, all were offering positions on what is colloquially called "soft money". In 1961 almost everybody had some sort of government-sponsored laboratory, and although the list of such establishments and the projects they were engaged in was impressive, none were what I was looking for. The exception, however, was Boston University, which was hiring faculty to replace the large fraction of the physics department that had disappeared to form the Mitre Corporation, taking with it, probably to the relief of those remaining behind, the high-security, government-lab-type optics division and its U-2 camera development. Thus it was that I found myself walking into the smallest suite on the floor and coming face to face with the little man with the bright eyes who, as chairman of Boston University's Physics Department, was presiding over the hiring desk.

My dissertation at MIT had been on the hyperfine structure of cesium and its attendant application in the atomic clock which became today's international time standard. Modern hyperfine structure investigations are

K. Gavroglu et al. (eds.), Physics, Philosophy and the Scientific Community, 47–72.

carried out using molecular beam techniques that are the outgrowth of the famous Stern–Gerlach experiment,[1] originally developed to disprove quantum theory, but which actually supported it to everyone's surprise. The original idea was to look for the supposed "space quantization" of atoms in an atomic beam by running the beam through an inhomogeneous magnetic field. The presence of the field would then cause each atom to have an energy that depended on the orientation of its dipole moment with respect to the field direction, and the inhomogeneity would result in a net push on the atom as a whole which would be stronger the closer the atom and the field were to being lined up. The beam would then be separated at the detector into components depending on the orientations of the various atoms, and the classical result would be that, since all orientations were possible, all beam deflections would be possible and the detector would simply show a smudged out spot. Quantum mechanically, however, each orientation corresponds to a different energy, i.e., the energy given by the angle of rotation of the atomic dipole out of the field, and so if energy is quantized, only certain orientations would be possible. The beam should then be split into discreet components and the pattern on the detector resolved into spots rather than just a broad smudge. An interesting sidelight is that silver was picked for the beam in the original Stern–Gerlach experiment not because of any special atomic properties but simply because one could detect it. We must remember that we are dealing with beams of neutral atoms, so that detection is not the simple problem of observing a small electric current. Because metallic silver is produced in the development of the photographic plates in use at that time, Stern and Gerlach's detector was simply a glass plate on which the silver beam impinged and on which the resulting silver deposit could be observed by the photographic techniques of the day. Quite fortuitously, and an example of a kind of luck that pervaded nineteenth century and early twentieth century physics, the atomic structure of silver is relatively simple, with its single valence electron and $^2S_{\frac{1}{2}}$ ground state. Modern quantum theory correctly predicts that the silver atomic beam should split into two components in a Stern–Gerlach experiment, but it is interesting to note that the "old" quantum theory of Bohr and Sommerfeld called for three components, so that Otto Stern's work, although establishing space quantization beyond a doubt, also showed that old quantum theory had flaws that led to the later work of Schrödinger and Heisenberg.

The trouble with the Stern-Gerlach experiment, however, was that quantitative results could not be obtained because, with a simple iron magnet having pole pieces shaped to produce as big a gradient in the field as possible, neither the value of the field itself nor of its gradient at the position of the beam could be readily measured. One way of solving this problem is by the use of the so-called two-wire field in which magnetic field B is produced by two parallel wires spaced by distance $2a$, shown in end view in Figure 1. The beam is to be a narrow ribbon proceeding parallel to the wires (i.e., perpendicular to the page) as shown schematically in Figure 1(a) and is thus located at a distance z from the y-axis. Ramsey[2] gives the field and its gradient at z as

$$(1) \qquad B = \frac{\mu_0 I}{2\pi} \frac{2a}{r_1 r_2} \quad \text{and} \quad \frac{\partial B}{\partial z} = - \frac{\mu_0 I}{2\pi} \frac{2a}{r_1^3 r_2^3} (r_1^2 + r_2^2)\, z.$$

Thus theoretically, given current I in the wires, both the field and its gradient can be calculated at the position of the beam and hence also the amount of deflection to be expected with a particular dipole moment inclined at a given angle to the field direction. However, because the beam position is hard to determine with any precision and, more importantly, because high fields from what amounts to a one-turn coil in

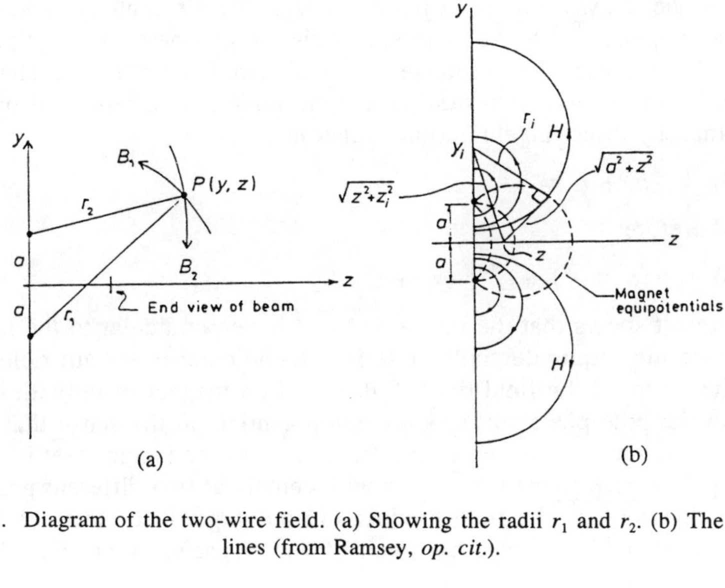

Fig. 1. Diagram of the two-wire field. (a) Showing the radii r_1 and r_2. (b) The field lines (from Ramsey, *op. cit.*).

vacuo require excessively high currents, Millman, Rabi, and Zacharias[3] sacrificed computation of the field and its gradient by going to iron magnets but retained the field configuration of the two-wire field (Figure 1(b)) by noting that both the lines of B and the perpendiculars thereto are circles. Edmonds and Corson[4] have shown this for the similar case of the electric field of two parallel wires carrying equal and opposite charges and obtain for the field lines circles whose centers are on the y-axis (Figure 1) at

$$(2) \qquad y_i = a \frac{\xi_i^2 + 1}{\xi_i^2 - 1},$$

and whose radii are

$$(3) \qquad r_i = \pm \frac{2a\xi_i}{\xi_i^2 - 1},$$

where ξ_i is any constant in the range $0 \leq \xi_i \leq \infty$. That the perpendiculars to these circles are also circles follows from the construction shown in Figure 1(b). Pick any point z on the z-axis and draw a circle of radius $\sqrt{a^2 + z^2}$, which must necessarily pass through the points representing the wire ends at points $y = \pm a$. Now pick any circular field line, say the ith one, draw a line from point z to that circle's center y_i, and draw the radii r_i and $\sqrt{a^2 + z^2}$ to intersect at the point where the "equipotential" circle drawn above intersects the ith circular field line. That the angle between these two radii is a right angle is demonstrated by the fact that, by direct algebraic manipulation

$$(4) \qquad a^2 + r_i^2 = y_i^2,$$

and therefore

$$(5) \qquad (a^2 + z^2) + r_i^2 = (y_i^2 + z^2).$$

This result shows that the radius $\sqrt{a^2 + z^2}$ is perpendicular to the radius r_i, hence the "equipotential" circle is perpendicular to the ith field line and hence to all the field lines. But in an iron magnet made with high-μ iron the pole-piece surfaces are equipotentials in the sense that they are perpendicular to the emerging field lines, hence if a magnet is made with pole pieces shaped as circles with centers at two different points z and radii $\sqrt{a^2 + z^2}$, the field produced in the gap will be identical with a two-wire field produced by a pair of wires spaced by distance $2a$. A diagram showing the end-on view of such a magnet appears in Figure 2, which also shows the ribbon-like beam, whose direction is perpendicular to the plane of the page.

Although in such a magnet the shape of the field and its gradient is known, the value is not, this having been given up in the interest of obtaining the large values of B available with an iron magnet. But subsequent development of the atomic beam method made such knowledge unnecessary. This development consisted of adding a second deflecting (i.e., inhomogeneous field) magnet following the first one, the two magnets being then referred to as A and B. One arrangement calls for the B-magnet to be oriented opposite to the A-magnet so that deflections due to the latter are cancelled by the former and the beam is refocused to the single point at the detector that it would strike in the absence of both fields. Now if between the two magnets a transition is stimulated from the initial state of a particular atom in the beam to one corresponding to a reorientation of that atom's dipole moment, then the deflection in the B-magnet will not cancel the deflection in the A-magnet, and the beam will not be refocused at the central point mentioned above,

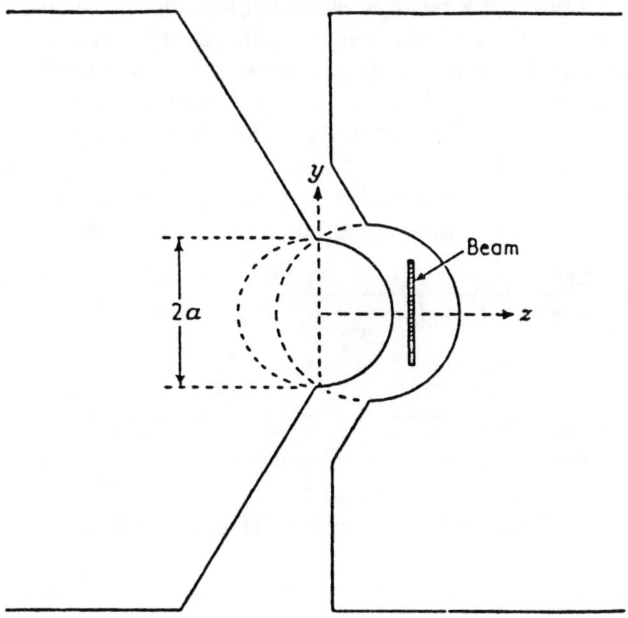

Fig. 2. Cross-section of a pair of magnet pole pieces shaped to produce a two-wire field.[5]

i.e., it will miss the detector. Such an arrangement is often called "flop-out", the word "flop" referring to the turning over of the dipole, and "out" indicating a decrease in the detected beam intensity when the transition is effected. Such a procedure has the disadvantage that only the state out of which a transition is made drops out of the beam, all atoms in other states getting through, so that the fall in intensity at resonance is often not large. A better method is not to reverse the B-magnet relative to the A, so that in the absence of a transition the detected intensity is zero. A transition that reverses the dipole of a particular state thus causes atoms originating in that state *and only those* to be refocused, so that a relatively large signal-to-noise ratio is obtained. This arrangement is called "flop-in" for obvious reasons, and is the preferred method in most cases.

A schematic diagram of a typical modern atomic beam apparatus appears in Figure 3, which shows the sort of beam geometry discussed above. Note that the beam is a stream of neutral atoms, so that the source (an oven in the case of a solid sample such as cesium) has no way of focussing it other than by collimation. The emerging beam thus has a spread through a range of small angles, and the diagram of Figure 3 shows one such component leaving the oven at angle θ. The usual layout calls for a collimator slit placed as shown so that the beam must be on the center line at three points – the oven, the collimator, and the detector – in order to be detected. Assuming equal and opposite deflecting accelerations a_z in the A- and B-magnets, we find that requiring the beam to be on the center line at the three points just mentioned imposes the following constraint on the apparatus dimensions given in Figure 3:

$$\frac{l_5(l_5 + 2l_6)}{l_4 + l_5 + l_6} = \frac{l_2(l_2 + 2l_1)}{l_1 + l_2 + l_3} . \tag{6}$$

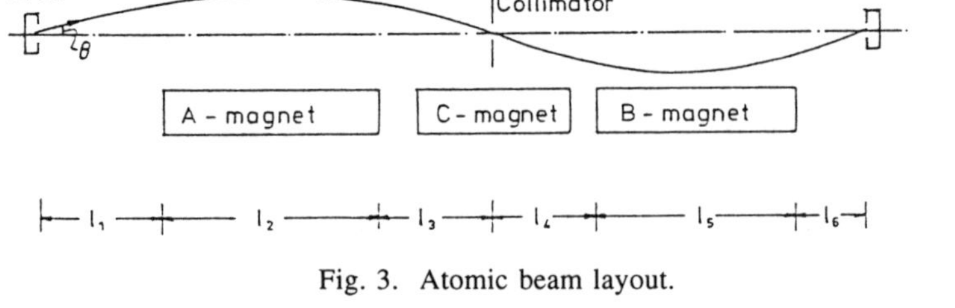

Fig. 3. Atomic beam layout.

Moreover, θ may be any angle within the small spread emitted by the source, and we expect the usual Maxwell–Boltzman distribution in the beam velocity v. We then find a relation between v and θ which must hold for a particular atom to get through the apparatus and reach the detector, given by

$$(7) \qquad \theta = -\frac{1}{2}\frac{a_z}{v^2}\frac{l_2(l_2 + l_3)}{l_1 + l_2 + l_3}\ .$$

Since at any θ there will be atoms at all velocities in the distribution, there will always be plenty to get through to the detector, but even more will fail to do so and thus be stopped at various points in the apparatus. In the cesium experiments this was of no particular consequence, although the famous Zacharias "crinkly foil" oven was designed to maximize beam collimation so as to minimize the range of θ and hence use as little cesium as possible (an important consideration for the sealed beam tube eventually designed for the commercial atomic clock). Such a consideration also becomes important when rare and/or radioactive samples are to be run.

Stern and Gerlach were interested only in the interaction of their sample's electronic dipole moment with an externally imposed field, but now the possibility exists of investigating the interaction of both the electronic and nuclear dipoles with this field and with each other. This leads to a most interesting analysis, first given by Breit and Rabi[6] in 1931. The so-called Breit–Rabi formula which they derived using simple perturbation theory describes the magnetic dipole coupling between an electronic angular momentum J and a nuclear spin I. In the absence of a magnetic field, the vector model gives the resulting hyperfine energy levels as

$$(8) \qquad W_F = W^0 + a\vec{I}\bullet\vec{J} = W^0 + \frac{a}{2}\,[F(F+1) + I(I+1) + J(J+1)]$$

where W^0 stands for the atom's energy with no hyperfine interaction, F is the total angular momentum quantum number, which can have the values $I + J$, $I + J - 1$, - - - $|I - J|$, and a represents the interaction energy. We now observe that Cs^{133}, the one stable isotope of cesium, was chosen for the clock development work for a number of reasons having no direct relation to its atomic structure. It has a vapor pressure sufficient to produce a beam at the very reasonable temperature of 80 °C and, most important, it is the heaviest of the stable alkalis and thus has the lowest ionization potential of any non-radioactive element in the periodic table. This makes for almost 100% efficiency in that simplest of all neutral atom detectors, the hot wire. This consists simply of a wire made of some material such as pure tungsten having as high a work function as possible and heated to a temperature not high enough to cause copious electron emission but sufficient to ionize low-ionization-potential atoms impinging upon it. In short, the hot-wire surface, instead of emitting electrons, preferentially steals the valence electron from the impinging cesium atoms and re-emits these latter as positive ions, which can then be detected by standard electrical methods. A small mass spectrometer is often used to remove spurious emissions from the hot wire and may be followed by an electron multiplier to facilitate detection of the small ion current. But another consequence of cesium's being an alkali is that $J = \frac{1}{2}$ so that F is limited to the values $I \pm \frac{1}{2}$ and W_F to $W^0 + \frac{a}{2} I$ and $W^0 - \frac{a}{2} (I + 1)$. There are thus only two hyperfine levels at zero external field, and the splitting between them, which is what is of interest in the present work, is given by

$$\Delta W = W^0 + \frac{a}{2} I - W^0 + \frac{a}{2} (I + 1) = a\left(I + \frac{1}{2} \right). \tag{9}$$

In a weak field, meaning one insufficiently strong to decouple I and J so that the resultant F precesses around the field direction, the overall magnetic moment μ_F is $g_F \mu_o F$, and the energy splitting due to this field $\vec{B}$ is given by

$$W_z = g_F \, \mu_0 \, \vec{F} \bullet \vec{B} = m_F g_F \mu_0 B, \tag{10}$$

where

$$g_F = g_J \frac{J(J + 1) + F(F + 1) - I(I + 1)}{2F(F + 1)} + g_I \frac{I(I + 1) + F(F + 1) - J(J + 1)}{2F(F + 1)}, \tag{11}$$

and m_F is the weak-field magnetic quantum number. Note that this is the g-factor for the general case of any two spins J and I coupled by a magnetic dipole interaction. If we put L for J, S for I, and J for F as in the case of electron spin-orbit coupling, then $g_J = 1$, $g_I = 2$, and we immediately obtain the famous Landé g-factor. Thus what we have been considering here is the Zeeman effect in hyperfine structure. We will also see a Paschen–Back effect in this structure (which is really the Zeeman effect in fine structure with some hyperfine splitting added), but we must remember that these effects, the only ones derivable from the vector model, are of necessity linear and are only approximations valid in the very low and very high field cases, respectively. What about the intermediate non-linear case, which must successfully go over to the low- and high-field cases when B is very small or very large? To solve this problem, Breit and Rabi considered the perturbing Hamiltonian

$$H' = a\,\vec{I} \cdot \vec{J} + m_I g_I \mu_0 B + m_J g_J \mu_0 B \tag{12}$$

which describes the three interaction energies to be considered. The usual perturbation procedure leads to an array of matrix elements H'_{ij} and the associated secular determinant, but the fact that in the case of cesium $J = \frac{1}{2}$ means that this determinant factors into sub-determinants which are at most 2×2 and thus allows the secular equation to be solved in closed form. The result is the Breit–Rabi formula,

$$W_{F,m_F} = W^0 - \frac{\Delta W}{2(2I+1)} + m_F g_I \mu_0 B \pm \frac{\Delta W}{2}\sqrt{1 + \frac{4m_F}{2I+1}x + x^2}, \tag{13}$$

where x is a dimensionless field factor given by

$$x = \frac{(g_J - g_I)\mu_0}{\Delta W} B. \tag{14}$$

Notice that Equation 13 expresses W_{F,m_F} directly in terms of the zero-field hyperfine splitting ΔW, which must thus be determined *a priori*, so that the interaction energy a does not appear, but the equation does show the positions of the two zero-field levels relative to the no-interaction level W^0.

In the case of cesium, the nuclear spin I is $\frac{7}{2}$, so that F has the values 3 and 4, and m_F can be 4,3,2, - - - –4 in one case, and 3,2,1, - - - –3 in the other, the two cases being distinguished in Equation 13 by taking the + or – sign respectively. The resulting plots of W_{F,m_F} against the field parameter x are shown in Figure 4. Note that the curves are straight lines near zero field ($x = 0$), corresponding to the Zeeman effect of hyperfine structure as discussed earlier, and again at high field corresponding to the Paschen–Back effect. Note also that the high-field plots fall into two groups of eight sublevels each, the two groups being the electronic fine-structure splitting of the $^2S_{\frac{1}{2}}$ ground state, with the sublevels representing the further hyperfine splitting due to the nucleus. Of further interest is the illustration of an important feature of perturbation theory,

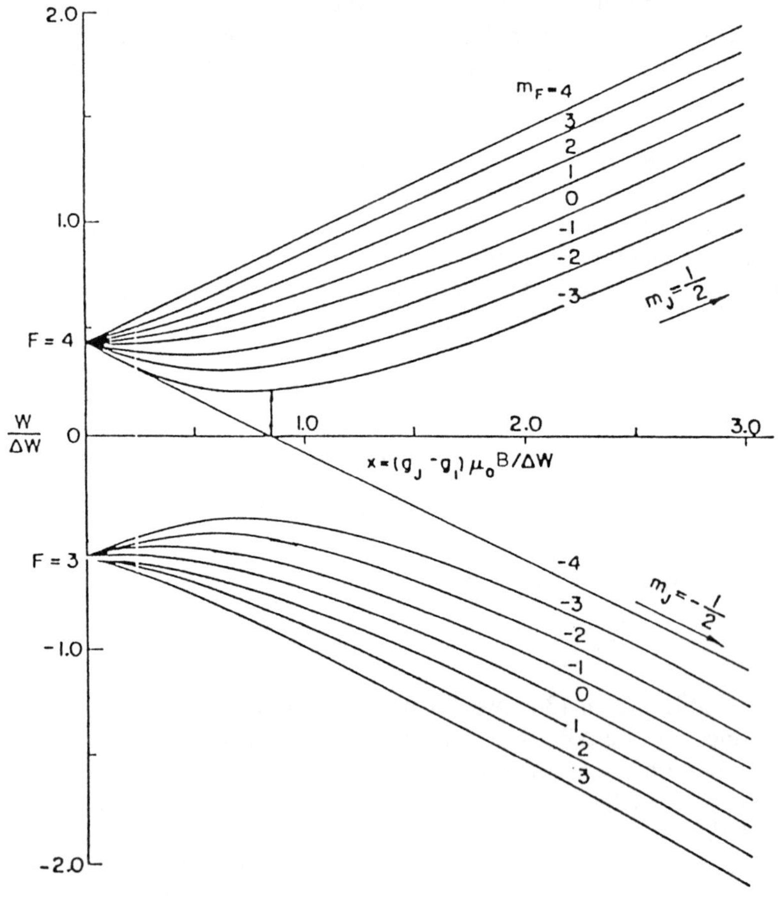

Fig. 4. Plot of the Breit–Rabi formula for cesium.

namely that states having the same magnetic quantum number "repel" each other so that they will not cross and become degenerate as B is increased. Thus the states with $m_F = 3,2,1,$ - - - -3 coming from the $F = 4$ level are repelled upwards, those from the $F = 3$ level are repelled downward, and thc $m_F = \pm 4$ levels, having no counterparts from the $F = 3$ level, plot as straight lines.

The so-called clock line is now the zero-field transition $F = 4 \leftrightarrow F = 3$, and since the selection rule for $\Delta F = 1$ is $\Delta m_F = 0,\pm 1$, transitions may be stimulated between any of the levels coming from the $F = 4$ level to levels from $F = 3$ satisfying this rule. If the A- and B-magnets are now set to some high field, Figure 4 shows that the upper group of levels have a fairly large positive slope whereas the lower group have the same slope except for its being negative. But we can show that *the slope of any Breit–Rabi curve is the dipole moment exhibited by that state at the field defining the point at which the slope is taken.* To do this, consider the simple case of an electric dipole consisting of charges $\pm q$ separated by distance b and therefore having dipole moment $p = bq$. If such a dipole is placed in an inhomogeneous electric field directed along the z-axis, the net force F_z exerted on the dipole is $p_{eff}\frac{\partial E}{\partial z}$ where p_{eff} is the component $p \cos \phi$ of the dipole moment p along the field direction. That this is true is easily seen from the fact that $p_{eff} = p \cos \phi = bq \cos \phi$ so that $b \cos \phi$ is the distance separating the charges as measured along the field direction, and $\frac{\partial E}{\partial z} b \cos \phi$ represents the difference in the field strength at the two charges. Hence $p_{eff}\frac{\partial E}{\partial z}$ represents the difference between the forces urging the charges in opposite directions, i.e., the net force F_z on the dipole. However, in any case, if W represents the energy of the dipole, F_z must in general be equal to $-\frac{\partial W}{\partial z}$ and hence to $-\frac{\partial W}{\partial E}\frac{\partial E}{\partial z}$. Therefore

$$(15) \qquad -\frac{\partial W}{\partial E}\frac{\partial E}{\partial z} = p_{eff}\frac{\partial E}{\partial z},$$

so that

$$(16) \qquad p_{eff} = -\frac{\partial W}{\partial E}.$$

Then by analogy, we have for our magnetic case $\mu_{eff} = -\frac{\partial W}{\partial B}$, where $\frac{\partial W}{\partial B}$ is of course the slope of any one of the curves in Figure 4. So important is this concept that we show a plot of $\frac{\partial W}{\partial B}$ in Figure 5, and now point

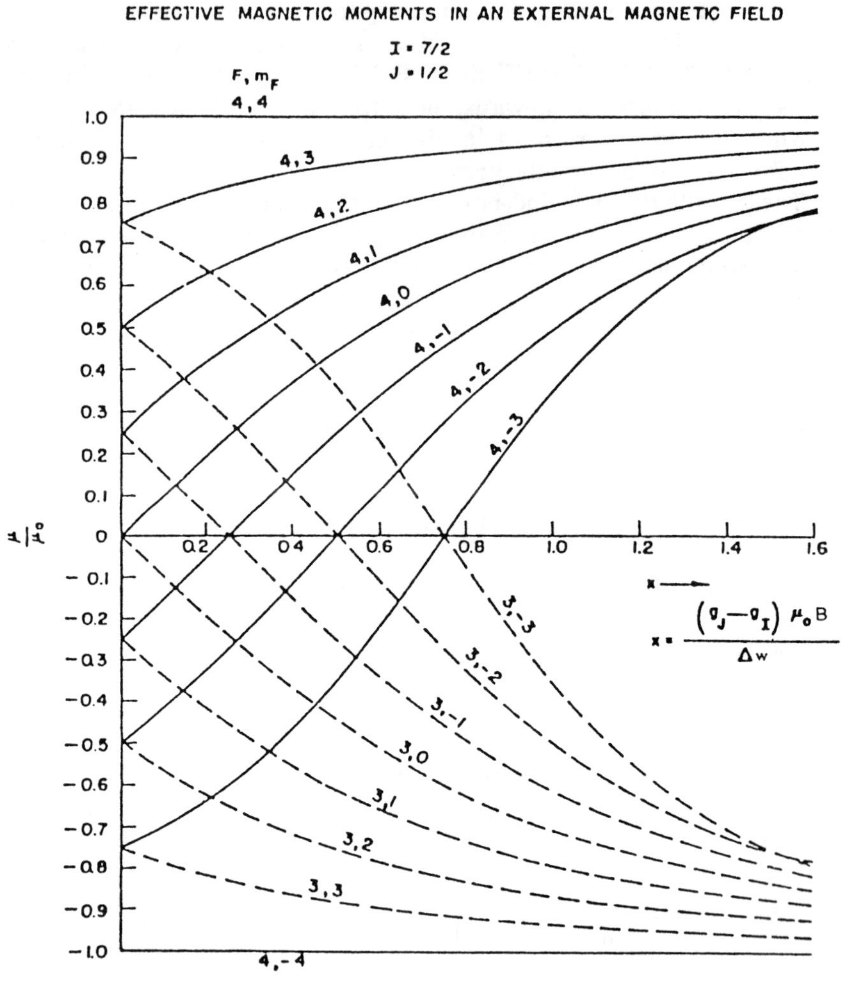

Fig. 5. Plots of μ_{eff} from the slopes of the curves of Figure 4.

out certain important features. One is that in the intermediate field region, many of the curves of Figure 4 go through minima or maxima, which means they have zero slope and hence zero moment. This is shown by the corresponding curves in Figure 5 going through zero μ. If the A-

and B-magnets are now set at the field giving zero moment for a particular state, then atoms in that state will go straight through the apparatus without deflection, while all others get thrown out of the beam. The apparatus thus functions as a state selector for that one state. However, if in the so-called C-region shown between the A- and B-magnets in Figure 3, a transition is made out of the selected state, then these atoms too will be thrown out of the beam in the B-magnet and the detected intensity will fall to essentially zero. This technique is called "zero-moment flop-out" and is characterized by a much better signal-to-noise ratio than other flop-out methods. A possible transition observable in zero-moment flop-out is indicated in Figure 4. Notice that zero moments corresponding to minima in the Figure 4 curves are to be used, as these are focussing, the deflecting magnets returning the beam to its central position if it drifts off in the gap to a field value on either side of that corresponding to zero moment. The maxima are defocussing, as excursions of the beam are augmented in this case until the beam is lost.

Another feature of Figure 5 of particular importance to the clock development is that, if at some field the slopes of two of the Breit–Rabi curves are equal (curves parallel), then a transition between these states will be field independent, at least to first order, as the spacing between the curves does not vary to first order with field in the neighborhood of that field value. Figure 5 shows the cases of equal slopes directly as intersections of the plots for the various states. In particular, at zero field the clock line $F = 4$, $m_F = 0 \leftrightarrow F = 3$, $m_F = 0$ is first-order field independent. Moreover, at high field these states are seen directly from Figure 5 to have equal and opposite moments, so that a flop-in experiment is possible with good deflections (hence good state separation) in the A- and B-magnets by reason of their being set to high field and hence high gradient. Close to zero field is maintained in the C-region by suitable shielding and/or coils to block out stray fields, and a highly homogenous magnet called the C-magnet may also be placed here as shown in Figure 3 if high-field transitions are to be studied. The C-magnet is designed to produce an extremely uniform field which takes no part in beam deflection (as it has no gradient) but serves merely to set the field and hence the energy levels at which some given high-field transition is to be observed. An interesting example of this is the pair of transitions $F = 4$, $m_F = -2 \leftrightarrow m_F = -3$ and $F = 3$, $m_F = -2 \leftrightarrow m_F = -3$, which are first-order field independent at fields of a little over

5000 gauss. If these values of F and m_F are inserted in Equation 13, we get

$$\text{(17)} \quad \frac{W_{3,-3}}{\Delta W} = \frac{W^0}{\Delta W} - \frac{1}{16} + 3Ax - \frac{1}{2}\sqrt{1 - \frac{3}{2}x + x^2},$$

$$\text{(18)} \quad \frac{W_{3,-2}}{\Delta W} = \frac{W^0}{\Delta W} - \frac{1}{16} + 2Ax - \frac{1}{2}\sqrt{1 - x + x^2},$$

$$\text{(19)} \quad \frac{W_{4,-3}}{\Delta W} = \frac{W^0}{\Delta W} - \frac{1}{16} + 3Ax + \frac{1}{2}\sqrt{1 - \frac{3}{2}x + x^2},$$

$$\text{(20)} \quad \frac{W_{4,-2}}{\Delta W} = \frac{W^0}{\Delta W} - \frac{1}{16} + 2Ax + \frac{1}{2}\sqrt{1 - x + x^2},$$

where x is the field parameter given by Equation 14 and for simplicity we have divided through by ΔW and introduced the constant $A \equiv -g_I/(g_J - g_I)$. The transition energy differences (and hence the transition frequencies when divided by Planck's constant h) are then

$$\text{(21)} \quad \nu_3 = +Ax + \frac{1}{2}\left[\sqrt{1 - x + x^2} - \sqrt{1 - \frac{3}{2}x + x^2}\right]$$

and

$$\text{(22)} \quad \nu_4 = -Ax + \frac{1}{2}\left[\sqrt{1 - x + x^2} - \sqrt{1 - \frac{3}{2}x + x^2}\right],$$

where the subscripts 3 and 4 refer to the $F = 3$ and $F = 4$ states, respectively. These transitions differ only by the term Ax which is just the small nuclear term $m_F g_I \mu_o B$ in the Breit–Rabi formula, and an interesting experiment carried out by the author is to scan these lines to find their field-independent points (where the transition frequencies go through minima) and compare the experimental results with theoretical ones obtained by setting $\frac{d\nu_3}{dx} = 0$ and $\frac{d\nu_4}{dx} = 0$. The results, which include values of the field at the field-independent points obtained without doing actual measurements in the magnet gap, are

$$\nu_{03} = 1487.659225 \text{ MHz}$$
$$\nu_{04} = 1481.95407 \text{ MHz}$$
$$x_3 = 1.562685750$$
$$x_4 = 1.553904964$$
$$A = 1.991351 \times 10^{-4}$$

where ν_{03} and ν_{04} are the frequencies of the two transitions at the respective field-independent points x_3 and x_4. Notice that in addition, by taking the accepted value 2(1.00126) for the electronic gyromagnetic ratio g_J, the nuclear gyromagnetic ratio and magnetic dipole moment may be found for Cs^{133}. The results are $g_J = -3.988514 \pm 0.000140 \times 10^{-4}$ referred to the Bohr magneton or -0.7323431 ± 0.0000220 referred to the nuclear magneton. We then find for the dipole moment μ_I the value 2.563200 $\pm$ 0.000077 nuclear magneton (uncorrected) and the values 5127.6 and 5098.8 gauss for the actual fields at the field-independent points. These calculations all depend on the determination of ΔW by measurement of the clock line, and original measurements by Hahn and Zacharias[7] gave 9192.631820 MHz. Today, and defining the international standard second as adopted by the Thirteenth General Conference of Weights and Measures meeting in Paris in 1967, the accepted value is 9192.631770 MHz.

The precision with which the above numbers are quoted seems surprising in view of the fact that in a stimulated transition the line width must be inversely proportional to the time during which the stimulating frequency is applied, according to the uncertainty principle $\Delta\nu\Delta t = \frac{1}{2\pi}$. As the atomic beam is proceeding at a fairly high speed (the most probable velocity for cesium atoms coming from an oven at 80 °C is about 22,000 cm/sec), a narrow line width depends on the C-region being quite long. This in turn imposes the unrealizable requirements of maintaining a highly uniform magnetic field (even with first-order field-independent transitions) and of applying the stimulating oscillatory field (which may be a radio frequency of any value from a few hundred kiloHertz to short microwaves) over this entire region. To his everlasting credit, Ramsey[8] solved these problems by inventing the separated oscillating field transition method, often referred to as Ramsey flop. Consider a long C-region with identical short regions at either end, in which the beam sees the stimulating RF field. Although Ramsey[9] has given the full quantum mechanical treatment of this situation, a classical approach is useful in understanding the results obtained. Suppose, then, that a magnetic dipole enters the C-region and then precesses about the C-magnet's field with the Larmor frequency. The length of the first transition region and the intensity of the RF therein are adjusted so that, with the RF set to the Larmor frequency so that a torque tending to invert the dipole is applied, the precessing dipole spirals down to the equatorial plane and is thus "half-flopped" (transition probability = $\frac{1}{2}$) when it leaves this region. Although the half-flopped state is virtual (i.e., unobservable), we may think of the dipole as continuing to precess at the Larmor frequency as it proceeds through the C-region. It will then reach the second transition region in phase with the RF, so that the transition to the inverted (observable) state is completed. Notice that not only do we have only to apply the RF in two small places but also that the only requirement on the C-field is that its *average* over its entire length be equal to its value at the transition regions. In between, where we are not looking, it doesn't matter whether the dipole speeds up or slows down provided only that it hits the second transition region in phase with the RF. Nevertheless, because phase information has accumulated over the entire interval during which the dipole was in the C-region, this is the interval that appears in the uncertainty principle to determine the line width.

It is interesting to note further that if the applied RF is shifted slightly from the Larmor frequency, the precessing dipole will slip in phase as it moves through the C-region, and if the shift is just right so that the phase slips by an integer multiple of 2π radians, the dipole will also enter the second transition region in phase so that the transition to the inverted state is completed. Thus the theory predicts a whole series of resonances on either side of the true one. However, we note that for the true resonance, the dipole stays in phase with the RF at all times so that the time it spends in the C-region and hence its velocity doesn't matter. The other peaks are velocity-dependent and thus wash out as we get further and further from the true resonance because of the velocity distribution in the beam. The pattern computed by Ramsey (and which closely follows the experimental results) appears in Figure 6. Notice the broad peak characteristic of the line width obtainable with one short transition region alone and the deep depressions on each side of the Ramsey peak indicating a phase slip of π radians that brings the dipole

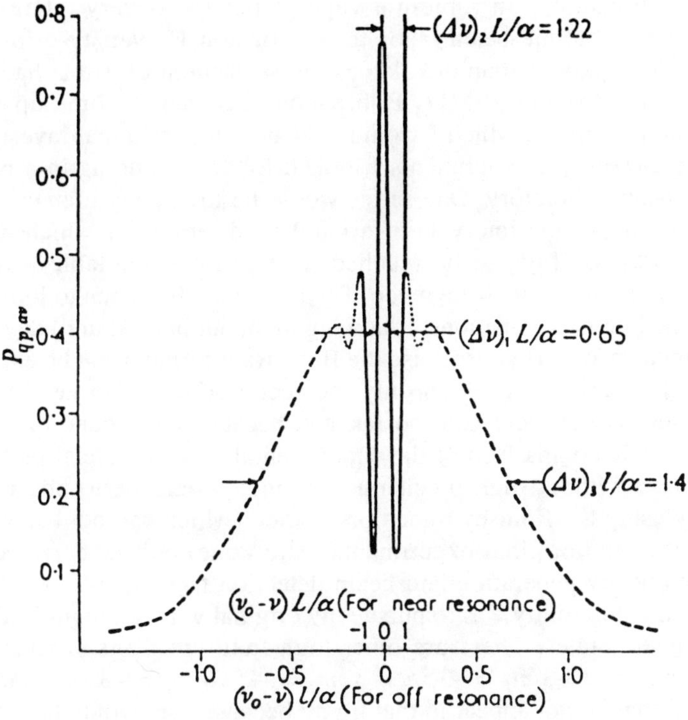

Fig. 6. Separate-field transition probabilty as a function of frequency. L is the length of the C region, l of each transition region, and α the most probable velocity in the beam. (From Ramsey, *op. cit.*)

into the second transition region exactly out of phase with the RF, which then undoes the half-flop of the first region resulting in no transition at all. The full curve is calculated for frequencies near resonance, the dashed one for off resonance, and the dotted one represents an interpolation between the two.

This was the sort of expertise I brought to Robert Cohen's department when I joined it in 1961. At first I was just a research professor, for the one tenure-track slot that had been available went to Wolfgang Franzen, whom I had known at Princeton University during my brief graduate career there and who had gone on to do research at Arthur D. Little. We worked together at Boston University on observing double-quantum transitions in rubidium vapor,[10] but I was very interested in setting up an atomic beam laboratory at Boston University using more up-to-date apparatus than the old cast brass vacuum cans that had given me so much trouble at MIT. Bob encouraged me to come up with a proposal, the first in which I would appear as the principal investigator. There were several experiments I looked forward to doing in a modern atomic beam laboratory. One series would be the remeasurement of the zero-field hyperfine intervals of a number of atoms for which the frequency was too high to be reached with sources available when the original measurements were made. The procedure had been to look at the hyperfine Zeeman transitions, following them out in field until deviations from linearity occurred. Because the Breit-Rabi formula can be expanded in a power series, for which only the second-order term need be kept if the field is not increased too far, and because the second-order term contains ΔW, comparison of the experimental and theoretical plots gave ΔW, albeit without high precision. The idea would be to observe ΔW directly using the Ramsey transition method (which had not been developed when the original experiments were done) and modern vacuum, high-frequency generation and beam detection techniques.

Another possibility was to pursue my original work on high-field field-independent states. There are other high-field crossings similar to the $F = 4$, $m_F = -2 \leftrightarrow m_F = -3$, $F = 3$, $m_F = -2 \leftrightarrow m_F = -3$ pair shown in Figure 5 that do not appear in that figure because they would be off-scale to the right. They are highly field independent, and the investigation would give interesting results in a modern apparatus equipped with a magnet that could reach these fields.

Also of particular interest would be the extension of the work on the Bohr–Weisskopf effect started by Stroke *et al.* in 1957.[11] This effect is the appearance of anomalies in the hyperfine structure as a result of the failure of the approximation that the nucleus is a point dipole. As Equation 8 shows, the hyperfine splitting for any atom with $J = \frac{1}{2}$ depends on the interaction energy a between the nuclear and electronic dipoles. The well known Fermi-Sergré formula,[12]

$$(23) \qquad a = \frac{16}{3} \pi \, g_I \, \mu_0 \, |\Psi(0)|^2 \, \frac{m}{M}$$

gives this energy in terms of the electronic wave function Ψ evaluated at the origin, thus inserting the assumption that the nucleus is a point dipole, i.e., that the electronic wave function does not penetrate a region that could be considered part of the nuclear volume. If this is true, then for two isotopes of the same $J = \frac{1}{2}$ atom, Equation 8 says

$$\text{(24)} \qquad \frac{\Delta W_1}{\Delta W_2} = \frac{a_1\left(I_1 + \frac{1}{2}\right)}{a_2\left(I_2 + \frac{1}{2}\right)} = \frac{\frac{16}{3}\pi\, g_{I1}\, \mu_0\, |\Psi(0)|^2\, \frac{m}{M}\left(I_1 + \frac{1}{2}\right)}{\frac{16}{3}\pi\, g_{I2}\, \mu_0\, |\Psi(0)|^2\, \frac{m}{M}\left(I_2 + \frac{1}{2}\right)}$$

in which, for isotopes of the same atom, many factors cancel out, leaving

$$\text{(25)} \qquad \frac{\Delta W_1}{\Delta W_2} = \frac{g_{I1}\left(I_1 + \frac{1}{2}\right)}{g_{I2}\left(I_2 + \frac{1}{2}\right)} = \frac{g_{I1}}{g_{I2}}\,\frac{2I_1 + 1}{2I_2 + 1}\,.$$

Deviations from this result have been attributed by Bohr and Weisskopf[13] to failure of the point-dipole approximation, and it would be interesting to look for such deviations among the isotopes of other $J = \frac{1}{2}$ atoms besides those of cesium studied by Stroke.

Finally, I have always wanted to look at the hyperfine structure of calcium 41 (Ca^{41}). This is a stable isotope of calcium and is interesting because the nucleus consists of the doubly-magic Ca^{40} nucleus with a single neutron tacked on. Because the ground electronic state of calcium is ^1S_o, there is no electronic dipole, and the doubly-magic nucleus of Ca^{40} is a tightly-bound structure also having no dipole moment. Thus Ca^{41} looks like a big, heavy, stable neutron, and its hyperfine structure consists merely of the interaction of that single neutron's dipole with an externally applied field. Measurement of this Zeeman effect would give the dipole moment of the neutron, presumably to a precision not available from experiments limited by the free neutron's half-life, and comparison with such measurements on free neutrons would give information on the effect of including a neutron outside the otherwise complete shells of a stable nucleus. Because there is no electronic dipole, the A- and B-magnets would have to push on the much smaller nuclear dipole, so that high fields and gradients would be needed, but an atomic beam apparatus with magnets external to the vacuum system could easily be designed to carry out this work.

II. BOB'S DEPARTMENT – THE MICROTRON

The atomic beam proposal was enthusiastically supported by Bob Cohen, who signed it as department chairman and wrote a strong accompanying letter of recommendation. However, at about this time the famous Mansfield Amendment, mandating that the government should only sponsor research having definite applications to national defense, came into effect, and it was difficult to claim that basic atomic-beam research had such an application. The proposal was therefore not funded, but Bob's support remained. His department was a pleasure to work in, and I still remember the weekly faculty lunches at 147 Bay State Road, which was then the very good and now long extinct Faculty Club. Far from being a drag, these meetings represented a high point of the week for me. The department was small, individual research was encouraged, and Bob was always ready with help and encouragement for everybody, especially neophytes like myself. There were department parties, spontaneous faculty and student get-togethers, and in general a level of "quality of life" that has never been equaled, even today with our present chairman, Larry Sulak, who places considerable emphasis on such matters.

It was during this time, following the collapse of plans for an atomic beam laboratory and partly as a result of my post-doctoral experience at the Cambridge Electron Accelerator, that I became interested in a type of accelerator called a microtron. This is essentially an electron cyclotron and, like the cyclotron, uses a fixed magnetic field. It therefore has the possibility of being a CW machine, and I wanted to exploit that possibility. The conventional microtron, first proposed by Veksler[14] in 1944 and subsequently developed by Redhead,[15] Froelich,[16] and Kapitza,[17] uses a fixed circular magnetic field similar to that of a cyclotron and therefore has the cyclotron's weak focusing characteristic. However, instead of the orbits forming a spiral as in the cyclotron, they are all circles tangent at one point and passing through an accelerating cavity placed there in lieu of the D's used in a cyclotron. It is interesting to note that cyclotron people work with heavy particles at relatively low energies and so can assume that the mass remains constant while acceleration means increasing velocity, whereas microtron (and synchrotron) people work with light particles and high energy and so can assume that speed remains constant (at essentially the speed of light) while acceleration means increasing mass. In a microtron, the electrons are fully relativistic upon injection (or at least after the first pass through the accelerating cavity) and may thereafter be taken to be at the speed of light. Thereafter each electron acquires energy ΔE as it returns to and passes through the cavity, so that successive orbits have larger and larger radii corresponding to the larger energy and hence larger mass characterizing each orbit. Since the electrons are assumed to be at the velocity of light c, the time required for each round trip thus increases for each successive orbit, and the distinguishing feature of a microtron is that the orbits are arranged so that the time around increases each time by an integer number of accelerating RF periods, thus allowing the electrons to enter the cavity in phase with the accelerating field after every revolution. Mathematically, the time for an electron to complete the kth orbit is

$$(26) \qquad T_k = \frac{2\pi}{eBc^2} E_k = \frac{2\pi}{eBc^2} (E_{\text{inj}} + k\Delta E)$$

where B is the value of the magnetic field and E_k is the total energy in the kth orbit, which is seen to be the sum of the injection energy E_{inj} (which includes the electron's rest energy) and k times the energy gained on each passage through the cavity. We now require that two successive orbit periods differ by an integer n times the RF period τ, so that

$$(27) \qquad T_{k+1} - T_k = n\tau.$$

Hence

$$(28) \qquad \frac{2\pi}{eBC^2} [E_{\text{inj}} + (k + 1)\, \Delta E] - \frac{2\pi}{eBc^2} (E_{\text{inj}} + k\Delta E) = n\tau,$$

or

$$(29) \qquad \frac{2\pi}{eBc^2} \Delta E = n\tau = \frac{2\pi n}{\omega} ,$$

where ω is the angular frequency of the radio frequency applied in the cavity. Equation 29 can then be solved for the magnetic field B required to satisfy the microtron resonance condition with a particular frequency ω and energy gain ΔE. The result is

$$(30) \qquad B = \frac{\omega \Delta E}{ec^2 n} .$$

This result, although apparently indicating that a practical microtron is feasible, nevertheless exhibits certain disadvantages. To begin with, if we put into Equation 30 values for ω and ΔE corresponding to an S-band RF source and reasonably attainable cavity fields, B turns out to be quite small (about 1000 gauss) even if we choose $n = 1$, as is usually done. A high-energy machine must then be inconveniently large. Although one could raise ω by going to K-band RF, and this would reduce the size of the cavity and hence the gap in the magnet in which the cavity has to fit, the smaller cavity would also not sustain high RF fields, so that ΔE would also be reduced. An interesting point here is that after injection an electron in the first orbit must traverse that orbit in some integer number, say n_o, of RF periods in order to enter the cavity in phase. Thus we also require $2\pi E_{inj}/eBc^2 = n_o\tau$, or, after substitution from Equation 30,

$$\frac{\Delta E}{E_{\text{inj}}} = \frac{n}{n_0}\,. \tag{31}$$

With $n = 1$, the minimum injection energy ($n_o = 1$) will be the same as the cavity energy gain, and in any case we want the electrons to be fully relativistic after injection, which means that both E_{inj} and ΔE should be at least 500 kV. As there is little room for a 500 kV electron gun in the conventional microtron setup, this poses a serious problem, although a solution has been proposed by Manca *et al.*[18] in the form of an annular electron gun placed at the entrance to the cavity and through which the circulating beam can pass. Electrons in the first orbit then have energy ΔE so that $E_{inj} = \Delta E$ and Equation 31 is satisfied with $n = n_o = 1$.

A final problem is producing a suitable ΔE in a CW machine, which thus requires a CW microwave source capable of a continuous power output sufficient to maintain at least a 500-kV amplitude in the cavity. A design addressing all these problems was nevertheless developed by a graduate student, Mr. Peter Wintersteiner, and myself[19] for a 15 MeV CW machine. With the help and support of both Bob Cohen and a senior member of the department, Prof. E. C. Booth (who hoped he could have an accelerator laboratory at Boston University rather than having to use MIT's medical Van der Graaf at night), a proposal was written for such a machine but unfortunately, in view of the presence of accelerators across the river at both Harvard and MIT, was not funded. It was at about this time that Bob relinquished the department chair and I went off on sabbatical leave to Canada, where a project developing a variation of the conventional microtron called a racetrack was being started at the University of Western Ontario in London.

A racetrack microtron is a development of the conventional machine in which the circular magnet is cut in half to form two D-shaped magnets each of which bend the beam through 180°. The two magnets are separated to produce a straight section between them, so that the orbits look like racetracks, hence the name. The insertion of straight sections in each orbit modifies Equation 30 to relieve the stringent requirement on B and also provides room for the cavity, which can now have several sections to provide an adequate ΔE. In fact, the multisection cavity can take the form of a small linear accelerator, and in designing the overall machine for a particular energy, one has to decide whether to have a long linac and few orbits or a short linac and many orbits. A fine distinction exists between a racetrack microtron, in which the predominant feature is the array of orbits, and a recirculating linac, in which the predominant feature is a big linac which is reused a couple of times by a beam recirculated in one or two orbits. However, the focusing feature of the conventional microtron is lost in the racetrack, as can be seen from the following: It can be shown that a 180° bending magnet produces weak focusing with the source point and focal point at zero distances from the entrance and exit, i.e., with zero object and image distances. Hence in the conventional microtron, whose circular magnet may be thought of as two 180°-bending magnets put together, the focusing condition is satisfied, the focal point after the first 180° bend being the source point for the next. The presence of the straight sections in each orbit destroys this situation, and special steps have to be taken to restore focusing. Froelich[20, 21] was able to shape the magnet pole pieces to achieve this goal, and Prof. J. William McGowan, then head of the Physics Department of the University of Western Ontario, sought to take advantage of the microtron expertise in his faculty by negotiating a contract for the development of a Froelich-type racetrack microtron with the Commercial Products Division of Atomic Energy of Canada, Ltd (AECL), Canada's counterpart of our Atomic Energy Commission. The idea was to develop a machine to be used for cancer therapy, and in fact Froelich and his group developed a variable energy racetrack microtron having a three-section accelerating cavity delivering a ΔE of 1 MeV and eighteen orbits for a maximum final energy of 18 MeV. A successful prototype was built, but because of various political machinations a recirculating linac designed by the AECL's own people at Chalk River was given preference. Ironically, this machine gave a lot of trouble, and AECL finally arranged with the French CSF company to import their Sagittaire and Neptune linacs for cancer therapy. A final irony is that, twenty years later, my late wife was treated on a Neptune linac that I had helped install.

III. TEACHING

That Robert Cohen was one of the best teachers at Boston University is a widely recognized fact, and he was an inspiration to those in physics instruction. He was instrumental in getting the physics part of the six-year medical program installed, having brought in the late Prof. J. Gordon Stipe, who had just published a combined physics and calculus text and was eager to organize the appropriate course. Gordon and I hit it off immediately, and I started my teaching career as his assistant.

In addition to the regular physics courses for science majors, there were also the "poetry physics" courses, the service courses, which included those for liberal arts majors, the School of Education students, and the physical therapists of Sargent College. With his abilities as both a teacher and a philosopher, Bob was the ideal man to take on such courses, and it was with the greatest pleasure that I served with him in this work and learned from him how to talk science to students who are not going to be scientists and are in a physics course only because they have to be. Lectures to such classes constituted my toughest teaching assignment, for I had to speak their language rather than the natural language of physics. It was also probably the most rewarding.

Another interesting feature of Bob's department was its response to the anti-science feeling prevalent in the late sixties. Then anyone admitting to being a chemist or a physicist was automatically scorned as a weapons maker, and the only way a student in a science course could hold his or her head up was to claim he or she was only there preparing to become a doctor. There were few physics majors, and even in the service courses enrollment dropped off seriously. Bob led his department in organizing a number of so-called "interdisciplinary" courses – physics and philosophy, physics and literature, physics and religion, physics and politics – to attract students. My contribution was a course called "How Things Work," which purported to show both how the laws of physics lay behind the working of various examples of everyday technology and how that technology was a fine demonstration of those laws. The course relied heavily on demonstration, and a devoted graduate assistant, Carl Russo, now a distinguished solid-state physicist, and I spent many hours developing them. We demonstrated Newton's laws, showed what to expect with electric and magnetic fields, built a model electric motor, and assembled a complete automobile engine in lecture, explaining each part as we went and finally starting the completed engine, somewhat to the discomfiture of neighboring classes. We also delved into electronics, showing what a vacuum tube and a transistor were all about and building breadboard-style a complete radio transmitter and receiver. Finally we showed the basic scheme for the transmission of television pictures. Each student was given a sheet of graph paper, and the lecturer had an identical sheet with a simple picture drawn on it. He would then call out the color of each square in order, starting in the upper left-hand corner, reading across the top row of squares, and then proceeding to the next row, until every square had been called out. At first, graph paper with large squares was used, so that the reproduced pictures were very rough but could be transmitted in a relatively short time. We then tried smaller squares, which gave higher definition for the reproduced picture but took longer to call out. Thus was illustrated the meaning of definition in television, so that the students would know what High Definition Television meant, and in addition they learned what a scanning line was and why synchronizing signals are necessary. Moreover, and perhaps most important, was the illustration of the compromise between time to transmit a picture and the definition obtained. This is one way of looking at the very basic engineering trade off between required bandwidth and rate of transmission of information. An interesting example of this is the fact that facsimile transmission uses essentially the same process as television but can be sent over ordinary telephone lines, whereas television signals require special coaxial cables to handle the high frequencies involved. Why? Because you can't send a movie by fax! It takes a couple of minutes to transmit a page, whereas you can televise a movie – the illusion of motion – because in television a complete picture is transmitted every 1/30 second. Carl and I had a lot of fun with this course, which was a great success, going from a small seminar of fifteen students to an enrollment of over 150 in subsequent years, but Bob deserves much of the credit for its existence. I might add that my own successful progress along the tenure track was largely due to his support and his idea that they also serve who only stand and teach. I hope that the Festschift for which this paper is written can suitably honor the man who laid the foundation for what is now a world-class department in both research and instruction.

Boston University

NOTES

[1] See, for example, W. Gerlach and O. Stern, *Zeitschrift für Physik* **8**, 110 (1921), and **9**, 349 (1922). Later papers by the same authors are *Annalen der Physik* **74**, 673 (1924) and **76**, 163 (1925).

[2] N. F. Ramsey, *Molecular Beams*, the Clarendon Press, Oxford (1956), pp. 398, 399.

[3] S. Millman, I. I. Rabi, and J. R. Zacharias, *Physical Review* **53**, 384 (1938).

[4] D. S. Edmonds, Jr. and M. R. Corson, *American Journal of Physics* **54**(9), 811 (1986).

[5] Taken from Ramsey, *op. cit.*, who adapted it from J. M. B. Kellogg and S. Millman, *Rev. Mod. Phys.* **18**, 323 (1946).

[6] G. Breit and I. I. Rabi, *Physical Review* **38**, 2082 (1931).

[7] R. D. Haun and J. R. Zacharias, *Physical Review* **107**, 107 (1957).

[8] N. F. Ramsey, *Physical Review* **78**, 695 (1950).

[9] N. F. Ramsey, *Molecular Beams*, *op. cit.*, pp. 124–134.

[10] W. Franzen, P. B. Newell, and D. S. Edmonds, Jr., *Physical Review* **170**(1), 17 (1968).

[11] H. H. Stroke, V. Jaccarino, D. S. Edmonds, Jr., and R. Weiss, *Physical Review* **105**, 590 (1957).

[12] E. Fermi and E. Sergré, *Zeitschrift für Physik* **82**, 729 (1933).

[13] A. Bohr and V. F. Weisskopf, *Physical Review* **77**, 94 (1950).

[14] V. I. Veksler, *Dokl. Akad. Nauk., SSSR* **43**(8), 329 (1944).

[15] P. A. Redhead, H. LeCain, and W. J. Henderson, *Nucleonics* **5**, 60 (1949).

[16] H. R. Froelich and E. Brannen, *Journal of Applied Physics* **32**, 1179 (1961).

[17] S. P. Kapitza, V. P. Bykov, and V. N. Melekhin, *Soviet Physics, JETP* **14**, 266 (1962).

[18] J. J. Manca, D. S. Edmonds, Jr., and H. R. Froelich, *Review of Scientific Instruments* **47**(9), 1148 (1976).

[19] P. P. Wintersteiner and D. S. Edmonds, Jr., *IEEE Transactions on Nuclear Science*, NS-14, No. 3, 749 (1967).

[20] H. R. Froelich, A. S. Thompson, D. S. Edmonds, Jr., and J. J. Manca, *IEEE Transactions on Nuclear Science*, NS-20, No. 3, 260 (1973).

[21] H. R. Froelich, A. S. Thompson, D. S. Edmonds, Jr., J. J. Manca, J. W. McGowan, J. C. F. MacDonald, J. Beard, and G. Bees, *Nuclear Instruments and Methods* **143** (1977).

AANT ELZINGA

EINSTEIN IN THE LAND OF NOBEL: AN EPISODE IN THE INTERPLAY OF SCIENCE, POLITICS, EPISTEMOLOGY AND POPULAR CULTURE

1923 was the year Einstein was in Sweden to give his Nobel lecture. Little has been written about this episode, and there is some confusion as to certain details.

Among the factors that easily contribute to misunderstandings concerning Einstein's sojourn in Sweden, one is the discrepancy between what one would normally expect in this connection and what actually occurred (the expected is that it was relativity that was being recognized, and that he went to Stockholm where the Nobel festivities take place each December). Another complicating factor is the symbolics which surrounds this prestigious prize-giving activity, something going far beyond the arena of pure science to touch other realms. In this case recognition in the scientific arena simultaneously had a significance in the domains of politics, epistemology and popular culture.

That award of a Nobel prize to Einstein was overlaid with political dimensions is immediately evident if one considers that it was finally the German Ambassador to Switzerland who delivered the diploma and medal to Berlin. Both Prussia and Switzerland claimed Einstein as "their" citizen, and in Stockholm the German and Swiss embassies contended with each other because both wanted to be the ones through whom the award would be sent to the prize winner. The solution arrived at was compromise, and in line with Einstein's own request.[1]

Einstein was a very popular figure at the time, much in demand in many countries, and not only in scientific circles. His inability to come to Stockholm at the time of the Nobel festivities was due to a lengthy trip during the winter of 1922/23 to Japan where he had been invited by the publishers Kaizo-Sha. On the boat trip back he stopped off in Palestine and Spain, meeting Zionists, Spanish royalty and anarchist students and workers. When he came to Sweden it was to participate in the 17th Meeting of the Scandinavian Association of Natural Scientists (*Skandinaviska naturforskarmötet*), which was held in Göteborg on the Swedish west coast 9–14 July 1923. This was in connection with the city's tercentenary celebration of its foundation. It was here in the midst

K. Gavroglu et al. (eds.), Physics, Philosophy and the Scientific Community, 73–103.

of a heady optimism and belief in the future of work and industry, underlined by a vast Exhibition and Trade Fair, that Einstein gave his Nobel lecture. The topic was one for which he had explicitly been refused the prize, relativity theory. Appropriately, the actual site was the Jubileum Hall on the premises of the currently still popular amusement park Liseberg, which had just been opened two months earlier.

The story of Einstein and the Nobel prize is full of compromises, symbolic meanings, surprises and curious anecdotes. My intent here is to unravel some of these. The point will be to show how some important events in the history of science can be understood simultaneously from two different points of view, one internalist, having to do with science, the other externalist, i.e., in relation to various contexts outside the process of peer review and recognition of achievement within science itself. It is not a question of the usual distinction between externalism and internalism, as for example between idea and context, or cognitive and social. Rather it is between science and politics, and a case of both/and, since some events associated with science may at the same time serve as a vehicle for (non-governmental) political action or opinion. It is found that scientific prestige and credibility-enhancing institutions generate symbolic capital that can be converted into political currency.[2]

In the case of Einstein and the Nobel prize some of this interplay of two kinds of credibility cycles, one scientific and one (non-governmental) political is also at work, but in a rather complex way.

In what follows I shall first depict a series of events in July 1923 as they appear at first glance. Thereafter I shall try to unpack various dimensions of symbolic significance related to a number of different contexts. In brief I shall touch upon the dominant ideal of science that was firmly entrenched in physics and philosophy departments in Sweden. Then something will be said about the symbolic significance of Einstein's appearance in Göteborg (not Stockholm) as a cultural event. Finally, the wider political significance of awarding the Nobel prize in the immediate post-World War I situation will be explored.

EINSTEIN AT THE AMUSEMENT PARK IN GÖTEBORG

On Sunday the 9th of July 1923 Albert Einstein arrived in Göteborg. He was invited by Svante Arrhenius who was on the program committee of the 17th Meeting of the *Skandinaviska naturforskarmötet*, which the city hosted that year. In all about 500 scientists from various parts of Denmark, Finland, Norway and Sweden attended. The different sections of the congress met in three main locations, zoologists near the botanical gardens at Slottskogen, physicists, chemists and physiologists on the premises of the Chalmers Technological Institute, and astronomers, geologists, botanists and meteorologists in the main building of the Gothenburg College, which today houses the central administration of the University of Göteborg and its Rector.

Tuesday afternoon the 11th of July the various sections cancelled their programs to allow everyone to join the city's commercial, industrial and political elites in the glass-walled Jubileum Hall in the Liseberg Amusement Park. The *Stockholm Dagblad* ran the story the next day, under the headline: "Einstein talks in stifling heat before an audience of 1000. King present during the celebrated lecture". The Swedish King, Gustaf V sat in a special seat between the two front rows facing the raised podium on which Einstein was standing to deliver his Nobel lecture. It was a popular presentation, in German, entitled "Basic ideas and problems in relativity theory".

During the course of an hour he presented some of the ideas that can be found in the work he completed in 1916 and published in the following year, *Über die Spezielle und die Allgemeine Relativitätstheorie, Gemeinverständlich.*[3] This, his most widely known book, brings together in popular form the special theory of relativity of 1905 with the general theory that was formulated ten years later. In the wake of the famous solar eclipse experiment of 1919, it was widely disseminated throughout the world at that time. An English translation came in 1920, followed during the next year in rapid succession by Spanish, Italian, Russian and French translations. The immediate motive for the translations was interest stirred by the event of 1919.

It should be remarked at this point that in Sweden the first translation was published only recently, in 1989.[4] This lag of almost 70 years behind other countries may be indicative of the fact that for a long time the educated public in this country had no difficulty reading German, a situation that changed after the Second World War (on the other hand one can sometimes wonder if this lag is not also a measure of the early unreceptiveness to Einstein's ideas to be found in established circles).

Photographs of the solar eclipse had been taken at two positions on the surface of the Earth, one on the island Principe in the Bay of Guinea on the African west coast, the other at Sobral, above Fortaleza in Brazil.

After a certain period of analysis of the results, the Royal Society in London announced that these were in accordance with Einstein's prediction that distant stars in the vicinity of the sun (as seen from the Earth) were displaced by an arc of 1.7 seconds compared to the position in the sky where they would otherwise be seen if the sun was somewhere else. The refraction was twice what one might expect on the basis of Newtonian physics.

As is well known this victory for Einsteinian relativity was portrayed as a spectacular event in the mass media. Sensationalism mingled with genuine desire to understand the new world picture. After the report in the *Times of London* the 7th of November 1919, newspapers throughout the world picked up the event and soon Einstein's name became a buzz word. The Reuter telegram from London referred to the "greatest discovery in connection with the law of gravitation since Newton's day and one of the greatest achievements in the history of modern human thinking" (*Dagens Nyheter*, 8 Nov. 1919). Newspapers called it the verification of the "Einstein effect" and the man himself the new Copernicus.

In Sweden things were as usual a little calmer. The Stockholm daily cited above remained cautious and one writer held that it was misleading to say that "light like matter is subject to the law of gravity and can therefore refract or bend" in a gravitational field. In Uppsala one speculated that it was the magnetic field of the sun that influenced the light. In the *Göteborgs Handels-och Sjöfartstidning* (11 Nov. 1919) there appeared a cautious article, which ended with the comment that one waited expectantly to see what scientific journals would write. We are also told that a professor Hugo von Zeipel in Uppsala already in 1914 had travelled to Strömsund in the north to investigate if there was any interaction between gravitation and light rays. The weather conditions however were poor, and the observations failed.

Dagens Nyheter of 1st of December 1919 carried an article signed by K.G. Dernby, who gave a popular presentation of Einstein's ideas, among them the notion that time is not independent of space but rather a "function" of space. Here reference was also made to a book by the Swedish scientist The Svedberg entitled *Arbetets dekadens* (The Decadence of Work) which had appeared in 1915.[5] The title is a play on words, where work is used in the scientific sense of the term in thermodynamics, with attention focused on the loss of entropy. This work contains one chapter in which the author gives a brief exposition of

Einstein's special theory. Later in December 1919 the popular weekly read by many of the educated public, *Veckojurnalen*, carried a little item on Einstein and his great accomplishment. Even though some Swedish philosophers were to remain sceptical until well into the 1940's, by 1923 Einstein and his relativity theory had certainly captured the popular imagination, also in Sweden.

The Göteborg daily *Handels-och Sjöfartstidningen*, which prided itself on being factual and fairly highbrow, presented a thorough account of Einstein's lecture, in which the main ideas were unfolded in the same order as we find them in his popular book of 1917. From the newspaper report it is evident how Einstein emphasized not the revolutionary nature of his ideas but rather their continuity with tradition, showing the scope of validity of the classical principle of relativity and how the apparent discrepancy between Newtonian mechanics and the electrodynamics flowing from Maxwell's equations could be resolved. Consistent with this ambition he was obviously loathe to call himself a revolutionary. Instead he underlined the continuity between his relativity theory and Galilean-Newtonian physics. By dissolving the theoretical contradiction between Maxwell and Newton, between electrodynamics and classical mechanics he saw himself as having perfected the latter. The newspaper reporter notes how "for anyone who has heard various presentations of the theory of relativity before the lecture hardly gave anything new. It was by and large a quick review of the elements of the theory" (GHS 12/7 1923). What impressed this reporter most was Einstein's humility. "Characteristically the speaker did not use expression like, for example, 'my theory' or some such. Instead he began to speak of the theory 'which one has come to call the theory of relativity' ".

Evidently Einstein felt that his theory had become the victim of a colossal mistake. The wrong name had been attached to it. It conjured up wrong associations, especially since people connected it with everything's relativity and the lack of secure values. An alternative had been to call it "Invariantentheorie", but now it was already too late, since "relativity theory" had become the commonly accepted term and a change of name would only add further to the confusion.

By "invariance" he meant the basic trait of the theory, that it gave prominence to symmetry, harmony and constancy. In an interview with another reporter in answer to the question, "Do you find the worlds of music an mathematics are similar?", he responded, "Naturally! They both instill in us exactly the same architectonic enjoyment of beauty" (DN 11/7 1923).

The reporter for the Stockholm paper *Dagens Nyheter* who wrote this differed from his Göteborg colleague in the account of the Nobel lecture. He seemed less intent on the scientific content than on the person and the popular media image of the magician, muse and cult figure. Thus he wrote how "the silence was endless, and when professor Einstein in order to concretize his lecture a little bit gestured elegantly up in the air towards the hall's voluminous arch of the ceiling when he spoke of different coordinate systems, or pointed to the corner and mid-point of the lecture to clarify his definition of the concept of simultaneity, many people in the audience with perhaps already overworked intellect stretched their necks as though they were witnessing a spiritist experiment or the transformation of spirits (from the other side of the grave) into matter" (DN 12/7 1923).

A POET AND A MUSICIAN

On the previous day the Nobel laureate had attended a smaller gathering in the new physics room of the Chalmers Institute to give a lecture to a select group of scientists in the physics section of the ongoing congress. The *Dagens Nyheter* man was also there. This time he wrote: "The face could have been that of a musician or a poet just as much as a scientist's . . . and what a voice the man has! Low, soft and soothing. One might think that he was reading a poem of Heine when in actual fact he talked about the most intricate problems" (DN 11/7 1923).

The topic this time was, "Zur Weyl-Eddingtonschen Theorie", in other words it dealt with contemporary attempts to develop a unified geometrized space-time theory. Arvid Hedvall, a lecturer at Chalmers was present and later in his memoirs recalled the discussion that broke out after Einstein's paper. "I remember especially a temperamental statement by Abraham Langlet, but Albert Einstein took this opposition, like the other not very well informed interventions, calmly. I did not understand much either, and it was indeed rather indifferent if Albert Einstein was right or wrong in all that he said, since the obvious thing was that, whatever the case, a new epoch of mathematical-physical-philosophical thought had been introduced".[6]

To complete the story of Einstein's visit to Göteborg as it meets us at face value a few more details extracted from Svante Arrhenius' correspondence with one of the city fathers and patron of science, Dr. Gustaf Ekman may be added.[7]

During his visit, it was Svante Arrhenius who was Einstein's host at the congress of Scandinavian scientists. For the city of Göteborg it was Gustaf Ekman, director of the Carnegie Sugar Works and Brewery who took care of the illustrious guest. Einstein visited the Ekman family's country estate *Råda säteri* in Molnlycke just outside the city,[8] and he lived at Dr. Ekman's city residence, the Carnegie House, now known as the Chalmers House in central Göteborg overlooking the inner canal at Södra Hamngatan 11 where the Gothenburg Royal Society for the Arts and Sciences[9] currently has rooms on the second floor above the exclusive old bank offices of *Skandinaviska Enskilde Banken.* According to still living members of the Ekman family who were there at the time, in the mornings just after breakfast and before he went out of the house to join colleagues at the congress, their rather unconventional guest used to play a few strokes on his violin.

Actually Einstein had been slated to speak at or before the banquet that was held on Sunday the 9th of July after the provincial governor Oscar von Sydow had officially opened the Congress. However the guest of honour had made a mistake in his calender. Since he was under the impression that he was not due to speak until Monday the 10th he came with a later train via Copenhagen. Thus he arrived at the banquet during Dr. Gustaf Ekman's speech and very discretely tried to sit down somewhere in the back amongst the audience. The Chalmers scientist Arvid Hedvall writes that he remembers the moment he saw Einstein wandering about "at the gala dinner in the Exhibition Hall and pointed out to him that he naturally should occupy the place of honour" at the head table, with the King and Oscar von Sydow.[10] Other eyewitness reports mention how an excited murmur went through the crowd when it became clear who it was that had just come to sit down at the head table with the other dignitaries.

Later in the week Einstein apparently participated in the boat cruise in Skagerak, which Arrhenius organized for congress participants. Folklore has it that he probably also travelled in the cable car between the Liseberg Amusement Park and the upper part of the Industrial Exhibition grounds where the Main Exhibition Restaurant, as it was called, lay – the area where the University of Göteborg today has its Central Library and a large part of the Humanities faculty. In that case, so the story goes, he must have reflected on how he was prolonging has life perhaps by a nanosecond as he swung high in the air over passers by on the road below.

There is another curious story concerning the Jubileum Hall on the Liseberg fairgrounds. Apparently the benches on which the 500 scientists and 400 others were seated to listen to Einstein deliver his Nobel lecture were freshly lacquered. In the stifling heat of this unusually hot summer day some of the audience found themselves glued in their places when they tried to leave. If it is true, it was befitting the occasion, as was the fact that the lecture was held in an amusement park. After all it was on relativity theory, a subject quite different from the one that was the ostensible motivation of his Nobel prize. He had been awarded the prize for the "wrong" accomplishment.

DELIBERATIONS AND DILEMMA AT THE ROYAL ACADEMY OF SCIENCES

Some scientists, like Wolfgang Pauli, have been amused by or derided the fact that Einstein was awarded the Nobel prize for his "discovery" of the law of the photoelectric effect rather than Einsteinian relativity. Indeed, even before 1922 whenever Einstein's name came up on the list of nominees for the prize it was mostly in connection with the theory of relativity. From 1916 to 1922 his was a top name, supported by leading physicists all over the world. Among the roughly fifty nominations he received since 1910, the photoelectric effect is only mentioned twice, by one and the same person, the Swede Carl Wilhelm Oseen, professor in mechanics and mathematical physics at Uppsala.

Oseen was, together with the mathematically inclined Henning Pleijel at the Royal Technical Institute in Stockholm, one of the younger generation who acted as leading spokesman for Einstein in Sweden. In 1922 Oseen was made an extra special member of the Nobel Committee for Physics, where he had his chance to push for Einstein's nomination. The moment was a propitious one.

The Nobel Committee for Physics had been unable to decide Einstein's case, neither in 1919, nor 1920, nor 1921. At the same time international peer pressure in favour of Einstein was steadily mounting. Therewith the Einstein case had gradually turned into a dilemma for the Swedish Royal Academy of Sciences, since the Nobel Committee within its ranks had a strong bias against relativity theory. It simply lacked the necessary competence to satisfactorily review the work in question.

Being of such a highly theoretical but fundamentally revolutionary and daring nature, it went against the grain of the experimentalist ideal of science that was so firmly entrenched, especially in Uppsala, where a majority of the Committee members had their affiliation. Consequently it is not surprising that relativity theory was regarded as too speculative, lacking in empirical foundations, and adherence to it more an article of faith than anything else. For the Nobel Committee to endorse a theory just to please foreign peer opinion, while a majority within its own ranks was convinced the theory was fundamentally flawed, would have been a breach of integrity.

Dependence on outside assessments moreover involved an element of risk. Who knows, maybe the general theory of relativity was only a passing fad that would fade; it had certainly not been validated to the satisfaction of the hard core experimentalists – thus, a veteran member of the Committee might reason. In addition philosophical colleagues in Uppsala were claiming that they had demolished the epistemological foundations of Einstein's theories of relativity, and therewith denounced the man's work as belonging to the realm of metaphysics, not science.

Finally, the daily newspapers were featuring the strongly diverging opinions, with arguments for and against Einstein, in some cases tending to fall back on ideological differences. Anarchists, social democrats and communists were speaking up for Einstein, while conservatives and the ultra-right were against both the man and his ideas. He had become a symbol for modernity and revolution and hence a kind of litmus paper test for political color.

This in short was the climate of opinion in which the Physics Committee deliberated over whether or not to recognize Einstein's achievements, and if so, on what scientific grounds. Given all the commotion and controversy there was every reason to proceed with caution. Unfortunately the mismatch between the reigning ideal of science and Einstein's revolutionary theory was so great that however many reviews the Committee undertook it would probably not have been sufficient to overcome the existing bias and negative verdict. It was here Oseen was able to come in and show a way out of the dilemma of inaction.

As noted, the Committee had already had the theories of relativity under review a couple of times. In 1919 the conclusion was that one should wait until the phenomenon of the red shift was clarified and the results of the solar eclipse had been worked through. The following year Arrhenius wrote a special report on the general theory. Here too the conclusion was negative: the red shift actually measured did not match with the theory, and criticisms had been levelled at the results of the solar eclipse experiment. In 1921 the Committee assigned the task of reviewing relativity theory to Alvar Gullstrand, a leading authority on optics. He too concentrated especially on the empirically testable consequences predicted by general relativity. His conclusion was also negative. "The effects that are measurable with physical means", he wrote, "are, however, so small, that in general they lie below the limits of experimental error".[11] Oseen for his part had come up with the proposal that Einstein should receive the prize for his work on the photoelectric effect instead, since this effect by now had experimentally been well verified, among others through some work by Millikan. Arrhenius was given the task of reviewing Einstein's contribution on the photoelectric effect. For his part he had no objection to this work, but when it came to awarding a prize he thought that in this case it would be better to give it to an experimentalist who had dealt with the phenomenon. Thus the Committee was unable to arrive at a decision in 1921. Unable to agree to give it to Einstein; the physics prize for that year was not awarded.

In 1921 Gullstrand published his critical analysis of the problem of the planet Mercury's perihelion and remarked that in the mathematical treatment there was an integration constant whose value could be arbitrarily chosen. Einstein had chosen to give it the value zero to get his result. Gullstrand maintained that this was a weakness, since it meant that the problem would have an infinite number of solutions. Against this Oseen replied in the same journal that there was nothing wrong with this, since the main thing was to use the observations to see if the theory could explain these. "Was man von der Theorie verlangen kann, ist nur, dass sie mit den Beobachtungen übereinstimmt".[12] This led to a sharp exchange in the following year between Gullstrand and another scientist who criticized the calculations the Uppsala authority on optics had made. Gullstrand brusquely replied that his critic was way off the mark and that there was no point in discussing the matter further. In his view acceptance of Einstein's relativity theories was simply "a matter of faith (*trossak*)". Privately he had stated that Einstein should never receive a prize.[13]

In 1922 the play was repeated. Gullstrand presented an update of his earlier review of Einstein's general theory, again with a negative verdict. B. Hasselberg another members of the Committee, who died later that same year, asserted from his sick bed that "it is highly improbable that Nobel considered such speculations as these [i.e., Einsteinian relativity] to be the object of his prize".[14] Oseen as a special extra member repeated his compromise proposal to award the prize to Einstein for, as he called it, the photoelectric law. This time the proposal was presented with more tactical polish, and the argumentation was reinforced by linking it to the positively received idea to recognize Bohr's work in atomic physics. Thus it was that the frozen award of 1921 went to Einstein for "the wrong" thing and the 1922 prize went to Niels Bohr for atomic theory. "Almost all confirmations of Bohr's atomic model", Oseen wrote in the general report for the Committee, "are also confirmations of Einstein's principle".[15]

In the official motivation for Einstein's prize, relativity theory is explicitly disavowed. It is not attributed any role in the recognition accorded him. On the contrary, in the covering letter accompanying the original diploma one can read in stately Swedish how the prize has been awarded "without taking into account the value which will be accorded your relativity and gravitation theories after these are confirmed in the future".[16] To be sure, the prize was explicitly disassociated from these theories that had caused the Nobel Committee such headaches. Also the motivation refers to Einstein's "discovery" of the law of the photoelectric effect. Note that even here it is not for his quantum hypothesis used to explain the effect, but rather the "effect" itself. In an earlier draft Oseen had included a reference to Einstein's theoretical explanation of his "discovery", but in the final report this was taken away. Instead the law of the photoelectric effect is now recorded as having contributed to the development of quantum theory in general.[17] Consequently in connection with the photoelectric effect which was finally settled on, it was apparently to be the "discovery" of a law that could be awarded a prize, but not an "explanation" of the same.

AT ODDS WITH THE REIGNING IDEAL OF SCIENCE

It will be noted how the Committee's verdict on Einsteinian relativity was entirely consistent with its ideal of science. As such it was also in line with the view of the experimental scientific establishment in Sweden. There was a strong aversion to what was considered pure theory and abstraction. This standpoint is perhaps most clearly reflected in an early article on relativity in 1918 by a younger researcher, Hilding Faxén (later professor of mathematics at Chalmers). He was not unsympathetic to Einstein and was at pains to give a just account of the general theory. However his conclusion was coloured by the reigning ideal of science within the research culture in which he had been brought up. "Einstein seems to have understood *theory* [as such] as the goal of physics and this is typical for the younger. Older persons, who have lived through many theories . . . are more inclined to be satisfied only with factual knowledge, i.e., experimentally demonstrated theories, and they are sceptical when it comes to theories which cannot be verified. A theory, which is not permeated by facts, lacks in beauty for this very reason, and one can feel that it is a simple and uncultivated taste that gets carried away by such [abstract] theories".[18]

The majority of expert voices in Sweden at this time were thus negative. This included the Nobel Physics Committee, at least until 1923, when the strikingly narrow instrumentalist experimental ideal of science was softened by the election of two younger generation physicists, Oseen and Manne Siegbahn, replacing two older members who died in 1922. Oseen referred to it as the start of "a new era", and envisioned the creation of a Nobel Institute of Theoretical Physics.[19] However in 1922, just as in 1920 and earlier, the strongly instrumentalist and experimental physics ideal continued to exert its grip. An indication of this is 1920, when Einstein, a top name on the nominations list was passed over, and the award was given to Charles Guillaume, head of the International Bureau of Weights and Standards in Paris, for his "services in connection with precision measurement in physics through the discovery of anomalies in nickel-steel alloys".[20] In an analysis of the nominations for the physics prize, 1907–1919, Bengt Nagel has classified eight of a total number of twelve prizes as experimental, two as theoretical and two as mixed. The two purely theoretical ones were for Max von Laue's work on X-ray diffraction in crystals (1914) and Planck's discovery of energy quanta (the prize of 1918 awarded in 1919).[21]

Apart from sub-specialties in physics, research achievements can also be grouped in terms of their function in relation to the development of fields of knowledge. Küppers, Ultizka and Weingart in a review of decision-making within the Nobel Committee for Physics suggest a categorization along the following lines, with reference to various examples:

- new observations that extend existing knowledge (X-rays, radioactivity, the Zeeman and Stark effects);

- explanations of observations which allow their integration into theoretical concepts which, in turn, are often changed in the process (classical theoretical physics, van der Waals, Boltzman and Heavyside, and also Planck's radiation law);
- new methods and instruments of measurement which open up new possibilities for observation and extend the realm of experience (Wilson, Hale and others);
- precision measurements and experiments that make it possible to decide between competing theories (von Laue, Michelson and others);
- new theoretical concepts which allow an integration of experimental anomalies and revolutionize traditional principles (Planck, Einstein).[22]

During the early period of the Nobel institution the first four categories were in total dominance. The first three relate to traditional areas of cognitive action in science: experiment, construction of theories and development of methods. Precision measurement is a sub-category of the first, but it is taken up as separate category here because it was much discussed by the Nobel Committee. Guillaume's work is an example. All four of these are areas the Nobel Committee was able to deal with. It was nominations in the fifth category that caused trouble. It was a question of unfamiliar territory, with strange concepts underlying measurement and what is counted as "experience". In other words, novelty in this case seemed basically counter-experiential.[23] For the traditional experimentalist relativity theory was necessarily at odds with experience in the accepted sense of the term. This is evident from Gullstrand's report of 1921, where he wrote that for him a theory "is a summing up of a number of descriptions of related phenomena".[24] The categories and concepts used have to be derived from experience, and the theory has to deduce the observed phenomena from earlier experience, which has been recorded in previous works of science. Thus something that makes a radical break with previous work will, for someone like Gullstrand, appear to totally contradict accepted "experience". The definition of what counted as experience was contextually dependent. In a sense it reflected the cultural practices of experimental science at a particular conjuncture in time. The cultural practices of laboratory life furthermore had a wider context in which the anti-theoretical bias was further reinforced.

It should be remembered that Sweden was an industrializing nation many of whose industries were built on mechanical invention. This was an important cultural factor underlying the mandate of the Nobel

Foundation. Alfred Nobel himself had made his fortune in munitions among other fields. In his will he explicitly stated that the prize should be to recognize tangible achievements in science – discovery and invention. Mathematics was definitely excluded.

In terms of its academic tribes, the culture of Swedish physics in Sweden as a whole was also strongly experimental, as distinct from theoretical. Mechanics and mathematical physics was a minority field. An indication of the degree to which this was the case may be found by looking at the topics of all doctoral monographs produced in physics and mechanics at Swedish universities during the years 1910–1925. In this period 46 doctorates were published, the majority of them (54%) at Uppsala, followed by Lund (35%) and least in Stockholm (11%). Almost all of the topics taken up are experimental ones; hardly a handful of dissertations deal with topics that might be said to have a predominantly theoretical slant.[25]

It is not so strange then that scientists in the Nobel Committee had difficulty with Einstein. They were incapable of coping with Einstein's theories of relativity, since these were perceived as having too weak an empirical base and too strong a mathematized flavour, with abstract concepts. In experimentalist circles an earlier rule of classical physics seemed to reign: that whatever one presents in concepts one should also be able to "grasp" physically, at least with the help of instruments.

Significantly enough, at the Nobel festivities in Stockholm on the 11th of December 1922, Arrhenius commented that the prize-awarding authority was of the opinion that relativity theory belonged to the theory of knowledge rather than physics (see below).

SOME FURTHER TWISTS

The irony of the story is that it was not Einstein who first discovered the photoelectric effect, but rather Hertz, and his pupil Halbwachs carried out the experiment that contributed to making the phenomenon more broadly known among researchers. Phillip Lenard had also worked in this area, and for his part he felt himself bypassed by Einstein when the latter published the theoretical explanation or hypothesis in 1905.

Oseen, who had come up with the compromise was probably unconcerned about this. It was a question of doing Einstein justice. Moreover, he had some important interests of his own at heart. He wanted to develop atomic research in Sweden, and therefore it was quite convenient if

Einstein's work on the photoelectric effect of 1905 was taken up instead of the unacceptable relativity theories. In his report, even with its tactical tone, he was able to point to an important link with early quantum physics. The two Nobel prizes in 1922 helped put the spotlight on this field, lent it glory, and put it squarely on the national research policy agenda. In this case the symbolic capital inherent in the Nobel prize was destined to translate into political good will and hard cash for Swedish atomic science.[26]

It has sometimes been suggested that because of its strong Germanic orientation the official Swedish science community did not want to irritate their German colleagues by recognizing Einstein for his work on relativity. It was after all this work that had become the target of a German nationalist campaign 1920, in which it was proclaimed that as a Jew and pacifist Einstein's ideas were being pushed into the limelight by an international conspiracy of Jews, Communists and key people in the mass media. Philip Lenard had explicitly branded relativity theory as a Jewish creation. The members of the Swedish Academy of Sciences certainly did not agree with this view, but it is possible, so the claim goes, that they were concerned not to antagonize their German colleagues by further fanning the flames of the controversy that had been generated in Germany. Moreover, it was not until 1922, when Arrhenius and von Euler personally were present at a conference where Lenard was carrying on, that they realized what was really happening. Thence they were keen to distance themselves by signalling positive recognition of Einstein. Thus it was quickly decided to award Einstein the prize, but not for relativity theory.[27]

Elisabeth Crawford, who has pored over the contents of the Nobel archives more than most maintains that there is no evidence that can be mustered to substantiate the foregoing reconstruction of what might have happened. On the other hand it is clear that the main thing that weighed against relativity theory is the ideal of science that was so firmly entrenched in Swedish research communities and in particular, at least until 1923, in the Nobel Committee for Physics. Also, as far as the politics of symbolic representation is concerned, the main thrust in the awarding of a prize to Einstein was if anything to bridge the gulf that had arisen during the world war between scientists in the Allied and the German camps. Scientists in the neutral countries like the Netherlands and Sweden did what they could to repair scientific internationalism.[28] In this respect the Swedish Nobel Institution had an important role to play. Before going on to the symbolic value attached to Einstein's sojourn in the land of Nobel, it remains to say a few words in connection with Arrhenius' comment to the effect that Einstein's work on relativity lay more in the realm of epistemology than physics.

PHILOSOPHICAL RESISTANCE

In his speech during the Nobel festivities already referred to, Svante Arrhenius stated: "There is certainly no other currently living physicist whose name has become so well known in such wide circles as Albert Einstein's. Most discussed is his relativity theory. It concerns to a large extent the realm of theory of knowledge and has therefore been vigorously discussed in philosophical circles".[29] To be sure, Einsteinian relativity evoked many critical comments from the leading gatekeepers of the Swedish philosophical establishment. In the main they clung to the classical concepts of space and time, rejected Einstein's innovations and insisted that his theories contained contradictory concepts and mixed fiction with fact.

The Finnish philosopher Georg Henrik von Wright, who is well acquainted with Swedish academic life, has suggested that Einstein did not get the Nobel prize for relativity theory because of the critique that came from philosophical quarters. He refers therewith to two "influential and sharp Swedish philosophers Hägerström and Phalén who could not accept Einstein's new view of space and time".[30] Now, this is to give too much credit to the philosophers, but certainly considering Arrhenius' comment above, their views may have had some influence in the overall deliberations that took place.

The stronghold of Swedish philosophy was the Uppsala school, which prided itself on its uncompromising stance of doing battle with metaphysics. The main figure in this school was Axel Hägerström who held the chair of practical philosophy from 1911 to 1933. His main comrade-in-arms was the chairholder in theoretical philosophy from 1916–1931, Adolf Phalén.[31]

Hägerström is best known for his theory of value-nihilism which is generally regarded as having had a strong impact in legal and social scientific thought. His book of 1908, *Das Princip der Wissenschaft* constitutes a rationalist critique of empiricism. In it he formulated a number of criteria to characterize his ideal of science. These include what he calls determinateness (*bestämdhet*), self identity, freedom from internal contradictions and presence of logical necessity. Truth was considered to be a function, not of agreement with an external reality, but absolute (immanent) knowledge which could be cast in an internally consistent and logical system.

His student Phalén adopted a more realistic epistemology, but agreed with his mentor that the task of philosophy was to fight error, not only in earlier philosophical systems but also in commonsense understanding of the world and science. His first independent work bears the telling title *Kritik av subjektivism* (1910). Anti-subjectivism was the banner under which the Uppsala school did battle. Anything that in their eyes looked like subjectivism and metaphysics became a likely target. Unable to fathom Einstein's ideas, they put them into that basket, and the man was accused of subjectivism and using contradictory concepts. Relativity theory in other words was easily defined as falling outside the domain of what could be considered science. It was metaphysics, pure and simple.

It was in a volume on philosophy published the year after Einstein had been in Sweden that Phalén maintained Einstein mixed fiction with reality.[32] Hägerström's and Phalén's students who came to occupy philosophy chairs in Uppsala, Lund and Stockholm echoed their masters' views as late as the latter part of the 1930's. One of them, Einar Tegen, wrote in 1936: "A contradictory theory can never be demonstrated – not even by any experiment whatsoever".[33] This was in defence of Phalén and the standpoint of Uppsala philosophy against an increasing number of scientists who were pointing out how history had already shown how wrong this school had been in their negative assessment of Einsteinian relativity.

Einstein did not fare much better in the eyes of other leading philosophers, at least not in the early 1920's. Allen Vannérus later in the same year Einstein visited Sweden wrote a newspaper article under the heading "Absolute time". With impulses coming from Høfding and Spencer, he had developed a kind of materialism, and saw himself as the defender of "material substance" and "objective and general fundamental time". He found that Einstein excelled in "the not altogether praiseworthy effort to relativize our conception of external reality. Nothing is allowed to be absolute: neither time nor space nor motion nor mass nor instruments of measurement; not even the most elaborate chronometer, without the qualification that it is thus and so, depending on different situations and points of view".[34] In Lund, Alf Nyman who was under the influence of neo-Kantianism wrote an article in 1924 where he took up a similar type of critique that came from Henri Bergson, who had questioned the relativization of "real" physical time. This then became an additional basis for the Swedish claim that Einstein mixed reality with fiction.[35] It was the culmination of a series of articles that had appeared in daily newspapers, both for and against Einstein during the years 1922 and 1923. Miller's repetition of the Michelson-Morley experiment during the years 1921–1925 drew considerable interest in Sweden. His finding of a difference in the velocity of light in the two directions instilled some hope in those who still believed in the ether. Anna Beckman, who was later to be the secretary of the Swedish Association of Physicists, noted in 1926, "if Miller's result is correct, the Einsteinian relativity theory falls".[36] At the same time she added that by virtue of the "logical mathematical truths Einstein has given us", he will not be forgotten.[37] Alf Nyman the philosopher was prompted to develop his Bergsonian critique of Einstein, now claiming that two methodological mistakes in Einstein's special theory had cancelled each other: "the relativization of time and length compensate for the contradiction originally introduced by the postulate of the constancy of light. It is a case of what Vaihinger calls an 'antagonistic operation' ".[38] It is granted that the theories are interesting and valuable "as if" constructions, but their claim to describe physical reality must be denied.

By 1930 physicists had come around to accepting the theories of relativity. Even Hilding Faxén, who in 1918 had declared that experimentalists had no use for Einstein's theory, now admitted his mistake, writing that it was beginning to show its fruitfulness as one of the fundamental principles of the new physics. It turns out that he actually became convinced of this already shortly after his earlier article, but it was not until now that he let this be known in print.[39] This may be an indication of the changing climate of opinion amongst natural scientists. Carl-Olov Stawström, whose interesting study I draw upon here, sees the early 1930's as the time of the breakthrough for special relativity in Sweden, and he points to two factors that helped it along. One is the experimental support the theory had gained by then; the second is the shift of generations in the physics community, where people like Oseen and Siegbahn now played a leading role. Oscar Klein, professor in Stockholm since 1930, was also instrumental in creating a better understanding of both special and general relativity in Sweden, and like Oseen he spent time dealing with the critique of the philosophers, showing they were off the mark.

In the mid-thirties the Uppsala philosophers were still attacking Einstein's theories, leading among others to a reply from Philipp Frank, who met Phalén's critique point by point in the newly established Swedish philosophical journal *Theoria* for 1937. A younger member of the Uppsala school replied that Frank had totally misunderstood Phalén, and that the latter's "Kritik der philosophischen Grundgedanken der Relativitätstheorie steht noch unwiderlegt".[40] In the same year Axel Hägerström, who had spent some 20 years formulating his critique of Einstein now published an article in a festschrift for Phalén. A more detailed presentation was made public in 1946, and finally in 1957 the complete dissertation appeared, in a collection of Hägerström's work brought out by his friend and student Martin Fries.[41] There one can read among other things that "the Einsteinian relativity theory . . . is a mathematical bastard of Humean sensualism in its present day form influenced by Mach, and therefore it falls into the same contradiction".[42] It was essentially the same tune that was played time and again, going back to the criteria and summary analysis Hägerström and Phalén had put forward at the very outset when Einstein's theories were still new.

When Tegen, a product of the Uppsala school, was installed as professor of philosophy in Stockholm in 1937, he also took the opportunity to try and demonstrate how Einstein's concept of time was wrong. This started a debate which lasted well into the early 1940's, with among others Oscar Klein and Victor Kraft coming out against the untenable position of the Swedish philosophers. The most polemical counter-attack however came from a philosophically interested engineer, Ragnar Liljeblad who for a time was the Chief Executive Officer of the company ASEA. He was so fed up with the backwardness of the philosophers that he felt compelled to speak out, and did so in the popular engineering sciences journal, *Teknisk Tidskrift*, under the heading "Swedish university philosophy and the physical world picture".[43] In a temperamental blast he asked, "what strange kind of clique existence is it that can render Swedish university philosophy alien to the greatest revolution in the field of theory of knowledge since the days of Kant?"[44] Nyman in Lund was also attacked, because "he shows how completely he has misunderstood the whole fundamental idea in relativity theory".[45] And further, "it is now 20 years since the general theory of relativity first saw light, and still in philosophical circles one has not yet understood its value (*sammanhanget*).[46]

The debate roused in *Teknisk Tidskrift* continued in several issues, with Oscar Klein supporting Liljeblad, although in somewhat milder tones. Liljeblad saw little hope for philosophy until a new generation had taken over. Especially singling out the Uppsala school, he wrote how "there is probably nothing more to do than to wait for a younger generation of philosophers to throw overboard their petrified masters and take on the challenge of really studying the philosophical significance of the decades past of revolutionary transformation of physical conceptualisations".[47]

In the debate that followed in *Theoria* Victor Kraft tried to mediate and dampen the intensity of feelings and invectives on both sides, calling for a rapprochement and mutual dialogue between physicists and philosophers. It was only with a new generation of philosophers that the conflict between modern physics and Swedish philosophy dissolved. The reorientation and refoundation of philosophy went along modern analytical lines with its extensive use of logical methods.

The only place spared the anti-Einsteinian turn in philosophy was Göteborg, where in the 1930's there were two philosophers with a sound understanding of modern science, Gunnar Aspelin and Ernst Cassirer. Cassirer found a haven at Gothenburg College between 1935 and 1941 after his flight from Hamburg and nazi-Germany. He came to Sweden via Oxford, and at the invitation of Malte Jacobsson, an active social democrat who had held the Gothenburg chair in philosophy, but left it to become the local governor of the region that includes Göteborg.[48] Cassirer had already written a book, *Zur Einsteinischen Relativitätstheorie* (1921), in which he gave a neo-Kantian interpretation of the problems raised by the theory of relativity. In the preface to this work he writes that one of his purposes was to work toward a "mutual understanding between the philosopher and the physicist on questions, concerning which they are still widely separated. That I was concerned, even in purely epistemological matters, to keep myself in close contact with scientific physics and that the writings of leading physicists of the past and present have essentially helped to determine the intellectual orientation of the following investigation, will be gathered from the exposition".[49] From the preface we learn also that Einstein himself read the essay in manuscript form and provided critical comments.

Here then one finds a completely different attitude to that of the

Uppsala school and other Swedish philosophers of the time. While in Göteborg Cassirer published his book, *Determinismus und Indeterminismus der modernen Physik* (1937), and the same year Gunnar Aspelin in Göteborg published his *Tankelinjer och trosformer* (Lines of Thought and Forms of Faith). The latter work inspired some young students, among them Håkan Törnebohm who studied in Göteborg under Cassirer and Aspelin. Later, after having come to logical empiricism via Hans Reichenbach's work, Törnebohm wrote a doctoral dissertation entitled *A Logical Analysis of the Theory of Relativity* (1952) (published in Lund; Törnebohm also spent a year at the University of California at Los Angeles where Reichenbach was). In the 1960's and 70's he made a number of other original contributions to the philosophical analysis of Einsteinian relativity, first with regard to the special theory, and later on the theory of gravitation.[50] The interest in relativity theory continues at the Theory of Science Department that was founded in Göteborg in 1963 when a personal chair for Törnebohm was created. The dissertation of 1952 was an important item in the review that preceded this decision. Thus, in Göteborg, Einstein got his revenge – in more ways than one.

To a reader unfamiliar with the Swedish academic system the events reviewed, and the lengthy resistance of a number of philosophers and their poorly informed argumentation may appear rather curious, anecdotal or even esoteric. It would be all too easy to dismiss it as a case of marginality at a time when the Vienna Circle and affiliated movements were beginning to set what would be the agendas for mainstream philosophy of science for many years to come. Swedish philosophy was obviously out of joint with this development.

However it would be a mistake just to dismiss the history I have just reviewed. In Swedish context, those cited were not marginal figures. For a long time their thinking dominated the official philosophical establishment. They acted as gatekeepers and their pronouncements were important in the oligarchic institution of Swedish philosophical life, the universities. A sociological explanation for why this was so to such a high degree must be sought in the professorial chair system somewhat similar to the German model. A discipline at a particular academic site revolved to a large degree around a single individual appointed by the King as the civil servant and the authorized spokesperson for the field in question. Thus a professor in this system was the intellectual baron of an epistemic fief in a realm of officially certified knowledge. This meant that struggles around appointments could be long, drawn out and acrimonious, especially if there were applicants to a chair who might challenge the orientation of the established field of research and education. Jealous boundary management was reflected in the many struggles over professorial appointments.[51] If a school, like the Uppsala one, was strong, well entrenched, with many high level hangers, on its chances to reproduce itself and expand into other academic sites were excellent. On the other hand, with the reorientation of Swedish philosophy, thanks to a younger generation during and after the Second World War, it also meant that the approach of the old school was immediately experienced as anachronistic and constraining, and it disappeared as if overnight, almost without a trace, while the new – this time modern analytical – orientation readily established its hegemony via the professorial chair system.[52]

SYMBOL FOR MODERNITY AND PEACE

During the early 1920's, the conservative Swedish newspaper *Nya Dagligt Allehanda* started an anti-Einstein campaign that was to last until at least the end of the decade. Into this were drawn some theoretically conservative physicists and engineers, and articles from other countries were also published. One of the main themes was that relativity theory had a bad influence on the moral and cultural life of society, since it tended to relativize moral values and knowledge claims. In addition there was a claim that it might contribute to an erosion of the credibility of science, as the general public is confronted time and again with these learned debates where exact mathematical figures are mixed with invectives between different sides.

Relativity theory was also opposed because it was associated with modern fashion. The lawyer Sten Lothigius who had an amateur interest in physics and helped found the Swedish Association of Physicists, serving as its first treasurer,[53] drew a parallel with modern art. "What cubism is in art, Einstein is in science. In old Hellas one knew better what had lasting value".[54] Lothigius dismissed both Einstein and cubism. In Sweden he was one of the very few who followed the example of the German ultra-right, attacking Einstein for ideological reasons. Among other things, he wrote: "More than once Einstein stood at a crossroads where he had to choose between scientific truth and his own fame and without doubt he chose the latter, which he sought to accelerate via an infamous, open and hidden advertisement in the daily newspapers".[55] Here he echoed a leaf out of the smear campaign launched by Phillip Lenard and company. It is significant to note that hc was not a physicist or philosopher. Carl-Olov Stawström, in his study of how Swedish physicists and philosophers understood Einstein's relativity theories, concludes that no one in either of these categories followed Lothigius. Even the sharpest of critics drew a line between Einstein's ideas and his person, homing in only on the former.[56]

While conservatives were inclined to oppose relativity theory, liberals and socialists often tended in the opposite direction. The historian of ideas Kjell Jonsson in an essay, recently, regarding science and metaphysics in the interwar years suggests this is because of the connection relativity theory was assumed to have with modernity, even though the content of the arguments in this direction varied considerably. The young anarchists writing in the paper *Brand* supported Einstein for ideological reasons. The world view he had outlined harmonized with their individualism and yearning for freedom. They found legitimation for their view that each society, each local community, etc., should be allowed to develop on its own premises. "The essence of Einstein's world picture", they said, "the meaning of his relativity theory is, that every system of the world, perhaps each solar system, encloses a final totality (*avslutat helt*), finite in its conception of space and independent in its activities. . . . The universe should be pluralist in character, many-sided, not monistic or unitary. . . . The worlds are thus like individuals who organize themselves in mutual relationships with each other according to their own sex, or isolate themselves from each other".[57]

Georg Brandes, who had met Einstein in Copenhagen in 1920, wrote a popular article five years later, where he tried to popularize relativity theory, making use of Bertrand Russell's writings on the subject. In his presentation Brandes concluded by comparing Newton's and Einstein's world systems. "Newton's system may be called monarchist, since the planets in it have to obey the sun. Einstein's system can be called anarchist, since the planets in it do not take account of the sun, but follow the way that is closest for them, as water finds its way down a hill".[58] The metaphorical significance of this counterposition of monarchy and authoritarianism versus democracy and individual freedom was not lost on the readers.

Brandes belonged to the progressive intellectual movement that worked for peace and international solidarity. Others in this movement were Romain Rolland, Stefan Zweig and Henri Barbusse who with his novel *Clarté* founded a movement with the same name. Einstein belonged to the oppositional spirits in this international network and as such he was committed to improving the relations between countries, among other ways through the League of Nations. In this connection he supported the proposal to create an international university, and he participated in the Committee on Intellectual Cooperation of the League of Nations, which also counted Marie Curie and Hendrik Antoon Lorentz amongst its members. Just before his visit to Göteborg he temporarily resigned from the Committee. This was in connection with the French and Belgian troops', occupation of the Ruhrgebeit.[59] The following year however he rejoined. In Berlin he also became one of the first members of the *Gesellschaft der Freunde des Neuen Russland in Deutschland*, founded in 1923 to promote mutual understanding and cultural exchange with the young socialist republic.[60] His activities as a proponent of pacifism, socialism and civil disobedience helped shape the symbolism whereby revolution in the realm of science became intimately associated with democratic, progressive trends and even revolutionary strivings in society.

When Einstein visited Göteborg in 1923, his presence as a symbol of a radical break with the past and a premonition of something new fit in well with the overall theme of the city's tercentennial celebrations. Here optimism, and with it a heady belief in the power of science and technology, dominated. This could be seen in many of the exhibits put on by shopkeepers, crafts guilds, industrial companies, sports associations, professional organizations, and the like. In contrast to Germany, where defeat in World War I had evoked a wave of cultural and even technological pessimism, in neutral Sweden the cultural critique of technology came much later – it emerged as a significant factor only after the economic crisis years of 1929/30, the Kreuger crash, and with the great depression. Oswald Spengler's pessimistic message in his two volumes on the decline of the West (appearing in German 1918–1922) was certainly discussed, but did not find much of a seed bed in Sweden of the 1920's.[61] Thus it was Einstein the muse and innovator that was celebrated by the city of Göteborg, while the organizers of the *Skandinaviska naturforskarmötet* (the scientific congress that had invited him) also saw him as an ambassador of peace. Svante Arrhenius, who had been involved in the deliberations of the Nobel Committee, was strongly committed to the role the Swedish research community could play to bring together scientists from the former warring countries. Like others in neutral Sweden he was anxious to bring the Germans back into the fold of scientific internationalism.[62]

Scientific internationalism, altruism, and other oft-quoted values are ideals. Paul Forman has characterized scientific internationalism as an ideology. Its tenets include the notion of the universality of scientific knowledge, transnational collaboration and competition, as well as a globalisation of standards and standard-setting. The latter has to do with measurement practices as well as convergence of mutual assessment and recognition amongst peers regarding standards of performance more generally.[63] Frequently lip service is paid to this ideology, while in actual practice scientists are mobilized within their own countries for national chauvinist policies. The rhetoric of transnationalism is therewith used to promote nationalist goals. Forman, in his analysis, shows how this functioned after World War I, when German proclamations of scientific internationalism served as a "Macht-ersatz". Germany had lost the war, and now scientific and scholarly achievement in its own right was at the same time a surrogate for the other, lost attributes of a great power.[64] In the words of one German politician at the time "German science is the one thing for which the world still envies us".[65]

Elisabeth Crawford in her study of internationalism in science during and after World War I has recently shown how, during and directly after the war, the frequency of German scientists nominating candidates for the Nobel prize from their own country was much greater than it had been before the war. This was part of the patriotic surge of rallying to the cause of the nation via recognition on the international arena.[66] In this, Einstein, as Crawford puts it, "doubtless was Germany's greatest asset in its efforts to regain a position in international science in the early 1920s".[67] Apart from German nominations he also more than anyone else in Germany at the time was supported by nominations from British, French, and American nominators, as well as from scientists in the neutral countries. This was not only because of his scientific stature and the admiration many felt for his theories of relativity, but also for political reasons: indirectly, his outspoken pacifism during the war was also being recognized.

In other words, even if the reasons were different, there was a desire to translate scientific recognition also into political opinion both inside and outside Germany. This convergence of interests was an important ground on which unofficial international scientific relations could be developed by the Germans, while they at the same time on the whole maintained a policy of intransigance vis a vis the official body of international cooperation, the International Research Council (IRC) and its unions.[68] This was in reciprocation to the exclusionist policies of the IRC. On both sides, science was being used as an instrument of international relations. In other words, international relations in science were intertwined with international relations between countries, with the neutral nations and their scientists finding themselves in the role of intermediaries. In this tradeoff between science and politics, the Nobel prize was an important symbol which could be mobilized as a much needed cultural resource to try and transcend the arenas of conflict and political tension. The use of science as the continuation of politics by other means thus took on different forms depending on geopolitical context and the domestic symbolic capital available to the different parties.

From the outset, directly after the war, the French and Belgian researchers within the IRC pressed to exclude German and Austrian scientists. The appeal of ninety-three German intellectuals (including a number of Nobel laureates) during the war, condoning German atrocities and great power chauvinism, was not forgotten. French and Belgian, but even other, scientists demanded from their German colleagues a recantation. Since this was not forthcoming, for a time after 1918 a cold war continued in science. The French boycott of German science continued until 1926.

Daniel Kevles has shown how George Ellory Hale, the distinguished astrophysicist, Director of the Mt. Wilson Observatory and the foreign secretary of the National Academy of Sciences in the U.S., expounded internationalism, but used its rhetoric to support the French and Belgian position of excluding the Germans from international collaboration.[69] Hale wanted to build up a system for protection against German revanchism in international science. Some of the neutrals, scientists from non-warring countries, such as the Netherlands and Sweden, were regarded with suspicion after the war. In some cases they were suspected of being crypto-Germans. However, the neutrals were drawn into the new regime of international cooperation by the IRC, for fear that they might otherwise side with the Germans, which would have strengthened the already strong position of German science, while weakening the Allied community's postwar scientific potential.

The Swedish scientists were not unaware of the role of the Nobel institution in this international context. Reflecting the neutralist ideology of a majority of their colleagues, the prize committees sought to use the prize as a political instrument, to pour oil on troubled waters, and seek a rapprochement between the hostile camps. Immediately after the war, this led to conciliatory gestures to the Germans, as, for example, in the awarding of the chemistry prize of 1919 to Fritz Haber. The prize ceremony was probably one of the first occasions for an official meeting between German and Allied scientists.[70] The process was not easy. Many of the Allied scientists found it inappropriate to award a prize to Haber and protested. Haber after all had fathered the German army's gas warfare program. In Sweden itself there was also controversy within the Academy of Science, as witnessed by a statement to the press by a minority of members in December 1919. The Social Democratic politician, Hjalmar Branting, in an open letter, wrote: "The heart of the matter is that Swedish science had lost all decency, tact, and neutrality by showing this conciliatory attitude toward the German scientists. The fact is here a prize has gone to honor destruction, at a time when great destruction is just behind us."

The Nobel prizes were meant to transcend politics. As Crawford puts it, they were to be the unassailable symbol of internationalism in science that many had felt was a casualty of the war. In Einstein's case it was obviously a much easier task to uphold this ideal. He himself was a symbol for peace and progress, so the symbolic capital of recognition was easily converted into political currency. Not only that, in the eyes of many people it also, in turn, helped reassert the apparent purity of the prize as cultural capital.

University of Gothenburg

NOTES

[1] Abraham Pais, *Subtle is the Lord. The Science and Life of Albert Einstein* (Oxford: University Press, 1982), p. 504 ff.

[2] A similar point is made by Paul Forman, "Scientific Internationalism and the Weimar Republic: The Ideology and its Manipulation in Germany after World War I', *ISIS* **64** (1973), pp. 151–180.

[3] The last word in the title was later dropped; see, e.g., the 23rd edition (Braunschweig/Weisbaden: Friedr. Vieweg & Sohn, 1988).

[4] Albert Einstein, *Den speciella och den allmänna relativitetsteorin* (Göteborg: Daidalos, 1989), with an Introduction by Aant Elzinga.

[5] The Svedberg, *Arbetets dekadens. Naturvetenskapliga essäer* (Stockholm: Hugo Gebers Förlag, 1915).

[6] J. Arvid Hedvall, *Göteborg i Centrum* (Göteborg: Rundqvists tryckerie, 1970), p. 57;

this book also contains the picture of Einstein on the podium of the Concert Hall in the Liseberg Amusement Park, delivering his Nobel lecture. The same picture is reproduced in Carl-Olow Stawström, 'Relative Acceptance. The Introduction and Reception of Einstein's Theories in Sweden 1905–1965', in Svante Lindqvist (ed.), *Center on the Periphery. Historical Aspects of 20th-Century Swedish Physics* (Canton MA: Science History Publications, 1993), p. 298.

[7] Mrs. Astrid Petzäl, a daughter of Dr. Gustaf Ekman is thanked for letting me consult a number of letters in her possession, including one by Albert Einstein to Mrs. Ekman.

[8] Einstein's initials are still there inscribed in a pane of a glass in a window of one of the buildings.

[9] *Kungl. Vetenskaps-och vitterhetssamhället, KVVS.*

[10] Hedvall, Göteborg i Centrum.

[11] Cited after Pais, *Subtle is the Lord*, p. 509.

[12] Carl-Olow Stawström, 'Synpunkter på hur svenska fysiker och filosofer uppfattat Einsteins relativitetsteorier', *Stockholm Papers in History and Philosophy of Technology*, no. 2018 (Stockholm: Royal Institute of Technology Library, 1986), p. 19. Some of the same territory is covered in Carl-Olov Stawström, 'Relative Acceptance', pp. 293–305; see also Carl-Olov Stawström, 'Idéhistoriska synpunkter på innehållet i vissa skriftliga arbeten av Carl Wilhelm Oseen', *Stockholm Papers in History and Philosophy of Technology*, no. 2019 (Stockholm: Royal Institute of Technology Library, 1986).

[13] Elisabeth Crawford and Robert Marc Friedman, 'The Prizes in Physics and Chemistry in the Context of Swedish Science: a working paper', in C. G. Bernhard, E. Crawford and P. Sörbom (eds.), *Science, Technology and Society in the Time of Alfred Nobel* (Oxford: Pergamon Press, 1982), p. 323.

[14] 'The Prizes in Physics and Chemistry', p. 323.

[15] Cited after G. Küppers, N. Ulitzka & P. Weingart 'Decisions about the Significance of Science', in C. G. Bernhard, E. Crawford and P. Sörbom (eds.), *Science, Technology and Society in the Time of Alfred Nobel* (Oxford: Pergamon Press, 1982), p. 349.

[16] Cited after Pais, *Subtle is the Lord*, p. 503. In accordance with a divorce settlement some years before, the proceeds of the Nobel prize were turned over to Einstein's ex-wife, his former fellow physics graduate student and companion in Zürich, Mileva Maríc. She bore him two sons in their marriage, and one daughter, in secret, before their marriage. The daughter was put up for adoption via Mileva's parents in Serbia. There has been some dispute amongst scholars about the extent of Mileva Maríc's role in special relativity theory; some even claim that her name appeared on the famous 1905 paper together with Albert Einstein's. See further Michael White and John Gribbin, *Einstein. A Life of Science* (London: Simon and Schuster, 1993), p. 125; and also Roger Highfield and Paul Carter, *The Private Lives of Albert Einstein* (London: Faber and Faber, 1993), pp. 108–117.

[17] See note 36 in Crawford and Friedman, 'The Prizes in Physics and Chemistry', p. 330, and note 36.

[18] Stawström, *Stockholm Papers* no. 2018, p. 12.

[19] Robert Marc Friedman, 'The Nobel Prizes and the Invigoration of Swedish Science: Some Considerations', in Tore Frängsmyr (ed.), *Solomon's House Revisited. The Organization and Institution of Science* (Canton MA: Science History Publications, 1990), p. 199.

[20] 'The Prizes in Physics and Chemistry', p. 322.

[21] Bernhard, Crawford and Sörbom (eds.), *Science, Technology and Society*, p. 358.
[22] 'Decisions about the Significance of Science', p. 345.
[23] 'Decisions about the Significance of Science', p. 345.
[24] Quoted in 'Decisions about the Significance of Science', p. 345.
[25] Estimations based on lists in John Tuneld, *Akademiska avhandlingar vid Sveriges universitet och högskolor 1910/11–1939/40* (Lund: Ohlssons boktrykerie, 1945).
[26] For a similar case where Oseen was also involved, using the renommé of the Nobel prize to advance Swedish science policy interests, see 'The Nobel Prizes and the Invigoration of Swedish Science', pp. 198–201; this was in connection with Manne Siegbahn's prize in 1925, reserved from 1924.
[27] Stawström, *Stockholm Papers* no. 2018, p. 21; cf. also Klaus Hentschel, 'The rise of antisemitic arguments against Einstein's theories of relativity in Germany since 1920', paper at the XVIIIth Congress of History of Science, Hamburg-Munich 1–9 Aug. 1989.
[28] Elisabeth Crawford, *Nationalism and Internationalism in Science, 1880–1939* (Cambridge: University Press, 1992), pp. 67–68; also Elisabeth Crawford, 'The Benefits of Nobel Prizes', in Tore Frängsmyr (ed.), *Science in Sweden. The Royal Swedish Academy of Sciences 1937–1989* (Canton MA: Science History Publications, 1989).
[29] Cited after Kjell Jonsson, 'Naturvetenskap, världsåskådning och metafysiskt patos i mellankrigstidens Sverige', *Lychnos* (1992), p. 128; see also Kjell Jonsson, 'Physics as Culture. Science and Weltanschauung in Inter-War Sweden', in Svante Lindqvist (ed.), *Center on the Periphery*, pp. 3–23.
[30] Georg Henrik von Wright, *Vetenskap och förnuftet* (Stockholm: MånPocket, 1988), p. 92.
[31] Svante Nordin, *Från Hägerström till Hedenius. Den moderna svenska filosofin* (Lund: Doxa, 1984).
[32] Kjell Jonsson, *Vid vetandets gräns* (Lund: Arkiv avhandlingsserie 26, 1987), p. 179 and corresponding note on p. 216.
[33] Nordin, *Från Hägerström till Hedenius*, pp. 76–77.
[34] 'Naturvetenskap, världsåskådning', p. 129.
[35] Jonsson, *Vid vetandets gräns* p. 176.
[36] Quoted in Stawström, *Stockholm Papers* no. 2018, p. 25.
[37] Stawström, *Stockholm Papers* no. 2018, p. 25.
[38] Stawström, *Stockholm Papers* no. 2018, p. 26.
[39] Stawström, *Stockholm Papers* no. 2018, p. 27; see also Suzanne Gieser, 'Philosophy and Modern Physics in Sweden. C.W. Oseen, Oskar Klein, and the Intellectual Traditions of Uppsala and Lund, 1920–1940', in Svante Lindqvist (ed.), *Center on the Periphery*, pp. 24–41, which examines, among other, the debates between physicists and philosophers during the period indicated.
[40] Quoted after Stawström, *Stockholm Papers* no. 2018, p. 29.
[41] See Stawström, *Stockholm Papers* no. 2018, p. 31; the final publication was Martin Fries (ed.), *Axel Hägerströms Filosofi och Vetenskap* (Stockholm: Ehlins, 1957), with commentary by Fries.
[42] Quoted after Stawström, *Stockholm Papers* no. 2018, p. 31.
[43] Stawström, *Stockholm Papers* no. 2018, p. 29.
[44] Stawström, *Stockholm Papers* no. 2018, pp. 29–30.
[45] Stawström, *Stockholm Papers* no. 2018, p. 30.

[46] Stawström, *Stockholm Papers* no. 2018, p. 30.
[47] Stawström, *Stockholm Papers* no. 2018, p. 30.
[48] It is interesting to note that Cassirer also wrote a book on Hägerström' philosophy, but his critical remarks in it deal only with value nihilism, not with Hägerström's views on relativity theory. E. Cassirer, *Alex Hägerström* (Göteborg: Högskolan, 1939); see also D. Gawronsky, *Ernst Cassirer: his life and his work. In The Philosophy of Ernst Cassirer* (Evanston Ill.: 1949). Malte Jacobsson's account of Cassirer's arrival and sojourn in Göteborg can be found in M. Jacobsson, *Minnesbilder* (Stockholm: 1964).
[49] Ernst Cassirer, *Einstein's Theory of Relativity*, publ. 1923, and again in 1953 together with *Substance and Function* (London: Dover Publ., 1953); the quote is from the latter edition, p. 349.
[50] Håkan Törnebohm, *A Logical Analysis of the Theory of Relativity* (Stockholm: Almqvist and Wiksell, 1952); Håkan Törnebohm, *Concepts and Principles in the space-time theory within Einstein's Special Theory of Relativity* (Göteborg: Acta Universitatis Gothoburgensis, 1963); Håkan Törnebohm, *A Paradigm Shift in Theories of Gravitation* (Gothenburg: Scandinavian University Books, 1977).
[51] For a review of some of these struggles in philosophy, see Svante Nordin, *Från Hägerström till Hedenius*.
[52] Cf. Sven-Eric Liedman, 'In search of Isis: general education in Germany and Sweden' in Sheldon Rothblatt and Björn Wittrock (eds.), *The European and American University since 1800. Historical and sociological essays* (Cambridge: University Press, 1993), pp. 74–106; also Aant Elzinga, 'Universities, research and the transformation of the State in Sweden', in the same volume, Rothblatt and Wittrock (eds.), pp. 191–233.
[53] This was in 1920, and Oseen was the Chairman; see Stawström, 'Relative Acceptance', p. 300.
[54] Quoted after Jonsson, 'Naturvetenskap, världsåskådning', p. 129.
[55] Quoted in Stawström, *Stockholm Papers* no. 2018, p. 15.
[56] Stawström, *Stockholm Papers* no. 2018, p. 15.
[57] Quoted in Jonsson, 'Naturvetenskap, världsåskådning', pp. 129–130 (my translation); also in Jonsson, 'Physics as Culture', p. 12.
[58] Georg Brandes, *Kulturbillider. Studier og Strejftog* (Kopenhavn: Hage & Clausens Forlag, 1932), p. 155; I am grateful to Bertil Nolin of the Literature Department at the University of Göteborg for drawing my attention to Brandes' references to Einstein and the general theory of relativity.
[59] Pais, *Subtle is the Lord*, p. 317.
[60] Kenji Sugimoto, *Albert Einstein. Die Kommentierte Bilddokumentationen* (München: Verlag Moos & Partner, 1987), p. 93.
[61] See Conny Mithander, 'Implications of Technology. The Debate in Context', in Catharina Landström (ed.), *Intellectuals Reading Technology*. Proceedings from the Nordic Symposium 'Technology-Ideology-Culture', *S.t.i.c. report* no. 4 (Göteborg: Theory of Science Department, 1991), pp. 3–25. For an interesting critique of Spengler's views see Otto Neurath, "Anti-Spengler", in Rudolf Haller und Heiner Rutte (eds.), *Otto Neurath Gesammelte philosophische und methodologische Schriften Band 1* (Wien: Verlag Hölder-Pichler-Tempsky, 1981), pp. 139–196.
[62] See Crawford, *Nationalism and Internationalism in Science*, also Forman, 'Scientific Internationalism and the Weimar Physicists'. In what follows I largely rely on these two authors.

[63] 'Scientific Internationalism and the Weimar Physicists', pp. 153–156.
[64] 'Scientific Internationalism and the Weimar Physicists', p. 163.
[65] Quoted after, 'Scientific Internationalism and the Weimar Physicists', p. 164.
[66] Crawford, *Nationalism and Internationalism in Science*, pp. 64–71.
[67] Crawford, *Nationalism and Internationalism in Science*, p. 70.
[68] 'Scientific Internationalism and the Weimar Physicists', pp. 177–179.
[69] Daniel J. Kevles, "Into Hostile Political Camps: The Reorganization of International Science in World War I", *ISIS* **62** (1973), pp. 151–180; for a discussion of Einstein's response to the "appeal of the 93", signing a pacifist counter appeal initiated by the physiology professor at Berlin University, Georg Nicolai, see White and Gribbin, *Einstein. A Life of Science*, pp. 119–120.
[70] Crawford, *Nationalism and Internationalism in Science*, pp. 67–68.

ULRICH RÖSEBERG*

DID THEY JUST MISUNDERSTOOD EACH OTHER? LOGICAL EMPIRICISTS AND BOHR'S COMPLEMENTARITY ARGUMENT

Einstein's theory of relativity provided a challenge to many physicists and philosophers. Some of the fathers of logical empiricism started their scientific careers with analyses of the revolutionary changes forced upon them by the theory of relativity in the concepts of space, time, mass, energy, and causality. Philipp Frank (1884–1966) published several articles on relativity before he acceded to Einstein's chair in Prague in 1912. Moritz Schlick (1882–1936) published a book on relativity (1917) and another one on general epistemological problems (1918) and discussed these subjects with Einstein who supported his move to Vienna in 1922. Hans Reichenbach (1891–1953) was among the few students in Einstein's first seminar on general relativity in Berlin and developed his own method of logical analysis of scientific theories by analyzing the theory of relativity (1920; 1924; 1928). Later, he declared this method to be the only legitimate one in scientific philosophy. Einstein liked Reichenbach's ability to analyze contemporary physical theories and supported his efforts to receive an academic position. Last but not least, Rudolf Carnap (1981–1970) – whose knowledge in physics should not be compared with Frank's, Schlick's and Reichenbach's – wrote a dissertation on space (1922) in which he prepared his later logical reconstructions of scientific language. Whereas the theory of relativity existed before the new philosophy of science developed in Vienna, Prague, Berlin, the development of this philosophy was immediately challenged by another physical theory, quantum mechanics. But quantum theory confronted physicists and philosophers with a situation different from relativity theory.

Both of Einstein's theories were developed from first principles. Moreover Einstein himself understood the main physical and epistemological messages of special as well as general relativity. The task of philosophers in this field was the systematization of the new theory, maybe its axiomatization, the analysis of relativistic concepts and their

* Deceased.

K. Gavroglu et al. (eds.), Physics, Philosophy and the Scientific Community, 105–123.

epistemological consequences, and the criticism of all kinds of physical and philosophical misinterpretations.

The quantum mechanical formalism, which was developed between 1925 and 1927 by Heisenberg, Born, Jordan, Dirac, and Schrödinger, was not based on first principles. There were different approaches for overcoming the older mixture of classical and quantum concepts in atomic physics. The new theory was first of all in need of a physical interpretation; whereas all the practical questions of how to manage the quantum mechanical formalism for solving physical problems could be answered relatively quickly the controversies over the epistemological questions raised are continuing today. Since the early 1930s, no essential differences have obtained in the use of quantum mechanics for all kind of practical purposes. But large differences did obtain in reflections about the place of quantum mechanics in the system of contemporary physical knowledge. Bohr's complementarity argument developed from 1927 onward, and Einstein's polemic against it has remained until today one of the most influential positions.

Philosophers are confronted with the problem that the epistemology of quantum theory and of relativity seem to be different. Some will go with Shimony's 'peaceful coexistence' between the non-locality of quantum mechanics and the relativistic structure of space – time (1985, 314), while others prefer definite decisions. Most of the discussions take note of the history of the problem, but whereas the Einstein – Bohr debate is analyzed in numerous books and papers the early debates between logical empiricists and Bohr have been neglected or ignored.[1] Depending on the author's position, summaries of the Einstein – Bohr debate stress either the positivistic or the anti-positivistic attitude in Bohr's thought. Both however are oversimplifications of what really happened between 1927 and 1945.

1. PHILOSOPHY COMES LATER

Ph. Frank in Prague, H. Reichenbach in Stuttgart (since 1926 in Berlin), M. Schlick in Vienna and other philosophers of the emerging logical empiricist philosophy of science became interested in the development of quantum mechanics as soon as the first papers were published in 1925. They communicated their astonishment to colleagues and students, invited physicists for discussions, and started to analyze the latest papers and monographs in this field, which was in a revolutionary phase of its development. Thus, that the main discussions in physics took place elsewhere was not objected to by the philosophers who had made their own revolution outside the traditional philosophical institutions. Their experience with relativity theory led them to the conviction: Philosophy comes later! The philosopher's task was to analyze existing scientific theories. Neopositivistic philosophers of science were ready to wait until the development of quantum theory was more or less completed.[2]

Among the young philosophers Hans Reichenbach obviously possessed the best prerequisites to follow the new physical development. In 1920 he was a colleague of Erwin Schrödinger in Stuttgart from where Schrödinger moved to Breslau and to Zürich. Schrödinger and Reichenbach stayed in touch.[3] In 1927 Schrödinger acceded to Planck's former chair in Berlin. Reichenbach wrote that he would take part in a quantum colloquium with Schrödinger, London and von Neumann.[4] He had more informal discussions with Fritz London (until 1926 in Stuttgart and after that in Berlin). London explained some mathematical details of the new theory and wrote that the physicists were waiting for a philosopher to tell them what they had accomplished.[5]

But the leading physicists did not wait for such a philosopher. They made their own philosophy and ignored what was done by philosophers analyzing physical theories. This at least was Reichenbach's experience. He expressed his frustration in a long letter to Max Born in 1928. Reichenbach thought that the correct interpretation for relativistic concepts was given only in his own analysis. Furthermore he criticized physicists who declared that only quantum mechanics made an essential use of probabilistic ideas and therefore destroyed the classical concept of causality. Already before quantum mechanics was discovered he had shown that the classical concept of causality could not be formulated without the concept of probability (Reichenbach, 1925). Even in classical physics the concept of probability possessed the importance which was now claimed for it in quantum mechanics.[6]

Reichenbach's conviction that there was no radical discontinuity between classical and quantum mechanics, especially not with respect to the use of probability, could hardly impress a man who had introduced the probability interpretation of Schrödinger's wave function. Max Born seems to have understood Reichenbach's criticism as one of many misinterpretations of the quantum revolution, which claimed that it was new wine in old bottles! He gave no direct response to this letter. But Reichenbach found other ways to learn something about the discussions in Göttingen and Copenhagen.

First of all, among Reichenbach's students in Berlin was Martin

Strauss (1970–1978), who had studied mathematics, physics and philosophy of science. He continued his studies under Born in Göttingen and stayed in touch with Reichenbach until the end of the war.[7] They exchanged long letters about the details of quantum mechanical formalism and its physical understanding. When Strauss went to Copenhagen for the first time in 1929 he reported on Bohr's discussion of ß-decay and the problems of relativistic quantum theory. At that time Bohr had published his first papers on complementarity and tried to find better formulations for his arguments in order to clarify the concepts he had employed. His aim was to understand what the quantum revolution in physics really meant and what the philosophical consequences of the new theory were. Bohr spoke of its "epistemological lesson" and challenged both the physicists and the philosophers.

2. BOHR'S COMPLEMENTARITY ARGUMENT

Since 1927, Bohr's philosophical message was his complementarity argument (the Como lecture and the Solvay paper). He introduced it to prevent contradictions in the interpretation of the quantum-mechanical formalism which on the one hand was developed to overcome classical physics and on the other hand employed even the classical physical concepts in a very special way. This was indicated by Heisenberg's indeterminacy relations. They are a direct consequence of the mathematical formalism of quantum mechanics. Under the condition of the completeness of this theory for every pair of conjugate variables (with q as the position, and p as the momentum) there exists the relation

$$\delta q \,.\, \delta p \geq h/2$$

This relation answered the question why there are tracks of electrons in the Wilson chamber whereas the matrix mechanics had eliminated the concept of trajectories as unobservable. Bohr understood this formula as an important step forward but he criticized Heisenberg's wording as misleading. According to Heisenberg's first formulations it was the observation of one variable which does not allow simultaneous observation of the other one. Bohr stressed that the observation of one variable excluded the definition of the other one and concluded that it made no sense to say that after observing a particle's position its momentum can not be measured simultaneously, for the momentum will not be defined at all when the position was measured sharply.[8]

With his complementarity argument Bohr stressed the interesting point that for specific pairs of variables (position and momentum, energy and time) or concepts (waves and particles; space-time and causality) the empirically determined possibilities of observation and the theoretical possibilities of definition are deeply connected with each other: "Indeed, in the description of atomic phenomena, the quantum postulate presents us with the task of developing a 'complementarity' theory the consistency of which can be judged only by weighing the possibilities of definition and observation." (Bohr, 1934, 55). Later Bohr did not speak about a 'complementarity theory' but used to speak of a 'mode of description'.[9]

This means that, for Bohr, in atomic physics the practice of physicists (experiments or observations under macrophysical conditions with measurement-instruments which should be described by classical physics) and the theoretical description of this practice (classical physical variables and concepts that are allowed under the empirically given conditions in quantum mechanics) are connected in a way unknown to classical physics. The physical theory was not to be misunderstood as sophisticated talk about a nature-in-itself. Rather we have to say explicitly what we are doing (or intending to do) by observing (or intending to observe) something.

According to Bohr we have to recognize

> that the description of the experimental arrangement and the recording of observations must be given in plain language, suitably refined by the usual physical terminology. This is a simple logical demand, since by the word 'experiment' we can only mean a procedure regarding which we are able to communicate to others what we have done and what we have learnt.
>
> In actual experimental arrangements, the fulfilment of such requirements is secured by the use, as measuring instruments, of rigid bodies sufficiently heavy to allow a completely classical account of their relative positions and velocities (Bohr, 1963, 3).[10]

Situations in which we cannot describe the behavior of objects without describing the conditions of their behavior must be described in terms of complementarity. This is an objective description. And, if the theory is a complete one, then it is the only possible objective description. In such situations ontological statements without reflection on the epistemological conditions will not be allowed. Any ontology isolated from epistemology must be declared senseless.[11]

This is the philosophical message of Bohr's complementarity argument. And it reminds us of Kant who, two hundred years ago, criticized the ontological illusion of speaking about nature without reflecting conditions of possibilities for perception. But whereas the German philosopher formulated his position as a consequence of philosophical arguments against his precursors, the philosophical position entailed by the complementarity argument was a consequence of a particular physical theory. Roughly speaking, this philosophical position followed directly from physics and it went further than the Kantian because

even in the great epoch of critical philosophy in the former century, there was only question to what extent *a priori* arguments could be given for the adequacy of space-time coordination and causal connection of experience, but never question of rational generalizations or inherent limitations of such categories of human thinking (Bohr 1958, 65).

According to my reconstruction, complementarity in quantum mechanics is the dialectical negation of correspondence in the former quasi-classical Bohr-Sommerfeld theory. In the new theory we can find the traces of Bohr's successful research program. The systematic question of the relation between two theories (classical mechanics and quantum mechanics) is in a non-trivial sense the consequence of the historical development of quantum mechanics. Physics has a history and it shows! (Röseberg 1991, forthcoming). Neither the limitation of ontology via complementarity nor the historical dimension of this limitation was discussed in the 30s. Logical empiricists as well as most of the physicists had difficulties in understanding what exactly Bohr meant by complementarity.

3. BE CAREFUL!

Hans Reichenbach who understood his own work as '*wissenschaftliche Naturphilosophie*' had published in his popular writings positions that were far from Bohr's. Reichenbach still believed that physical theories should formulate ontological statements about nature-in-itself. In this respect he followed Einstein. But whereas Einstein thought that Bohr's position must be wrong, Reichenbach had the feeling – at least until 1930 – that maybe he did not understood Bohr well enough.

This was quite different from what was thought in Vienna and Prague at this time. In the name of '*wissenschaftliche Weltauffassung*' Carnap, Neurath, Schlick, and Frank criticized some physicists for a renewal of metaphysics in science. They found even in Bohr's argumentation elements of mysticism and irrationalism foreign to science. And in agreement with the neo-positivistic elimination of philosophy they demanded that the concepts of space, time, and causality be analyzed by the same methods as all other scientific concepts. The theory of relativity and quantum mechanics opened possibilities to learn something new and the logical empiricists tried to pursue their investigations without reverting to the old metaphysics. But what they really did was often not yet in agreement with their own methodological demands. For example, Frank understood that Heisenberg's uncertainty relation did not allow a causal picture for single electrons in terms of position and momentum. But the use of statistical calculations was not essentially new for him, and he thought there could be other concepts in which the ideal of a causal description could be fulfilled in the future (1929, 993).[12]

Neurath criticized any metaphysical interpretation of wave mechanics (1930/31, 313) and learned from a letter that Reichenbach as editor of *Erkenntnis* wanted to delete this criticism.[13] The other editor, Carnap, thought this criticism should be published: Even if we are not able to analyze quantum mechanics as a whole we have to criticize any metaphysical interpretation![14] Reichenbach did not change his mind,[15] but nevertheless published Neurath's criticism.

Reichenbach had already analyzed the problem of probability in 1916 in his dissertation. Now he returned to this subject. The *Wahrscheinlichkeitslehre* and the troubles of his emigration to Turkey did not leave him enough time to analyze quantum mechanics just as carefully as he had analyzed relativity theory earlier. But there was a different reason too. Since 1927 most philosophical discussions in this field started with Heisenberg's uncertainty relations, but Reichenbach thought that in this respect the importance of the physicist's work was overestimated. According to Reichenbach, in 1927 Heisenberg had simply shown empirically what he himself had proven conceptually already two years earlier. Causality should be understood as probabilistic, prediction with certainty being the exception (Reichenbach 1935, 11). Heisenberg's idea of uncertainty did not amount to a physical discovery but simply transformed a philosopher's idea into empirical knowledge. An interpretation of the situation in physics like Bohr's complementarity argument[16] must be definitely wrong;[17] the failure of classical causality could not legitimate the mixed use of incompatible concepts like wave and particle.

Reichenbach did not publish this inadequate criticism of Bohr's argument. In 1935/36 Strauss communicated from Copenhagen more about Bohr's views. In a first letter he informed Reichenbach about his plans for a paper on the logical reconstruction of quantum mechanics. Strauss was happy to discuss the details with Bohr himself.[18] According to Bohr – Strauss wrote – the quantum theory is first of all a theory of measurement as a real physical process.[19] The possibilities for the measurement of the kinematical variables like position and time on one hand, and for the measurement of dynamical variables like momentum and energy (legitimated by conservation laws) on the other hand, are given only in agreement with the uncertainty relation, and this is called by Bohr "the complementary mode of quantum-theoretical description". Strauss was very critical of Jordan's reconstruction of Bohr's complementarity whereas Reichenbach had published Jordan's paper with great pleasure a year earlier in *Erkenntnis*.[20]

Reichenbach encouraged Strauss to analyze quantum mechanics. "Unfortunately until now no constructive axiomatization of quantum theory exists; and therefore we remain in the dark."[21] Reichenbach asked Strauss for a response to Frank's, Schlick's, and Zilsel's "dogmatic criticism" of complementarity (all 1935). He wrote that he was much more in agreement with Bohr and Jordan. But the whole discussion in *Erkenntnis* as well as the reactions to Einstein, Podolsky, Rosen (1935) had shown that it was necessary to talk with Bohr himself. The Second International Congress for the Unity of Science in Copenhagen offered an opportunity for this.

The Vienna circle knew that Bohr was interested in discussing his ideas even with logical empiricists who criticized him as a metaphysician. Neurath had reported about his experiences in Copenhagen in 1934:

> Bohr. Idiosyncratic. An intense man. Came to two lectures and joined the discussion enthusiastically . . . Basic line: he does not want to be considered a metaphysician. And he is able to express himself relatively non-metaphysically, when he is careful. Yet obviously, there lies a certain tendency in the selection of problems, insofar as the question of life etc. is discussed, and in the stress on uncertainty. In addition, his printed remarks are full of crass metaphysics. But he possesses certain basic attitudes which agree with mine, e.g., that in science one cannot clear up everything at once, but that the individual scientific-logical actions have to pay a price, as it were. An idea of compensation, which with him now naturally tends to be connected with the uncertainty relation. Obviously tries to come into agreement with us. But since his circle confirms him in his habit to express himself somewhat unclearly, one would have to be able to work on him for a long time, which he would be prepared to do."[22]

This report was in agreement with what Bohr had expressed himself. He had sent Neurath his booklet (1934) and asked him if the spirit in which the introduction was written is really so bad as its reputation. Bohr was glad to learn that Neurath's and his own attitude were not as far apart as seemed indicated by their different terminologies.[23] Bohr's statements reflected his own research experiences, and Neurath understood this. But he nevertheless could not refrain from saying that some of Bohr's statements "shall and really will give under all circumstances occasion for misunderstandings."[24] And Neurath had the feeling, expressed also by many other readers of Bohr, that behind his texts was hidden an attitude which should be made clear. *Horribile dictu*: Would there be any hidden metaphysics? But Neurath understood that even the Viennese philosophers "could not make a magic sieve with which one can eliminate metaphysics."[25]

He and his friends[26] hoped for a '*Nettodebatte*' in Copenhagen, i.e., a debate without trivial misunderstandings that could be prevented by an exchange of papers in advance. On 3 March 1936 Neurath wrote a long letter to Bohr in which he raised the main points of the expected discussions in June. He agreed with Bohr that one should not simulate undue exactness.[27] The logical and mathematical instruments employed should be permanently re-adapted to the aim of science, namely prognoses controlled by statements expressed in everyday language. Neurath also asked how it would be possible to formulate terms that can be used in all scientific disciplines in the same way. For this reason Neurath proposed to Bohr that he should give up words like 'free choice', 'teleological argumentation', 'irrationality', and criticized some of Bohr's specific formulations.[28] But Bohr did not respond to this and other challenges with a paper that could be distributed in advance, and even after the congress the organizers had to wait a long time before they received his manuscript.

4. COPENHAGEN 1936

Niels Bohr used his talk as an opportunity to show that his attitude did not support any mysticism foreign to the spirit of science and attempted to put an end to some misunderstandings. His aim was to explain that the complementarity argument was "an endeavor to clarify the conditions, in each field of knowledge, for the analysis and synthesis of experience." (1958, 63). More than ten years later he remembered:

Yet, I am afraid that I had in this respect only little success in convincing my listeners, for whom the dissent among the physicists themselves was naturally a cause of skepticism about the necessity of going so far in renouncing customary demands as regards the explanation of natural phenomena (1958, 63).

A listener like Strauss who knew physics very well was more optimistic about Bohr's success, but he was also very critical of the philosophers' achievements in Copenhagen (with the exception of Schlick's fine manuscript on quantum-mechanical and biological complementarity, and Frank's talk). Strauss reported in a 12 page handwritten letter to Reichenbach in Istanbul:

Also *Frank* who in Königsberg still misunderstood quantum mechanics in a completely positivistic manner converted to complementarity.[29]

In 1936 Strauss finished two papers on the logical reconstruction of complementarity in which he introduced a non-classical logic (partial Boolean algebra). Nearly at the same time Birkhoff and von Neumann developed another 'quantum logic' (non-distributive lattice algebra).[30] Strauss expressed the hope that logical analyses would advance the understanding of quantum problems. But neither Bohr nor Heisenberg shared this view. According to Bohr, complementarity was a consequence of quantum mechanics, whereas Strauss introduced it as a first principle for the axiomatization of the theory. Bohr recommended the publication of these papers, but accepted no responsibility for their logical part. His understanding of philosophy was far from that of a logician.[31] The same holds for Heisenberg who attended the 1936 physics conference in Copenhagen.

Strauss found Heisenberg's view very strange:

He swears a lot at the positivists . . . philosophy is simply = the great personality who has something to say, like Heidegger (who nevertheless is not great enough).[32]

In Copenhagen Strauss got the feeling that Heisenberg no longer would come to a conference like the one in Königsberg in 1930, but he was ready to make excuses for him (at least in 1936):

Yes, indeed productive scientific research takes precedence in any respect over the logic of science, and such a creative thinker should have the ideology which seems most suitable for himself.[33]

5. REICHENBACH'S ANALYSIS OF QUANTUM MECHANICS

Because of his isolation in Turkcy, Reichenbach did not do any research of his own on the foundations of quantum mechanics. He started this only in Los Angeles and finished the manuscript of his book (1944) in 1942.[34]

Hans Reichenbach analyzed quantum mechanics by thc same method by which he had analyzed relativity in his previous monographs. In agreement with his own understanding of science, the history of physical theories should be irrelevant for this logical analysis. Even though quantum mechanics was not – like Einstein's theory – developed from first principles, Reichenbach discussed it in an axiomatized form. In order to do so he introduced his own axiomatization.

Reichenbach's main epistemological concern was the problem of unobserved objects. He defined normal systems such that for them the behavior of the unobserved objects could be assumed to be the same as for the observed ones, and that there occurred no causal anomalies. This is the same type of a convention allowed in classical physics. Of course, we could reformulate classical physics without using the convention of normal systems. Our way of speaking about the macroworld is only the easiest way, not the only possible one. In quantum mechanics we are confronted with a quite different situation. There are causal anomalies which make it impossible to speak about unobserved objects of the microcosm in the same sense as of the macrocosm. You cannot find a

> normal system for the interpretation of the unobservables, and we cannot speak of unobservables in the same sense as is implied for the world of everyday life. We can regard the elementary constituents of matter as particles or waves; both interpretations fit the observations equally well and equally badly (Reichenbach 1951, 186).[35]

Reichenbach introduced his own definition of phenomena into quantum mechanics.

> The phenomena are connected with macroscopic occurrences by rather short causal chains; we therefore say that they can be 'directly' verified by such devices as the Geiger counter, a photographic film, a Wilson cloud chamber, etc. (Reichenbach 1944, 21).

This was different from Bohr who did not speak about causal chains and who used the word 'phenomena' for all possible experimental arrangements. But whereas Bohr did not allow any division of phenomena Reichenbach introduced 'interphenomena':

> We then shall consider as unobservable all those occurrences which happen between the coincidences, such as the movement of an electron, or of a light ray from its source to a collision with matter. We call this class of occurrences the *interphenomena* (Reichenbach 1944, 21).

According to Bohr, interphenomena were excluded in quantum theory as a consequence of Planck's discovery of the elementary quantum of action which revealed the feature of wholeness in the description of nature. The complementarity argument was Bohr's way to prevent all the contradictions that follow if one tries – at least in thought experiments – to divide atomic processes into parts without disturbing them (EPR). Reichenbach did not care about this. He allowed such thought experiments and contrary to Bohr proposed a revision of logic.[36] This was the way to return to speculative ontology.

In his previous books on space, time, and causality Reichenbach had modified the ontology of physics. He understood that even in physics it was not so easy to speak about the world in itself, but he did not criticize the idea of an ontological understanding of nature. In particular, he could not see what forced Bohr to give up this concept in his complementarity argument. Bohr's was a systematic reason, namely that all our experiences of atoms and molecules depend on macrophysical instruments. And this systematic reason has to do with the way quantum mechanics was formulated. Bohr's statements after 1927 reflected, at least implicitly, the genesis of the new theory, whereas Reichenbach saw no sense in doing this in his logical analysis.

Bohr read Reichenbach's book and saw in it a step forward in the discussions of fundamental questions. But he could hardly agree with an ontological position like Reichenbach's.[37] The main point of the observation problem of atomic physics, he stressed, was the impossibility of sharp separation between the behavior of an independent object and its interaction with measurement instruments. Especially in his 1939 paper he had shown that conceptual confusion could be prevented if the word 'phenomena' was preserved for a complete description of the whole experimental arrangement (including the report of the outcome of the experiment). Whereas such a phenomenon may be described completely in classical physical concepts, the measurement results can only be predicted by means of the generalized mathematical symbols of the quantum-mechanical formalism (which, incidentally is free from any contradiction). Bohr's complementarity mode of description excludes contradictions and prevents metaphysical speculations like Reichenbach's assumption of interphenomena and the behavior of unobserved objects. It was not the first time that a philosopher in his fight against all metaphysics transformed one piece of metaphysics into another one.

6. CONCLUDING REMARKS

Many physicists and nearly all philosophers of his time disliked Bohr's complementarity argument. The day before he died, in his last interview, Bohr said with some frustration: "No man who is called a philosopher really understands what one means by complementarity description" (1962). He criticized the philosophers because "they did not see that it was an objective description, and that it was the only possible objective description." Concluding this part of the interview, Bohr saw in philosophers "odd people who really were lost, because they have not the instinct that is important to learn something and that we must be prepared really to learn something of very great importance."

In the light of my research Bohr's criticism seems to be an overreaction. Some of the logical empiricists did have the instinct and were prepared "really to learn something of very great importance", but they were unable to suffer the views of a man whose understanding of clarification of scientific terminology seemed incompatible with the standards of formal logic. Still more important was that none of the logical empiricists could grasp why complementarity should be the "only possible objective description." In this respect they followed Einstein even if their argumentation was different from his in many details. Furthermore there were Einstein's and the philosophers' misunderstandings and misinterpretations of Bohr's ideas. As a result the philosophy of physics developed in logical empiricism tended to be along the lines of Einstein's epistemology. I believe it is necessary to develop a philosophy of physics along the lines of Bohr's epistemology, at least as a complementary mode.

NOTES

[1] One of the main reasons for this lies in the fact that reconstructions of these discussions must use unpublished materials. R. S. Cohen supported my idea to consult archives of physicists and philosophers for this purpose and opened several doors for my doing so. Thus I wish to dedicate to him a paper that may have some surprising news even for him who had the privilege to talk with most of the 'heroes' in this story. This paper is based on research supported by the former Academy of Science of the GDR, the

Center for Philosophy of Science at the University of Pittsburgh and the Center of History and Philosophy of Science in Berlin. I would like to thank T. Uebel for his valuable suggestions.

[2] In this respect it should be noted that the quasi-classical Bohr-Sommerfeld theory of atoms was already a subject of philosophical analysis. Yet analytic philosophers long neglected the development of theories. After quantum mechanics was found, the previous theory simply seemed to be only a wrong theory. Bohr's attitude was different. Quantum mechanics, which "may be regarded as a natural generalization of the classical mechanics" (Bohr, 1934, 110), should be understood as a fulfilment of Bohr's research program of 1913 (Röseberg 1991, forthcoming).

[3] Schrödinger was asked by a journal to illustrate an article on the new atomic theory and he learned that the author was Reichenbach. In a letter dated 25 Oct. 1926, E. Schrödinger made some suggestions and thanked H. Reichenbach, who tried to explain the new ideas in a wider *naturphilosophischen* context. This letter is held in the Archives of Scientific Philosophy (HR-015-18-12). These materials (described in the following by 'HR . . .') is quoted by permission of the University of Pittsburgh. All rights reserved.

[4] "In Berlin it nun die Quantenmechanik eingezogen mit Schrödinger, London und v. Neumann. Wir machen ein Quanten-Kolloquium . . ." H. Reichenbach, letter to P. Ewald, 26 Nov. 1927, HR-015-29-17.

[5] "Es besteht tatsächlich die Notwendigkeit, auch in der Quanten mechanik die *Bahn*umlaufsfrequenz existieren zu lassen. Wenn es Ihre Zeit gestattet, schauen Sie sich doch einmal die schrullige Sache an. Es geht mir mit ihr so, wie allen Physikern jetzt wohl, daß sie einen Philosophen brauchen, der ihnen sagt, was sie eigentlich angerichtet haben." F. London, letter to H. Reichenbach, 5 May 1926, HR-016-11-24.

[6] "So wird die Kritik des Kausalbegriffs jetzt als Ausfluß der Quantenmechanik hingestellt; aber meine Arbeiten haben schon seit langem gezeigt, daß bereits der klassische Kausalbegriff ohne den Wahrscheinlichkeitsbegriff garnicht formulierbar it, daß der Wahrscheinlichkeitsbegriff, die fundamentale Stellung, die ihm die Quantenmechanik jetzt einräumen will, bereits in der klassischen Physik besitzt. Der Sprung it also nicht so groß." H. Reichenbach, letter to M. Born, 12 April 1928, HR-044-03-40.

[7] Strauss was a member of a Communist student organization in Göttingen. The Nazis sentenced him to 18 months imprisonment. Then he had to leave Germany and went to Copenhagen where he stayed in 1935/36. With a grant organized by Bohr he was able to finish his dissertation in Prague where he worked with Frank. In 1939 he flew, via Holland and Neurath in The Hague, to England. Reichenbach, Carnap, and others tried to find a position for Strauss in the USA, but they did not succeed. Since 1953 Strauss lived in the GDR, where he had a permanent position in the Academy of Science and lectured at the Humboldt-Universität in Berlin.

[8] In one of Bohr's unpublished manuscripts the situation is described in the following way: "In this connection, it should not be out of place to emphasize the danger which lies in a formulation of the uncertainty principle such as: it is impossible to know accurately both the position and the momentum of a particle, which is liable to convey the impression that we are here dealing with a quite mysterious mutual exclusion between two elements of reality. The actual situation may rather be expressed by saying that in the field of quantum theory there exists a complementarity relationship between any two measuring procedures." This quotation is taken from Ph. Frank's letter to N. Bohr, 9 Jan. 1936

(Niels Bohr Correspondence 19.3, Copenhagen; in the following NBC). Frank was worried why Bohr did not publish these formulations and he learned: "Insbesondere war mein einziger Anlaß, den in Ihrem Brief zitierten Satz meines Manuskriptes in der Korrektur wegfallen zu lassen, nur eine Folge der durchgehenden kleinen stilistischen Änderungen der ganzen Arbeit." N. Bohr, letter to Ph. Frank, 14 Jan. 1936, NBC 19.3. These 'minor stylistic changes' in Bohr's texts made difficulties not only for philosophers who were trying to standardize the language of science.

9 "The fundamental postulate of the indivisibility of the quantum of action . . . inevitably requires us to forgo a causal mode of description and . . . forces us to adopt a new mode of description designated as *complementarity* in the sense that any given application of classical concepts precludes the simultaneous use of other classical concepts which in a different connection are equally necessary for the elucidation of the phenomena" (Bohr 1934, 10).

10 The formulation quoted here is taken from one of Bohr's later papers, but the idea itself was published even in his first papers on this subject.

11 Bohr himself did not use the word 'ontology' in this context.

12 This was exactly the same idea that Einstein had in mind, and it was clearly based on a metaphysical assumption.

13 "Ich denke Sie werden damit einverstanden sein; ich habe Ihre Attacke gegen die metaphysischen Ausdeutungen der Wellenmechanik lieber gestrichen, da wir doch diese Probleme noch viel zu wenig übersehen können . . ." H. Reichenbach, letter to O. Neurath, 24 April 1930, HR-013-41-70.

14 R. Carnap, letter to H. Reichenbach, 26.4.1930, HR-013-41-68.

15 "Die Quantenattacke von Neurath halte ich für sehr gefährlich. Man soll nicht über etwas urteilen, was man noch nicht durchschaut . . ." H. Reichenbach, letter to R. Carnap, 29 April 1930, HR-013-41-67. "Meine Bemerkungen zur Quantenmechanik waren nicht taktisch sondern sachlich gemeint. Ich halte es nicht für richtig, über Probleme, die man noch nicht endgültig durchschaut, vorwegnehmend eine Ansicht zu äußern." H. Reichenbach, letter to O. Neurath, 17 June 1930, HR-013-41-62.

16 "We know now that for material particles as well as for light different conceptual pictures are necessary to account completely for the phenomena and to furnish a unique formulation of the statistical Laws which govern the data of observation." (Bohr 1934, 94).

17 Cf. H. Reichenbach, letter to M. Boll, 5 April 1932, HR-014-44-27.

18 Cf. M. Strauss, letter to H. Reichenbach, 25 Sep. 1935, HR-013-35-25. In his letter of 2 Nov. 1935 (HR-013-35-20) Strauss wrote that he now had an agreement for a colloquium with Jörgensen, and on 5 Dec. 1935 (HR-013-35-16) he wrote how useful the discussions were, in which even the mathematician Fenichel and the physicist Kalckar took part.

19 Reichenbach and Strauss discussed mainly the interaction between the object and the instrument of observation. This was an important point in Heisenberg's argumentation. Bohr later understood that this was – at least in part – misleading.

20 "Es ist mir von großem Interesse, zu sehen, wie die Physiker jetzt immer mehr in die Gedankengänge hineinwachsen, die von unserem philosophischen Kreis schon seit längerer Zeit ausgearbeitet wurden. Zuerst ging es so mit der Kausalität: nachdem ich gezeigt hatte, daß die Kausalität nur als Limesbehauptung über Wahrscheinlich-

keitsbeziehungen sinnvoll ausgesprochen werden kann und zugleich die Möglichkeit entwickelt hatte, daß für die Steigerung der Voraussagewahrscheinlichkeit ein limes unterhalb der Gewißheit besteht, im Sinne einer Tatsache des Naturgeschehens, wurde dieser Gedanke von Heisenberg in seiner Ungenauigkeitsrelation als empirisch vorliegend nachgewiesen. Und jetzt mündet ihre neue Arbeit in eine Fassung des Existenzproblems der Außenwelt ein, wie sie wörtlich schon seit vielen Jahren von Carnap vertreten wird. Übrigens entsprechen auch Ihre Bemerkungen zur Willensfreiheit ganz den Gedankengängen, die ich schon seit langem darüber vertreten habe." H. Reichenbach, letter to P. Jordan, 16 June 1934, HR-013-17-16.

[21] H. Reichenbach, letter to M. Strauss, 26.10.1935, Archiv der Akademie der Wissenschaften, Berlin, 293 (my translation).

[22] O. Neurath, letter to R. Carnap, 14 Nov. 1934, RC-029-10-10. Quoted by permission of the University of Pittsburgh. All rights reserved. I thank The. Uebel who brought this document in his translation to my attention.

[23] Cf. N. Bohr, letter to O. Neurath, 24 Oct. 1934, *Archief Wiener Kreis*, inv. nr. 198. *Quoted by permission of Wiener Kreis Stichting*, Dr. Anne Kox.

[24] O. Neurath, letter to N. Bohr, 19 Nov. 1934. See previous footnote.

[25] O. Neurath, letter to N. Bohr, 25 Nov. 1935. See previous footnote.

[26] Carnap, Frank, and Schlick had planned to come to Copenhagen. Schlick then did not get permission to leave Austria. He was killed on 22 June 1936 in Vienna University, and his death was announced in Copenhagen at the Congress. Reichenbach could not pay for the long trip from Istanbul. In his letter to Bohr (8 June 1936, HR-013-40-34) he wrote: "Nachdem ich selbst mich mit den Fragen von Kausalität und Wahrscheinlichkeit viel beschäftigt habe, wäre ich sehr gern auch zu diesen Diskussionen gekommen. Meine Auffassung dieser Dinge ist wesentlich verschieden von der meiner Wiener Freunde, und der Ihren näher stehend."

[27] The same was expressed even by Ph. Frank: "Ich weiß ganz gut, daß der schöpferische Forscher zunächst alles im Werden sieht, noch ohne ganz scharfe Umrisse und daß die Forderung einer ganz scharfen Terminologie, wenn man sie an den Anfang stellte, die Forschung hemmen würde und nur das bestehende ordnen könnte. Aber wenn einmal ein Resultat gewonnen it, so muß es auch möglich sein, es so auszudrücken, daß keine mißbräuchlichen Anwendungen mehr möglich sind." letter to N. Bohr (undated), probably 1935, BSC 19.3.

[28] Ph. Frank explained in a letter to N. Bohr (undated, probably 1935) the political background of why the Viennese philosophers tried to develop a 'scientific language without any metaphysics': "Heute sind überall Kräfte am Werk, die an die Stelle der modernen Naturwissenschaft etwas von der Art der mittelalterlichen Scholastik setzen wollen, um damit auch die verschiedenen mittelalterlich-barbarischen politischen Systeme geistig zu untermauern." (BSC 19.3). Therefore ,according to Frank, the physicist should use a language that guarantees that his view may not be misused. For this purpose he needs a consistent positivistic or physicalistic terminology which may be used in physics as well as in psychology. Frank discussed Bohr's response to Einstein, Podolsky, Rosen (1935) in this context and understood his intention not to use the word 'physical reality' as a positivistic attitude. (Ph. Frank, letter N. Bohr, 9 Jan. 1936) Bohr agreed with Frank in his normal vague form: "Ich glaube auch, daß Sie ganz den Sinn meiner Bestrebungen getroffen haben." (N. Bohr, letter to Ph. Frank, 14 Jan. 1936) But to conclude from a sentence like this that Bohr understood himself as a positivist would be misleading.

[29] M. Strauss, letter to H. Reichenbach, July 29, 1936, HR-013-35-07 (my translation).
[30] H. Weyl told this to Strauss in Copenhagen. He assumed that von Neumann could be motivated to go this way by discussions with Einstein.
[31] Two years later, in Warsaw, he discussed with von Neumann the possibility of quantum logic. Bohr's position was published in the protocol: "Personally, he compelled himself to keep the logical forms of daily life to which actual experiments were necessarily confined. The aim of the idea of complementarity was to allow of keeping the usual logical forms while procuring the extension necessary for including the new situation relative to the problem of observation in atomic physics." (Bohr 1939, pp. 38–39).
[32] M. Strauss, letter to H. Reichenbach, July 1936, HR-013-35-07 (my translation).
[33] Same letter.
[34] There are lots of drafts and correspondence (with V. Bargmann, R. Carnap, N. Dalkey, P. Epstein, C. G. Hempel, S. Hook, A. Landé, V. Lenzen, F. London, W. Pauli, F. Regener, M. Strauss) in the Hans Reichenbach Collection in Pittsburgh that allow a reconstruction of the genesis of Reichenbach's ideas. This will be done elsewhere.
[35] Reichenbach followed "a formulation of the ideas of Bohr and Heisenberg which has been developed by Landé," who stated that all experiments can be explained in wave and in particle interpretation. (Reichenbach 1944, 23) But this was just the opposite of what had been stated in Bohr's complementarity argument. Within an experimental arrangement you make a choice between wave and particle interpretation, and once you have made such a choice you may not use the other interpretation. There is no ontolgical understanding of waves and particles in the classical sense. Complementarity reminds us of the epistemological conditions.
[36] I will not discuss Reichenbach's specific three-valued logic. This has been done by many other authors. My point is to show that Reichenbach missed the main point in Bohr's complementarity argument, whereas he thought himself to be in agreement with Bohr.
[37] The following argumentation is an extract from Bohr's letter to Strauss, 14 May 1946, BSC 32.3.

BIBLIOGRAPHY

Birkhoff, G. and von Neumann, J.: 1936, 'The Logic of quantum mechanics', in: *Annals of Mathematics* **37**, 823–843.

Bohr, N.: 1934, *Atomic Theory and Description of Nature*, Cambridge.

Bohr, N.: 1939, 'The causality problem in atomic physics', in: *New Theories in Physics*, Paris, pp. 11–46.

Bohr, N.: 1958, *Atomic Physics and Human Knowledge*, New York.

Bohr, N.: 1962, Interview with Th. S. Kuhn, A. Petersen and E. Rüdinger, 17 Nov. in: Archive for History of Quantum Physics (unpublished), Copenhagen.

Bohr, N.: 1963, *Essays 1958–1962 on Atomic Physics and Human Knowledge*, New York.

Einstein, A., Podolsky, B., and Rosen. N.: 1935 'Can quantum-mechanical description of physical reality be considered complete?', in: *Physical Review* **47**, 777–780.

Frank, Ph.: 1929, 'Was bedeuten die gegenwärtigen physikalischen Theorien für die allgemeine Erkenntnislehre?' in: *Die Naturwissenschaften* **17**, 971–977, 987–994.

Frank, Ph.: 1932, *Das Kausalgesetz und seine Grenzen*, Wien.

Frank, Ph.: 1935 a, 'Zeigt sich in der modernen Physik ein Zug zu einer spiritualistischen Auffassung?', in: *Erkenntnis* **5**, 65–79.
Frank, Ph.: 1935 b, 'Jordan und der radikale Positivismus', in: *Erkenntnis* **5**, 184.
Frank, Ph.: 1936/37, 'Philosophische Deutungen und Mißdeutungen der Quantentheorie', in: *Erkenntnis* **6**, 303–317.
Frank, Ph.: 1937/38, 'Physik und logischer Empirismus', in: *Erkenntnis* **7**, 297–300.
Frank, Ph.: 1941, *Between Physics and Philosophy*, Cambridge, MA.
Frank. Ph.: 1949, *Modern Science and Its Philosophy*, Cambridge, MA.
Heisenberg, W.: 1927, 'Über den anschaulichen Inhalt der quantentheoretischen Kinematik und Mechanik', in: *Zeitschrift für Physik* **43**, 172–198.
Heisenberg, W.: 1931, 'Kausalgesetz und Quantenmechanik', in: *Erkenntnis* **2**, 172–182.
Jordan, P.: 1934, 'Quantenphysikalische Bemerkungen zur Biologie und Psychologie', in: *Erkenntnis* **4**, 215–252.
Jordan, P.: 1935, 'Ergänzende Bemerkungen über Biologie und Quantenmechanik', in: *Erkenntnis* **5**, 348–352.
Neurath, O.: 1930/31, 'Historische Anmerkungen', in: *Erkenntnis*, pp. 312–314.
Neurath, O.: 1935, Jordan, 'Quantentheorie und Willensfreiheit', in: *Erkenntnis* **5**, 179–180.
Reichenbach, H.: 1920, *Relativitätstheorie und Erkenntnis Apriori*, Berlin.
Reichenbach, H.: 1924, *Axiomatik der Relativistischen Raum-Zeit-Lehre*, Braunschweig.
Reichenbach, H.: 1925, 'Die Kausalstruktur der Welt und der Unterschied von Vergangenheit und Zukunft' in: *Sitzungsberichtce der Bayerischen Akademie der Wissenschaften, Mathematisch-Naturwissenschaftliche Abteilung* 1925, 133–175.
Reichenbach, H.: 1928, *Philosophie der Raum-Zeit-Lehre*, Berlin.
Reichenbach, H.: 1935, 'Metaphysik bei Jordan?' in: *Erkenntnis* **5**, 178.
Reichenbach, H.: 1944, *Philosophic Foundations of Quantum Mechanics*, Berkeley and Los Angeles.
Reichenbach, H.: 1951, *The Rise of Scientific Philosophy*, Berkeley and Los Angeles.
Reichenbach, H.: 1989, *Philosophische Grundlagen der Quantenmechanik und Wahrscheinlichkeit* (Gesammelte Werke Band 5), Braunschweig/ Wiesbaden.
Röseberg, U.: 1991, 'Historical Explanations in Modern Physics: The Lesson of Modern Quantum Mechanics', in: *World Views and Scientific Discipline Formation*, ed. by W.R. Woodward and R.S. Cohen, Kluwer, Dordrecht, Boston, London.
Röseberg, U.: 'Hidden Historicity – the Challenge of Bohr's Philosophical Thought', in: *Niels Bohr and Contemporary Philosophy*, ed. by J. Faye and H. Folse.
Schlick, M.: 1917, *Raum und Zeit in der gegenwärtigen Physik. Zur Einführung in das Verständnis der allgemeinen Relativitätstheorie*, Berlin (there are other editions 1919, 1922, and 1922).
Schlick, M.: 1918, *Allgemeine Erkenntnislehre*, Berlin (2 ed. 1925).
Schlick, M.: 1935, 'Einige Bemerkungen über P. Jordans Versuch einer quantentheoretischen Deutung der Lebenserscheinungen', in: *Erkenntis* **5**, 181–183.
Shimony, A.: 1985, Reflections on the Philosophy of Bohr, Heisenberg, and Schrödinger, in: *A Portrait of Twenty-five Years*, ed. by R.S. Cohen and M.W. Wartofsky, Kluwer, Dordrecht, Boston, London.
Strauss, M.: 1936, 'Zur Begründung der Statistischen Transformationstheorie der Quantenphysik', in: *Sitzungsberichte der Preussischer Akademie der Wissenschaften, Phys.-math. Klasse* **27**, 90–113.

Strauss, M.: 1936/37, 'Komplementarität und Kausalität im Lichte der logischen Syntax', in: *Erkenntnis* **6**.
Strauss, M.: 1972, *Modern Physics and Its Philosophy*, Dordrecht.
Zilsel, E.: 1935, 'Jordans Versuch, den Vitalismus quantenmechanisch zu retten', in: *Erkenntnis* **5**, 56–64.

SYLVAN S. SCHWEBER

PHYSICS, COMMUNITY AND THE CRISIS IN PHYSICAL THEORY

For Bob:
A builder of communities.
In friendship, respect and admiration.

INTRODUCTION

In 1929, in the wake of the enormous success of nonrelativistic quantum mechanics in explaining atomic and molecular structure and interactions, Dirac, one of the main contributor to these developments, in a now famous quotation asserted that "The general theory of quantum mechanics is now almost complete." Whatever imperfections still remained were connected with the synthesis of the theory with the special theory of relativity. But these were ". . . of no importance in the consideration of atomic and molecular structure and ordinary chemical reactions . . . The underlying physical laws necessary for the mathematical theory of a large part of physics and the whole of chemistry are thus completely known, and the difficulty is only that exact application of these laws lead to equations much too complicated to be soluble . . ."[1]

Forty years later, in an after dinner speech entitled "The changing face of physics", delivered during a Batelle Colloquium on phase transitions and critical phenomena held in Geneva in September 1970, Léon van Hove, then the head of the Theory Division at CERN, commented that it seemed to him

> that physics now looks more like chemistry in the sense that, in percentage, a much larger fraction of the total research activity deals with complex systems, structure and processes, as against a smaller fraction concerned with the fundamental laws of motion and interaction. This colloquium is a good example. Surely, we all believe that the fundamentals of classical mechanics, of the electromagnetic interaction, and of statistical mechanics dominate the multivarious transitions and phenomena you discuss this week; and I presume that none of you expects his work on such problems to lead to modifications of these laws. You know the equations better than the phenomena . . . You are after the missing link between them, i.e., the intermediate concepts . . . which should allow a quantitative understanding and prediction of phenomena . . . .[2]

K. Gavroglu et al. (eds.), Physics, Philosophy and the Scientific Community, 125–152.

Van Hove was pointing to an important transformation that had taken place in physics: as in chemistry an ever larger fraction of the efforts of physicists were being devoted to the study of novelty rather than with the elucidation of fundamental laws and interactions. The picture of chemistry adumbrated by van Hove could have been elaborated further by pointing to its predominantly applied and utilitarian concerns and to the close ties that academic chemistry has traditionally had with industry. The same was becoming true for physics, and in fact for most of the sciences.

We are in the midst of a restructuring of the sciences: they have become or are becoming more like chemistry. My essay is about the internal reasons for this transformation within the physical sciences. It is suggestive rather than exhaustive for I have taken as my guideline Alexander von Humboldt's assertion to the young Charles Darwin that: "Works are of value only if they give rise to better ones."

THE ARGUMENT BRIEFLY STATED

It is my contention that the present situation is primarily the consequence of the enormous success of quantum mechanics – and more precisely of quantum field theory – and of the theoretical physics based on it.

The generation of physicists that matured with the birth of quantum mechanics was a "population"[3] in possession of traits that allowed it not only to adapt to existing conditions but in fact to dominate and transform the scientific and technological environment. The traits the physicists possessed were unusual versatility, outstanding mathematical abilities, and remarkable experimental and technical skills. These enabled the best among them to invent quantum mechanics, construct cyclotrons, develop atomic beams apparatus, design rhubatrons and klystrons – and thus acquire mastery over the atomic and nuclear domains. During the thirties a new generation was selected and trained – primarily on American soil.[4] The depression exerted a rigid selection pressure on the community: only the very best survived and these were a remarkable lot.

World War II gave physicists the opportunity to display their powers. They became inventors and creators of new gadgets. And whereas the *discovery* of the secrets of nature could be deemed a value-free enterprise, *invention and creation* carried with it moral responsibility. They emerged from the war transformed, with the state recognizing the value of their powers for its security and its power.[5]

The enormous expenditures for research and development during World War II and the lucrative support of science by the government after the war brought about a revolution in the physical sciences. The revolution was principally due to the availability of novel techniques and instruments. Microwave generators and detectors; nuclear reactors; cyclotrons, synchrotrons, and linear accelerators; transistorized electronics and computers; NMR; cryostats; masers and lasers – all these changed the practice of physics and of chemistry.

Building on the base laid by quantum mechanics, the theoretical foundations were likewise transformed. These culminated in a new understanding of condensed matter physics. In high energy physics it led to the establishment of what is now called the standard model, the description of matter in terms of quarks, gluons, leptons and weak and Higgs bosons – a representation that unifies the weak and electromagnetic forces and gives a particularly attractive depiction of the strong interactions. These conceptual developments in fundamental physics[6] have revealed a hierarchical structure of the physical world and have yielded successful representations that ascribe a quasi-decoupled relation to the different layers. These advances have supported the notion of the existence of objective emergent properties and have set an intrinsic limit to the reductionist approach that has been the hallmark of theoretical physics during the twentieth century. They have also given credence to the notion that to a high degree of accuracy our theoretical understanding of some of these domains have been stabilized. The activities in these fields are therefore driven by interests shaped by communities besides the scientific community. For this and other reasons, the moral sphere can no longer be considered as disjoint from the scientific one.

Let me amplify these remarks.

By the end of the nineteenth century it had become clear that theoretical physics, by elucidating the laws of nature, offered the possibility of the "intellectual mastery over nature' " a phrase coined by Hermann von Helmholtz, and that from the intellectual mastery of nature followed the possibility of the material mastery of nature, by letting "the natural forces work for us, after our will and wishes." In a lecture delivered just before the turn of the century, Helmholtz adduced as evidence for this the fact that "the whole development of industry in recent times, and the whole change in the form of human life and activity connected with it, depend in an essential way on our mastery over the forces of nature."[7] All the members of the theoretical physics community share this Helmholtzian vision of intellectual mastery over nature – though they are not necessarily committed to the Baconian one of material mastery. For some, for example Max Planck, the certainties of physics were to be the basis of their Weltanschauung and the foundation for their quest for understanding and meaning. In particular, the Leibnitzian notion of a preestablished harmony found wide currency in the discourse and rhetoric of the physicists and mathematicians of Göttingen. Indeed, many of them saw themselves as members of a priestly caste who were deciphering the secrets of the book of nature, and in fact, some of the theorists were characterized as such by members of the community. And faithful to their western legacy, the explanation these priests sought was to be an all-encompassing, global, unitary one. Quantum mechanics provided the framework upon which the subsequent unifications were to be based.

The revolutionary achievements in physics during the period from 1925 to 1927 stemmed from the confluence of a theoretical understanding, the representation of the dynamics of microscopic particles by quantum mechanics, and the apperception of an approximately stable ontology, electrons and nuclei. Approximately stable meant that these particles (electrons, nuclei), the building blocks of the entities – atoms, molecules, simple solids – that populated the domain that was being carved out could be treated as *ahistoric* objects, whose physical characteristics were seemingly independent of their mode of production and whose lifetimes could be considered as essentially infinite. These entities could be assumed to be essentially "elementary" pointlike objects, that were specified by their mass, spin, and statistics (whether bosons or fermions), and electromagnetic properties such as their charge and magnetic moment.

Quantum mechanics marks a turning point in the metaphysics of physics. The underlying metaphysics of physics until then predicated the existence of a *pre-existing, given* ontology at the microscopic level. Theoretical physics dealt with objects given in advance and these objects were seen as the objective correlates of the propositions of the theory. Although it was clear that the representations of theoretical physics considered very special experimental processes – in which a system is severed from the complex of its connections to the rest of the world – still it was assumed that the objects had an independent reality, that they already had been constituted.[8] Quantum mechanics indicated that this was not the case. Moreover the success of quantum mechanics at the atomic level, and more generally at the microscopic level, immediately made it clear to the more perspicacious theorists – as different in their outlook as Desmond Bernal and John Slater – that the laws behind the phenomena had been apprehended and that they could therefore control the behavior of simple microscopic systems, and more importantly, that they could on the basis of that knowledge create new structures, new objects, and new phenomena.[9] Condensed matter physics has indeed become the study of systems that have never existed before. Phenomena such as superconductivity, are genuine novelty in the universe.

Quantum mechanics reasserted that *the physical world presented itself hierarchically*. The world was not carved up into terrestrial, planetary and celestial spheres,[10] but layered by virtue of certain constants of nature.[11] As Dirac emphasized in the first edition of *Principles of Quantum Mechanics*, Planck's constant allows the world to be parsed into microscopic and macroscopic realms, or more precisely into the atomic and molecular domains and the macroscopic domains composed of these entities. The story repeated itself with the carving out of the nuclear domain: quasi-stable entities – neutrons and protons – could be regarded as the building blocks of nuclei and phenomenological theories could account for many of their properties and interactions.

During the 1970s the establishment of what has become known as the standard model description of the entities that populate the subnuclear worlds – quarks, gluons and leptons – marked the attainment of another stage in the attempt to give a unified description of the forces of nature. The program traces its origins to Newton and his realization of the universality of gravitation: gravitational interaction governs the fall of objects on the surface of the earth as well as the motions of the planets around the sun. Different aspects of the program were anchored during the nineteenth century. Oersted and Faraday – "natural philosophers" influenced by Naturphilosophie – gave credibility to the quest by providing the first experimental indication that the program of unification had validity. Thereafter, Maxwell constructed a model for a unified theory of electricity and magnetism and provided a mathematical formulation that was able to explain much of the observed phenomena and to make predictions of new ones. Fraunhofer and others demonstrated that the laws of physics discovered here on earth also apply to stellar objects. With Einstein the vision became totalitarian. Einstein advocated unification coupled to a radical form of theory reductionism. In 1918 he stated

> The supreme test of the physicist is to arrive at those universal elementary laws from which the cosmos can be built up by pure deduction[12]

This hope for unification animated the quest for the understanding of the molecular, atomic, nuclear and subnuclear domains – and success in obtaining a successful representation required a commitment to that vision.

Reduction and unification have been two tenets that have dominated fundamental theoretical physics during the present century. One characterizes the hope of giving a unified description for all physical phenomena; the other the aspiration to reduce the numbers of independent concepts necessary to formulate the fundamental laws. The conceptual – idealistic – component in the reductionism takes precedence over the materialistic one of reducing the number of so-called elementary particles. The impressive success of the enterprise since the beginning of the century has deeply affected the evolution of all the physical sciences as well as molecular biology.[13]

But ironically, the very concepts that helped achieve the important advances that led to the standard model were eroding the foundational tenets upon which the program is based. The ideas and concepts of symmetry breaking, renormalization group and decoupling adumbrate a picture of the physical world that is hierarchically layered into quasi-autonomous domains, with the ontology and dynamics of each layer essentially quasi-stable and virtually immune to whatever happens in other layers.

In fact at the height of its success, the privileged standing of high energy physics and the reductionism that permeated the field were attacked.

ANDERSON

In 1972 Philip Anderson, one of the foremost "condensed matter" physicist, challenged the radical theoretical reductionist view held by the majority of elementary particle physicists. His was not only an attack on the philosophical position of the high energy physicists, it also challenged their dominance within the physics community and within the councils of state. Anderson asserted that

> . . . the reductionist hypothesis does not by any means imply a "constructionist" one: The ability to reduce everything to simple fundamental laws does not imply the ability to start from those laws and reconstruct the universe. In fact, the more the elementary particle physicists tell us about the nature of the fundamental laws, the less relevance they seem to have to the very real problems of the rest of science, much less to those of society. The constructionist hypothesis breaks down when confronted with the twin difficulties of scale and complexity.[14]

Anderson believes in emergent laws. He holds the view that each level has its own "fundamental" laws and its own ontology.[15] But it is not enough to know the "fundamental" laws at a given level. It is the solutions to equations, not the equations themselves, that provide a mathematical description of the physical phenomena. Emergence refers to properties of the solutions. The properties of solutions are not readily apparent from the equations. Moreover, the behavior of a large and complex aggregate of "elementary" entities is not to be understood "in terms of a simple extrapolation of the properties of a few particles." Although there may be suggestive indications of how to relate one level to another, it is next to impossible to *deduce* the complexity and novelty that can emerge through composition.[16] The study of the new behavior at each level of complexity requires research that Anderson believes "to be as fundamental in its nature as any other." Although one may array the sciences in a roughly linear hierarchy, according to the notion: the elementary entities of the science X obey the laws of the science Y one step lower, it does not follow that science X is "just applied science Y." The elementary entities of condensed matter physics obey the laws of elementary particle physics, but condensed matter physics is not just applied "elementary particle physics", nor is chemistry applied many-body physics, *pace* Dirac. In his article Anderson sketched how the theory of "broken symmetry" helped explain the shift from quantitative to qualitative differentiation in condensed matter physics and why the constructionist converse of reductionism broke down.

Developments in quantum field theory and in the use of renormalization group methods have given strong support to Anderson's views. The insights from the renormalization group approach and from the effective field theory method have also greatly clarified why the description at any one level is so stable. In fact, renormalization group methods have changed Anderson's remark that: "The more the elementary particle physicists tell us about the nature of fundamental laws, the less the relevance they seem to have to the very real problems of the rest of science" from as folk theorem into an almost rigorously proved assertion.

THE CHANGED INTELLECTUAL SCENE: EFFECTIVE FIELD THEORIES

When renormalization was first developed after World War II it was regarded as a technical device to get rid of the divergences in perturbation theory, and seemed to be a remarkably effective way of sweeping the problems under the rug. Renormalizability became a criterion for theory selection.[17] Only renormalizable local relativistic quantum field theories were adopted for the representation of the interactions of what were then thought to be the "elementary particles". Note the cunning of reason at work: the divergences that had previously been considered a disastrous liability now became a valuable asset. However, the question as to why nature should be described by renormalizable theories was not addressed. Such theories were simply the only ones in which calculations could be done. The impressive advances of the late sixties and early seventies that culminated with the standard model were a triumph of renormalization theory. In 1976, Steven Weinberg, one of the principal architects of the unified theory of weak and electromagnetic interactions, declared that the fact there existed a limited class of theories that were renormalizable was for him "the most exciting thing" about quantum field theory. In the lecture he delivered in 1979 in Stockholm when he received the Nobel prize in physics, he indicated that

> To a remarkable degree, our present detailed theories of elementary particle interactions can be understood deductively, as consequences of symmetry principles and of the principle of renormalizability which is invoked to deal with the infinities.[18]

Relativistic local fields were again the "fundamental" entities; but now gauge symmetry and renormalizability, together with the insights derived from spontaneously broken symmetries of the Goldstone-Higgs variety, were the guiding principles.

But by the end of the seventies the cunning of reason once again had partially undermined the triumph. In the early fifties Stückelberg and Peterman, and Gell-Mann and Low[19] had made a fundamental observation whose importance was not realized by the community until much later. In the early 1970s, the implications of Gell-Mann and Low's paper were made manifest by Kenneth Wilson. Wilson had derived important insights from the work of condensed matter physicists attempting to explain phase transitions – in particular the researches of Lars Onsager, David Widom, Leon Kadanoff and Michael Fisher – and from the work of axiomatic field theorists who had elucidated the mathematical problems encountered in analyzing the properties of products of field operators. Wilson's work made clear that renormalization was not a technical device to eliminate the divergences "but rather is an expression of the variation of the structure of physical interactions with changes in the scale of the phenomena being probed."[20]

The renormalization group method made it clear that the cut-off, Λ, that is introduced in the regularization method that renders a relativistic quantum theory meaningful must be taken seriously. It was then realized that the physics that is observed at accessible energies (of order E) can be described, neglecting terms of $O[(E/\Lambda)^2]$, by an "effective" field theory, in which the interaction terms are finite in number and limited in complexity. In the effective field theory – the one that essentially any theory reduces to at sufficiently low energies[21] – the non-renormalizable interaction terms (that are consistent with the symmetries of the interactions) are all suppressed by powers of one over the cut-off, and hence negligible. Now once one accepts nonrenormalizable interactions, the key issue in addressing a theory like quantum electrondynamics "is not whether it is or is not renormalizable, but rather how renormalizable is it?" How large are the nonrenormalizable interactions in the "effective theory?" What range of energy scales are well described by the degrees of freedom that one keeps in the theory? Renormalizability has to do with the range in energy over which theory is valid.[22]

There are of course many theorists who do not accept the positivistic limitation entailed by the effective field theory approach.[23] At the present time they can be identified by their commitment to string theory.[24]

In condensed matter physics renormalization group methods were initially used to show that the detailed physics of matter at microscopic length scales and high energy is *irrelevant* for critical phenomena. What is important is the symmetry involved, the conservation laws that hold, the dimensionality of space and the range of the interactions. The method is not restricted to critical phenomena. It has been extended to show that for a many body system, integrating out the short wave length, high frequency modes (which are associated with the atomic and molecular constitution) one can arrive at a hydrodynamical description that is valid for a large class of fluids, and which is insensitive to the details of the atomic composition of the fluid. The particulars of the short wave (atomic) physics are amalgamated into the parameters – density and viscosity – that appear in the hydrodynamic description.[25] The physics at atomic lengths – and *a fortiori* high energy physics – has become decoupled.[26]

High energy physics and condensed matter physics have become essentially decoupled in the sense that the existence of a top and bottom quark, or any new heavy mass particle discovered at CERN or elsewhere, are irrelevant to the concerns of condensed physics, no matter how great their intellectual interest in them may be. The same is true in some measure of nuclear physics. This fragmentation has resulted in the exploration and conceptualization of the novelty capable of expression in the aggregation of entities in each level – novelty that is evidently contained in the "fundamental laws"[27] at that level and that does not challenge their "fundamental" character. The challenge is how to conceptualize this novelty – that is what Anderson meant by the statement "More is different".

In the statement that the condensed matter and high energy physics have become decoupled – the decoupling refers to the "ontology" that is used. Electrons and nuclei are the "elementary particles" of condensed matter physics, and the internal constitution of the nuclei is encapsulated in the parameters stating their spin, their magnetic moment, their electric quadrupole moment etc. More detailed descriptions are irrelevant for the description of the phenomena probed by condensed matter physics. There is however a great deal of exchange of ideas between the two fields. More than anything else, it was the importance of symmetry breaking, and in particular the beauty of the Higgs-Anderson mechanism of generating masses in massless gauge theories that made the particle physicists recognize the significance of the insights and techniques of condensed matter physicists. Then Wilson's work on the renormalization group brought home to both communities the importance of mutual education.

And indeed the cross fertilization has been of great value and mutual benefit to both sub-disciplines. The common pastures include such topics as: scaling and renormalization, topological defects, two dimensional models, Mote-Carlo techniques, non-linear σ models and much else.

The commonality of theoretical techniques used to address problems in what were different fields is a phenomenon more general than what is seen in high energy physics and condensed matter physics. Something similar is happening in areas of chemistry, meteorology, ecology, where the mathematics of dynamical systems has given deep new insights into such diverse phenomena as the onset of turbulence, oscillating chemical reactions, population dynamics . . . It is surely the case that the computer has been central in this development. It has allowed modelling and "visualizing" systems where the non-linearity is essential, and to deal with phenomena that were not amenable to the analytical methods fashioned until World War II. It is interesting to note that fixed points, domains of attractions, eigenvalues of relevant and irrelevant operators are part of the language common to all these fields.

What is also striking is the interdisciplinary nature of these new communities. What hold the communities together are not paradigms but tools, renormalization group methods, NMR machines, lasers, neural networks, computers, sequencing techniques, etc. What is also noticeable is that despite the apparent increase in specialization the interconnectedness of science is becoming more prominent. Tools and concepts are constantly being carried from one field to another in ways that are difficult to anticipate by any logical and structural process.[28]

It is my contention that the Kuhnian model will no longer do, and that we need to reconceptualize the growth of scientific knowledge. The new model will have to take into account that something important has happened. A hierarchical arraying of parts of the physical universe has been *stabilized*, each with its quasi stable ontology and quasi stable effective theory and the partitioning is fairly well understood. For the energy scales that are experimentally probed in atomic, molecular and condensed matter physics the irrelevance (to a very high degree of accuracy) of the domains at much shorter wavelengths has been justified. Effectively some kind of stabilization – one might call it a modified form of finalization – has taken place in these domains.[29]

Before turning to a discussion of some of the consequences of this stabilization let me make one further observation on the changing character of the physics.

GAUGE THEORIES, COSMOLOGY AND EVOLUTION

Advances in particle physics, and in particular, the establishment of the standard model and the subsequent attempts to unify the electroweak and strong interactions in grand unified theories, resulted in exciting interdisciplinary activities in cosmology and elementary particle physics. Thus the cosmological abundance of helium-4 fixes an upper bound on the number of flavors (quark varieties) in models that have symmetry between quarks and leptons. The striking success of the inflationary models in accounting for the flatness and horizon problems gives some credence to the approach – but a word of caution is in order since the validity of the Higgs mechanism will remain open until a Higgs particle is discovered and its properties determined.

I want to focus on one aspect of these developments to suggest that these advances have been responsible for yet another important transformation of physics.

The introduction of the idea of broken symmetry into elementary particle physics can be compared to the breaking of the spheres by Copernicus and Kepler: It transformed the way of describing and understanding elementary particle interactions.[30] When the dynamics of symmetry breaking at zero temperature were understood, it was natural to ask whether the broken symmetries of elementary particle physics would be restored by heating the system to a sufficiently high temperature, in the same way as the rotational invariance of a ferromagnet is restored by heating it above its Curie temperature. Kirzhnits and Linde[31] indicated that this was the case in field theories with broken global symmetries: starting from a spontaneously broken ground state, heating produces a phase transition and at high temperature the effects of the spontaneously broken symmetry disappear. Dolan and Jackiw, and Weinberg calculated the critical temperature in both gauge and nongauge field theories[32] and showed how finite temperature effects in renormalizable quantum field theories can restore a symmetry that is broken at zero temperature. Furthermore, Weinberg noted that the existence of such phase transitions had important cosmological implications in "big-bang cosmologies." These were elaborated by Sidney Coleman, Andrei Linde, Alan Guth and others.[33] The qualitative features of even the simplest model are sufficient for the point I want to make. At the very high temperatures, such as occur in the very early universe in big-bang cosmologies, for most gauge theories one is in the symmetric phase with the expectation value of the Higgs field equal to zero, corresponding to the only minimum of the effective potential for the Higgs field. As the universe expands and the temperature decreases the effective potential develops additional minima. At first these correspond to false vacua but at still lower temperature they become the true vacua. The state with the expectation value of the Higgs field equal to zero is now the false vacuum. Coleman showed that quantum fluctuations could create bubbles with one of the true vacuum on their inside. Such a bubble would grow and upon thermalization it would result in a large region of heated true vacuum. That region will be described by the theory with spontaneously broken symmetry.

If there is validity to this picture then an interesting shift has taken place in accounting for the regularities of the physical world – much closer in spirit to the way lawfulness is interpreted in biology. It would imply that evolution, i.e., history, would be a component in physical explanation. There are obvious caveats in that the scenario hinges on the Higgs mechanism, and the inflationary cosmology is not without its problems, the cosmological constant being one of them.[34]

It is of course the case that astrophysics brought a historical perspective to physics starting with Immanuel Kant's, William Herschel's and Simon Laplace's nebular hypothesis. Modern astrophysics deals with the life histories of stars and galaxies, nucleosynthesis and other such evolutionary process. Although we have come to appreciate that at some stage in the development of the universe, Mendeleev's table consisted of only two elements – rather than the present 100 or so – nucleosynthesis and stellar evolution were explained in terms of laws of nature that were thought fixed and immutable. It is the structures that evolved, not the "fundamental" laws. Normally, we don't think of Maxwell's equations or Schrödinger's equation as having evolved or having a history.

To state that QED is the low energy effective field theory approximation of the Weinberg-Glashow-Salam theory of electroweak interactions still casts the result within the traditional framework. To state that the Weinberg-Glashow-Salam theory of electroweak interaction is also the result of an evolutionary, historical process clearly adds a new dimension to the explanatory scheme.[35]

If there is legitimacy to this view, it would resonate with a position taken towards the end of last century by Charles Saunders Peirce, the great American philosopher who was one of the founders of pragmatism. Peirce had as deep an insight into Darwin's theory of evolution as anyone in the nineteenth century and was profoundly affected by it. Like Darwin himself he asked: How did natural selection evolve? The best Darwin could do was to suggest that perhaps maximalizing the amount of life on the surface of the earth is the global principle that drives the evolution of life on the planet, and gives rise to natural selection. To give an example of how regularity "evolved", in the *Descent of Man* Darwin traced the "evolution" of sexual selection from the lowest sexually reproducing life forms to human beings. Had he read the reprint Mendel sent him, he could have asked: "How did Mendel's laws evolve?" Clearly there was a time in the history of life on earth when there were no Mendelian laws.

Peirce took the position that to accept absolute, immutable, invariable laws makes the regularities ultimate and thus closes the door to the possibility of ever explaining them or explaining how there came to be as much regularity in the universe as there is.

> To suppose universal laws of nature capable of being apprehended by the mind and yet having no reason for their special forms, but standing inexplicable and irrational, is hardly a justifiable position. Uniformities are precisely the sort of facts that need to be accounted for. That a pitched coin should sometimes turn up heads and sometimes tails calls for no particular explanation; but if it shows up heads every time, we wish to know how this result has been brought about. Law is *par excellence* the thing that wants a reason.
>
> Now the only possible way of accounting for the laws of nature and for uniformity in general is to suppose them results of evolution. This supposes them not to be absolute, not to be obeyed precisely. It makes an element of indeterminacy, spontaneity, or absolute chance in nature.[36]

For Peirce evolution and "absolute" chance were fundamental. I want to stress the historical – evolutionary – component.[37]

I have indicated above that the advances in physics have suggested a picture of the world that is layered and differentiated into quasi autonomous domains. The standard model-based cosmology explains the occurrence of the upper quasi autonomous layers as a result of evolution. And as in the Darwinian account, the layering can be conceived of as diversification and speciation. This historicization of physical laws is one of the most striking implications of the revolutionary change in our world picture.

STABILIZATION AND COMMUNITY

The internal advances within physics have altered the traditional relation between the various branches of physics and have created a crisis for fundamental physics,[38] by which I mean primarily particle physics and cosmology. These fields claim to provide, in the reductionist sense, an ultimate basis for our understanding of the physical world. The crisis manifests itself both at the cognitive and at the social level.

Fundamental physics, as a rational pursuit, has been conducted principally within a certain world view, within a certain cultural tradition and within a certain economic framework. The world view is characterized by its atomistic thinking and reductionist methodology, the cultural tradition is distinguished by its commitment to individualism and to an instrumentalist conception of rationality, and the economic framework and the social conditions can be identified with capitalism. Yaron Ezrahi's *The Descent of Icarus* is a particularly insightful exposition of that thesis.

Fundamental research in physics is driven by a passion to reveal the secrets of nature by probing deeper and deeper into the physical world. This passionate internal dynamic, which has constantly propelled physics towards an intellectual mastery beyond the creation of technologically exploitable knowledge, is a deeply imbedded social practice that has ancient roots in the religious sphere. But the modern pursuit is conducted within a society whose dominant conception of rationality has become instrumental in nature. The justification of a reductionist pursuit within such a society depends, by and large, on the relevance of such resource-expensive research – both in tem of capital and talent – to the goals set by the society.[39]

The advances that I have outlined above have attributed a layering to the world. The stability of the layering and of the levels, i.e., the approximate quasi-independence of the levels, makes progress in fundamental physics seem to be irrelevant to the advance of other branches of physics and other sciences. In addition, in so far as the experimental high energy physics community is concerned, a convincing argument regarding irrelevance can also be made with respect to the development of technology, since the presumed benefits of building large accelerators – such as the improvement of large scale computing and the development of high T_c superconducting magnets – could be generated directly rather than as second order spin-off effects. Additionally, many of the assumptions regarding the connection between technological innovation and economic growth that were prevalent in the post World War II era have come to be challenged.

Both the conceptual and the social aspects of the crisis are attributable in part to the seeming failure of the reductionist approach that is essential to fundamental physics. The failure is rooted in the difficulties encountered by the reductionist approach in accounting for the existence of epistemological and *a fortiori* objective emergent properties.

The expectations that had seemed firmly established by previous successes, namely further reductionism and further unification, have not been met. In fact, the failure has called into question unification and reductionism as a strategy for the community – and thus undermined an important component of the value system shared by the community. At the social level, the cutbacks in federal funding, the contraction of job opportunities in both universities and national laboratories, are demanding a deep and painful restructuring of the community.

This and other factors, particular other factors peculiar to physics, have resonated with a wider cultural crisis that has been labelled the postmodern crisis. Fredric Jameson has suggested that taken together, the senses of crisis that permeate the epoch since the sixties – the crisis of social democracy, the crisis of the welfare state, . . . , and the constant refrains of the end of this or that – the end of ideology, the end of history, the end of philosophy, the end of subject, the end of social class, the end of art . . . – that echo throughout that timeframe, are some of the characteristic features of postmodernism.[40] My paper evidently adds another crisis to the list.

The so-called postmodern crisis has its roots in the critical confrontation with the legacy of the Enlightenment and its enshrinement of Reason that was initiated by Nietzsche at the turn of the century. For us in the United States the reassessment comes at a time when we have to face the cost of the waging of the cold war – a conflict that has left us almost bankrupt economically and in need of finding new bonds to hold the nation together since military expenditures and national security can no longer be relied upon to do so.

If one of the paths of the Enlightenment was to give rise to constant progress and to the perfectibility of human beings, another was molded by what Weber called *Zweckrationalität*, instrumental rationality, the rationality that is purposive and practical and aspires to master and to control. This path led us to the "scientific" waging of war and the rationalized, industrialized mass extermination that is the hallmark of the Holocaust.[41] When considered at any point along this path in the twentieth century, the hopes and aspirations of the Enlightenment philosophies appear to be a bitter and ironic illusion. They had envisioned and had expected a strong coupling between the growth of science, rationality, and human freedom. But as charted by Weber the main highway that carried the legacy of the Enlightenment was littered with the excesses of *Zweckrationalität*, whose growth during the nineteenth century had led to the creation of an "iron cage" of bureaucratic rationality, from which there seemed to be no escape. The heartless systems engineering that was expounded by RAND, adopted by McNamara's Defense Department and refined by the young Jasonites at IDA, and against which students rebelled in the late sixties, is part of that same rationality. Instrumental rationality shares the passion to control and master with certain political offshoots of the cult of rationality. Their common feature in the attempt to master reality is the banishing of particularized perceptions and the arrangement of such personalized insights into overarching, global, comprehensible regularities. The totalitarian states of the 20th century were the extreme political manifestations of this creed. They were explicit about their intentions and instrumental rationality helped build the gas chambers and the gulags.

Another legacy of the Enlightenment and of industrial capitalism – and the two have tended to march hand in hand – has been the cult of objectivity and the concomitant enthusiasm for quantification – particularly in the social sciences. But we have come to appreciate that quantitative knowledge is especially useful to coordinate the activities of diverse actors, and to lend credibility to forms of belief and action "when personal trust is in short supply." A fellow historian of science, Ted Porter, has stated the matter succinctly: "Objectivity has prestige in a pluralistic society mainly as a substitute for trust." And the logic of objectivity is most readily expressed by the language of numbers, quantification and mathematics. To appreciate how profoundly we have come to use numbers as the mechanism of replacing trust, think of present day medicine, and in particular how we certify death.[42]

The Enlightenment had as one of its progenitors the scientific revolution that had been wrought by Galileo, Bacon, Descartes, Harvey, Boyle, Newton, and others during the seventeenth century. Science became its emblem. Scientists and the institutions they spawned became the primary beneficiaries of enlightened states that depended ever more heavily on rationalized bureaucracies for their efficient and effective operations. It is therefore not surprising that the present attack upon the agenda of the Enlightenment should also manifest a sharp criticism of the scientific enterprise. That science could be the buttress of instrumental reason had already become clear to many at the beginning of the twentieth century. The power of science to propel technology had become evident with the successes of the electrical and chemical industries, and its potential in war was made plain during World War I.

How did sensitive and perceptive people react then? I have already alluded to the response of Max Weber. Let me briefly recall how Albert Einstein and Niels Bohr grappled with the legacy of the Enlightenment.

Einstein and Bohr are the towering figures of theoretical physics during the first half of the twentieth century. They laid the foundations for our quantum-mechanical understanding of the microscopic world. They grew up in a post-Darwinian environment and neither one of them could accept revelation or traditional religion as providing adequate answers. For both of them science embodied the ideals of the Enlightenment. They both lived for physics. For both, physics was the vehicle for their search for meaning, but their individual temperaments and characters shaped their particular ways to secure meaning. Abraham Pais, who knew both Einstein and Bohr personally and has written outstanding biographies of them, characterizes the most striking of all their differences as follows: "Einstein's apartness; Bohr's conjointness".[43]

Einstein always was the outsider, Bohr the builder of communities. Einstein had no real base. From 1913 to 1933 he was a Swiss in the country of his birth – Germany; and he remained an outsider in the United States. At one stage Einstein considered giving up his Jewishness. Bohr, on the other hand, as deeply committed as he was to internationalism – always was a Dane, and Copenhagen was always his anchor. Bohr and Einstein were rivals both in their scientific endeavors and in their Weltanschauung.

For Einstein physics became a retreat from the world, and he found a monastery at the Institute for Advanced Study in Princeton.[44] Meaning was given by his faith in the comprehensibility and the comprehension of nature. More specifically, Einstein believed "in the simplicity, i.e., the intelligibility of nature."[45]

Although Bohr did appreciate Einstein's position, for him science could not be an escape from the world. In fact for him there is no escape from the world. There is no way to step outside of it and obtain a God's eye view of it. We are always on the inside and whatever we know of the world we have mediated and constructed. He believed that all scientific understanding – all understanding – required that one test and risk one's convictions and prejudgments in and through encounters with others and in particular in encounters with what is radically new and alien. For Bohr the practice of science exhibited a commitment to an underlying moral order.[46] Scientists form an ideal society that relies on the virtues of honesty, tolerance, trust, truthfulness, and cooperation.

Without their moral commitment to be truthful and trustworthy members of this society that guarantees tolerance for the views of others and the verification of their claims, no important statements of fact could see the light of day and be stabilized. It was always self-evident to Bohr that science is a human activity, and constitutes a collaborative and cooperative enterprise. In fact, Bohr had a vision that the scientific community could serve as a paradigm for a possible future international order. Einstein would have mostly agreed – but he also believed that the communal aspect of science emphasized by Bohr could be transcended by individual acts of transcendence.

Ultimately, the difference between the two is the following: Deep down – although a child of the Enlightenment, Einstein was really a gnostic priest. He believed that he could read the mind of his Spinozan God and discover the laws that govern the objective physical world out there.[47] He believed in Truth with a capital T.[48] For Bohr science does not determine the truth, truth with a capital T – rather there is convergence toward truth and the understanding of physical reality. On the basis of their experiments – their queries to nature – scientists constantly reorder and reinterpret external reality, and these imaginative reorderings must be presented to the members of the community for their critical assessment by means of language. And because language is a uniquely human activity – our distinctive (though imprecise) means to communicate with one another and convince one another – and a unique means to stabilize communities without recourse to physical violence – language is a central and constant concern in Bohr's philosophical gropings.

Bohr's conception of community, as well as those of truth and reality have close affinity to the views of Peirce. Peirce's notion of reality combines a social conception, that "the real is the idea in which the community ultimately settles down" with the notion that the real is what it is regardless of what anyone may think of it.[49] For Peirce, the scientific enterprise is the model for how a society of free inquiring individuals can generate discourse and cooperation without chaos. What gives the actions of individuals their coherence is their decision as agents to aim toward a definite goal. We are all at work building and maintaining a scaffolding that will give stability and meaning to our lives and to the communities in which we share our lives. Science is part of that scaffolding; just as are literature, music, the visual arts, and the other activities we engage in. Science is a linkage between individualism and coordinated public choices and actions.

So too for Bohr. Bohr's conception of scientific activities in fact nurtures a wider vision of human interactions. Understanding the origin of the universe, its possible demise, unravelling the constitution of matter in its simplest manifestations and comprehending the rich and complex diversity matter can assume, understanding the limitations of mathematical theorizing and gaining insights into why it has been so effective in describing nature – are all part of our collective scaffolding – irrespective of the kind of scientific practitioners we are. Equally importantly, science is a piece of the scaffolding that makes for bridges to the members of other civilizations and cultures. Science, in fact, is that piece of what may be called the collective mind of the human race that is independent of place or individuality. And only in building together and in maintaining together the scaffolding can we guarantee stability.

WHERE WE ARE: THE ARGUMENT RECAPITULATED

For the most part the scientific enterprise is presently involved in the creation of novelty – of objects that never existed before in the universe. And precisely because we have created them we must assume moral responsibility for them. The huge success and the stability of the theoretical descriptions given by quantum mechanics and quantum field theory – which together with the quasi-stability of their related ontology suggests a quasi-finalization – also implies that the goals of most of the scientific enterprise are no longer determined internally; other interests come into play. And this is again an argument negating the separation of the cultural sphere of morality and that of science.

My emphasis on the act of creation is to make it clear that science as a social practice has indeed much in common with other human practices. We have invented priesthoods and monarchies, judiciary systems, forms of commerce and monetary exchanges, varieties of forms of government, trade unions, cheap newspapers . . . and we must assume moral responsibility for these. But there is one very important contrast: Nature plays the role of the ultimate arbiter in scientific practice. Or stated in more modest terms, the stabilities of the natural order studied by the natural sciences have a time scale that is "quasi-infinite" as compared to most other time scales.

I believe that in the reconstruction we are engaged in we must accept that the separation between the moral sphere and the scientific sphere cannot be maintained. We must reject instrumental rationality. The confrontation with the fact of the Holocaust allows the construction of a convincing case that deconstruction is not the answer, and makes explicit the grave dangers inherent in the relativism and the cynicism of the postmodernist position. Yet the stubborn question remains: How shall we determine the universalistic criteria that I believe we must have and commit ourselves to, criteria that transcendental philosophy had postulated as a priori? Part of the answer surely must come from the lessons we have learned from Darwin. We have a dual responsibility. On the one hand we have a moral responsibility for the future of our species – and in fact for all life forms on earth – and at the same time we have a moral responsibility to leave the future open – this in the face of having acquired the knowledge affecting our own evolution, and evolution in general.

Let me conclude by coming back to fundamental physics. I have argued that we should ground our philosophical reconstruction on taking prominently into account the creative component of our activities – and by virtue of creation our moral responsibilities. Fundamental physics has a privileged position in that the ontology of that domain and the other that it manifests is discovered rather than invented or created. But – though privileged – the community undertaking these investigations is part of the collective human enterprise. It must confront the implications of totalitarian visions such as Einstein's and recognize their danger. Yet it must also recognize the potency of that vision and accommodate it to a more human scale.

Fundamental physics has a special role to play – precisely because of what the proponents of these fields after WW II called their purity, their remoteness from everyday phenomena and their seeming lack of relevance in utilitarian matters. There ought to be a part of the scientific enterprise that does not respond easily to the demand for relevance. It has become clear that the demand for relevance can easily become a source for corruption of the scientific process. Elementary particle physics, astrophysics and cosmology are among the few remaining areas of science where the advancement of the field is determined internally, on the basis of its own experimental findings and its own intrinsic conceptual structure. Scientists engaged in fundamental physics have a special role – and a special responsibility – as a COMMUNITY committed to the Bohrian and Peircian vision. It is also a model of a Habermasian communicative community that exists under the constraints of cooperation, trust and truthfulness, one that is uncoerced in setting its goals and agenda, one that recognises that human emancipatory interests must be addressed. That community is the guarantor that one of the most exalted of human aspirations – "to be a member of a society which is free but not anarchical" – can indeed be satisfied.[50]

Brandeis University

NOTES

[1] Dirac, P. A. M., "Quantum mechanics of many-electron systems," *Proc. Royal Soc. (London) A* **126** (1929), 714–723.

[2] In fact, a year later Kenneth Wilson and Michael Fisher provided the missing link using renormalization group methods and 4-ε expansions. See K. G. Wilson and J. Kogut, "The renormalization group and the ε expansion," *Physics Reports C* **12** (1974), 131–268; K. G. Wilson, "Problems in Physics with many scales of length," *Scientific American* **241** (1979), 158–179; K. G. Wilson, "The renormalization group and critical phenomena," *Rev. Modern Physics* **55** (1983), 583–600.

Van Hove's speech is quoted in C. Domb, "Critical Phenomena: a brief historical survey," *Contemporary Physics* **26**/1 (1985), 49–72.

[3] Among the theorists: Hans Bethe, Lev Landau, Robert Oppenheimer, Rudolf Peierls, John Slater, Edward Teller, Sin-Itiro Tomonaga, John van Vleck, Victor Weisskopf, Hideki Yukawa, Eugene Wigner; among the experimenters: Enrico Fermi, Ernest Lawrence, Isidore Rabi, Felix Bloch, Emilio Segre, Bruno Rossi.

[4] E.g.: Robert Serber, John Wheeler, Julian Schwinger, Richard Feynman, Edward Purcell, Charles Lauritsen, William Fowler, Luis Alvarez, Norman Ramsey, Edwin MacMillan, Robert Wilson . . . .

[5] There exists by now an enormous literature on the physicists and World War II. This literature is quoted in M. Fortun and S. S. Schweber, "Scientists and the Legacy of World War II," *Social Studies of Science* **23** (1993), 595–642.

[6] In particular: symmetry breaking, renormalization group methods, decoupling, effective theories, etc. See T. Y. Cao and S. S. Schweber, "The conceptual foundations and philosophical aspects of renormalization theory," *Synthese* **97** (1993), 33–108.

[7] Russell MacCormmach and K. Jungnickel. *The Intellectual Mastery of Nature* (Chicago: The University of Chicago Press, 1987).

[8] See the suggestive article by Kurt Bayertz, "History of Science as a Natural Process?" *Scientia* **116** (1981), 285–293. See also the essays reprinted in Part I of E. P. Wigner, *Symmetries and Reflections: Scientific Essays* (Bloomington: Indiana University Press, 1967).

[9] Thus already in 1928 J.C. Slater had a vision of a molecular engineering practice based on quantum mechanics. See S. S. Schweber," J. C. Slater and the development of quantum chemistry," *Historical Studies in the Physical and Biological Sciences* **20**/2 (1990), 339–406. For J. D. Bernal see his *The World, the Devil and the Flesh.*

[10] According to Alexandre Koyré the most profound consequences of the scientific revolution of the 17th century stemmed from the fact it replaced the Aristotelian world picture – in which the cosmos was a finite, hierarchically ordered structure – by one in

which the cosmos is an abstract, infinite, homogeneous and void space populated by isolated bodies built of mutually separated and immutable smallest particles. It provided the metaphysical framework within which could emerge the Newtonian concept of fundamental laws of nature that aimed at a unified, reductionist representation of the physical world. Alexandre Koyré, *From the Closed World to the Infinite Universe* (Baltimore: The Johns Hopkins Press, 1957).

11 Thus non-relativistic quantum mechanics came to be seen as correctly describing that domain of nature delineated by Planck's constant h: Any system whose characteristic length (l), mass (m) and time (t) were such that the product ml^2/t was of the order of h, and such that l/t was much smaller than c, the velocity of light, was to be described by the new non-relativistic quantum mechanics.

12 Einstein 1918, In his "Autobiographical Notes" Einstein was even more explicit: "I would like to state a principle, which cannot be based upon anything more than a faith in the simplicity, i.e., intelligibility, of nature; that is to say, nature is so constituted that it is possible logically to lay down such strongly determined laws that within these laws only rationally, completely determined constants occur (not constants, therefore, whose numerical values could be changed without destroying the theory)." (Einstein 1949).

13 Reductionism together with its atomistic component is what Ian Hacking has called a style of scientific reasoning. "A style of reasoning makes it possible to reason toward certain kinds of propositions, but does not of itself determine their truth value." It determines what may be true or false and indicates what has the status of evidence. Similarly unification, the use of symmetry, the use of renormalization group methods, are examples styles of reasoning. Moreover, the fact these styles of reasoning are useful in both particle physics and in condensed matter physics – and in point of fact cross-fertilized these fields – illustrates the (nonlinear) additive properties of styles of reasoning. Styles of reasoning, as Hacking has noted, tend to be slow in evolution and are vastly more widespread than paradigms. They are not the exclusive property of a single disciplinary matrix [Hacking (1986)].

14 P. W. Anderson, "More is Different," *Science* **177** (1972), 393–396.

15 Translated into the language of particle physicists, Anderson would say each level has its effective Lagrangian and its sets of (quasi-stable) particles.

Anderson would say that in each level the effective Lagrangians – the "fundamental" description at that level – are the best we can do. When particle theorists explore the consequences of the $SU_3 \times SU_2 \times U_1$ standard model, their work is not more fundamental than what Anderson does exploring the consequences of the Schrodinger theory for condensed matter physics. Note in this case positivism leads to intellectual democracy.

16 The formulation of the strong interactions as a non-Abelian gauge theory can be considered a "fundamental" description. But to prove starting from QCD quark confinement in four dimensional space-time, or to go from that representation to an effective chiral Lagrangian to describe low energy pion-nucleon scattering, or to deduce from QCD the binding energy of the deuteron and explain why it is so small present enormous difficulties that have not as yet been overcome. Judging from the results obtained in the mathematical description of phase transitions in various dimensions, ascertaining the properties of the solutions of the "fundamental" equations – to say nothing of obtaining actual solutions – is an extremely difficult mathematical task involving delicate limiting procedures.

17 F. J. Dyson, Report to the Oldstone Conference held in April 1949.

[18] Steven Weinberg, "Conceptual foundations of the unified theory of weak and electromagnetic interactions", *Rev. Mod. Physics* **52** (1980), 515–524. See also S. Weinberg, "The search for unity: Notes for a history of quantum field theory," *Daedalus* (Fall 1977), Vol. II of *Discoveries and Interpretations in Contemporary Scholarship*.

[19] E. C. G. Stückelberg and A. Peterman, "La normalisation des constantes dans la theorie des quanta," *Helv. Phys. Acta.* **26** (1953), 499–520. M. Gell-Mann and F. Low, "Quantum Electrodynamics at Small Distances." *Phys. Rev.* **95** (1954), 1300–12.

See also Steven Weinberg, "Why the Renormalization Group is a Good Thing," in A. H. Guth, K. Huang, and R. L. Jaffe, eds., *Asymptotic Realms of Physics: Essays in Honor of Francis E. Low* (Cambridge, Mass: MIT Press, 1983), pp. 1–19.

[20] David Gross, "Beyond Quantum Field Theory," in J. Ambjorn, B. J. Durhuus, and J. L. Petersen, eds., *Recent Developments in Quantum Field Theory* (New York: Elsevier Science Publishers, 1985). For an overview of Kenneth Wilson's contributions see K. G. Wilson and J. Kogut, "The renormalization group and the ε expansion," *Physics Reports* **C12** (1974), 131–264; K. G. Wilson, "The renormalization group: Critical phenomena and the Kondo problem," *Rev. Modern Physics* **47** (1975), 773–840; K. G. Wilson, "Problems in Physics with many scales of length," *Scientific American* **241** (1979), 158–179; K. G. Wilson, "The renormalization group and critical phenomena," *Rev. Modern Physics* **55** (1983), 583–600.

[21] Steven Weinberg, "Phenomenological Lagrangians", *Physica* **96A** (1979), 327–340. Howard Georgi, "Effective quantum field theories," in Paul Davies, ed., *The New Physics* (Cambridge: Cambridge University Press, 1989) pp. 446–457.

[22] Peter G. Lepage, "What is renormalization?" in T. DeGrand and T. Toussaint, eds., *From Action to Answers* (Singapore: World Scientific, 1990), pp. 483–504.

[23] There is an interesting parallel to be made between Forman's thesis about quantum mechanics and Weimar Germany, and the development of effective field theories. Forman claimed that the Weimar physicists and mathematicians accommodated themselves to a Spenglerian point of view and rejected causality – even though this could not be related to any *internal* developments of physics, and particularly quantum theory. He thus wants to correlate the acausal features of quantum mechanics with *external* factors. The thesis however is much less prone to being challenged if stated as referring to the *reception* of quantum mechanics. In the present situation the question to be answered is the following: Were there external factors that made the acceptance of the effective field theory point of view particularly attractive to large segment of the high energy theoretical physics community? Was it an accommodation with the Zeitgeist? How did the proposal for building the SSC and large collider facilities impact on the reception and adoption of this point of view? This revised version of the Forman thesis does not claim that context determines the physics – but rather it suggests that there always exists a range of possibilities of view points – and that the milieu helps select among these. Paul Forman, "Weimar culture, causality and quantum theory, 1918–1927: Adaptation by German physicists and mathematicians to a hostile intellectual environment," *Historical Studies of Phys. Sciences* **3** (1971), 1–115; Paul Forman, "The reception of an Acausal Quantum Mechanics in Germany and Britain," in Seymor Mauskopf, ed., *The Reception of Unconventional Science*, AAAS Selected Symposium 25 (Westview Press, 1979); Paul Forman, "Kausalität, Anschaulichkeit, and Individualität, or how cultural values prescribed the character and the lessons ascribed to quantum mechanics," in Nico Stehr and Volker Meja, eds., *Society and Knowledge* (New Brunswick: Transaction Books, 1984).

[24] Thus in 1984 David Gross explicitly embraced Einstein's views regarding the possibility of ascertaining and formulating unique laws of nature and added that "It is a rather arrogant goal and a recent one in the history of physics." He indicated that he found support for this view from the uniqueness of QCD, referring "not to the fact that it seems to be uniquely singled out by experiment but rather to the fact it contains essentially no adjustable parameters." David Gross, "On the Uniqueness of Physical Theories," in *A Passion for Physics*, C. DeTar, J. Finkelstein, and Chug-I Tan, eds. (Singapore: World Scientific, 1985).

[25] The parameters, such as density and viscosity, that enter the hydrodynamical level of description encapsulate the ignorance of the short distance behavior.

[26] The differences between the non-relativistic (and essentially finite) field theoretic descriptions of condensed matter physics and the relativistic and divergent quantum field theories of high-energy physics ought not to be overlooked. The situation in the latter is clearly more ambiguous and more tentative – and the claims for the effective field theory methods ought to be tempered by that fact.

[27] The effective Lagrangian for that level.

[28] This point was made by Harvey Brooks already in 1973.

[29] The date when "finalization" occurred for non-relativistic quantum mechanics can be taken to be when Bardeen, Cooper and Schrieffer explained superconductivity. Prior to that there was always the nagging possibility that quantum mechanics broke down at distances of the order of 100–200 Angstrom. Thus Feynman wrote Landau in 1954: "It is a remarkable achievement in less than 30 years after quantum theory was discovered we can say that, as far as we know, the qualitative explanation of all extra-nuclear phenomena are understood as a consequence of that theory with two exceptions. One is gravity, which we believe cannot be understood in terms of the simple Schrodinger equation. The other is superconductivity. There remains that one point still resisting siege! (I omit the phenomena of life because we don't know enough about the atomic arrangement to know if quantum theory will hold up.)"

R. P. Feynman to L. Landau, November 22, 1954. Feynman Papers. Archives, C.I.T.

A few years later BCS removed that obstacle. BCS's theory of superconductivity can indeed be taken as successfully completing the non-relativistic quantum mechanical agenda.

The notion of finalization and decoupling clearly have important ramifications for policy issues.

The concept of a quasi stable theory has some affinity with the notion of a "closed theory" that Heisenberg advanced in the late twenties and that he published in *Dialectica* in 1948. See "The Notion of a 'Closed Theory' in Modern Science" in Heisenberg (1974). For a historical account of the formulation of theses ideas see C. Chevalley (1991). The notion of a quasi-stable theory was also used in Schweber (1989). But the conceptualization there emphasizes the stability of the theory against small changes in the formalism. Thus it seems extremely difficult to make small changes in the formalism of quantum mechanics that maintain logical consistency, empirical validity and the usual notions of causality.

[30] As is well known, dynamical symmetry breaking was brought into elementary particle physics from many-body physics. See L. M. Brown and Tian Yu Cao, "Spontaneous breakdown of symmetry: Its rediscovery and integration into quantum field theory," *Historical Studies in the Physical and Biological Sciences* **21** (1991), 211–235.

[31] D. A. Kirzhnitz and A. D. Linde, "Macroscopic consequences of the Weinberg model," *Phys. Letters* **42B** (1972), 471–4.
[32] L. Dolan and R. Jackiw, "Symmetry behavior at finite temperature," *Phys. Rev.* **D9** (1974), 3320–41. S. Weinberg, "Gauge and global symmetries at high temperature," *Phys. Rev.* **D9** (1974), 3357–77.
[33] See for example L. F. Abbott and So-Young Pi, eds., *Inflationary Cosmology* (Singapore: World Scientific, 1986); Adnrei Linde, *Particle Physics and Inflationary Cosmology* (New York: Harwood Academic Press, 1990) and Alan Guth and Paul Steinhardt, "The Inflationary Universe," in Paul Davies, ed., *The New Physics* (Cambridge: Cambridge University Press, 1989), pp. 34–60.
[34] See for example Larry Abbott, "The Mystery of the Cosmological Constant," *Scientific American* **256**:5 (1988), 106–113.
[35] Note that the viewpoint posited does not exclude the possibility that there exists a "fundamental" unified theory, which can be represented at various energy scales by "effective" field theories.

The present situation with respect to symmetries may also be analogous to the relation between special and general relativity. In the special theory of relativity space-time is a rigid stage unaffected by the matter and energy present – as was the case in Newtonian physics. General relativity dynamicized space-time, with the geometry of space-time becoming a response to the energy-momentum density present. The analogy is the following: before the standard model, symmetries were likewise thought of as "rigid." Gauge field theories have allowed the possibility to "dynamicize" the symmetries.
[36] C. S. Peirce, "The architecture of theories", in *Chance, Love and Logic. Philosophical Essays* (New York: Harcourt, Brace, 1923), p. 162. See also Henri Poincaré, "L'Évolution des Lois" in *Dernières Pensées* (Paris: Flammarion, 1913), pp. 48–67.
[37] Interestingly, in the late 1960's Cocconi had arrived at a somewhat similar position as a result of his confrontation with the diversity and complexity encountered in the world of elementary particles. G. Cocconi, "The Role of Complexity in Nature," in M. Conversi, ed., *Evolution of Particle Physics* (New York: Academic Press, 1970), pp. 81–87.
[38] The division between the various subfields of physics were sharpened during the eighties, partly because of shrinking budgets. The disparity in the time scales for constructing new accelerators and detectors and that for creating novel theories have brought a sense of crisis to the high energy community. There have essentially been no new experimental results since the early eighties when the electroweak theory was fully corroborated with the discovery of the W's and the Z°. Particle physics theorists are divided into various camps – with string theorists representing the tradition of trying to seek unifying (unitary) theories and with many of them resonating to Dirac's contention that theories ought to be "beautiful". But, since their efforts thus far are so removed from experimental relevance, they are branded mathematicians by the more phenomenologically inclined theorists committed to effective field theory approaches.

And everyone seems to agree that some of the most exciting aspects of high energy physics are to be found in astrophysics.

Nor has the condensed matter physics community been immune to the sense of crisis. The excitement generated by the solution of the problem of phase transitions has abated. The problems of explaining high temperature superconductivity have proven more refractory than initially anticipated. Nor has the departure of a number of distinguished practitioners to such fields as biophysics and neural networks escaped notice.

The end of the cold war in the late eighties has exacerbated the sense of crisis. The *New York Times* reported that some 800 applicants had applied for a single tenure track position at Amherst College. *Physics Today* in March of 1992 reported on the sense of despair that characterizes the atmosphere at the meetings of the AIP. These suggest that the discipline is facing a situation as difficult as that during the early thirties at the height of the depression. Furthermore the demographics of the physics community has brought these issues to the nitty gritty of departmental politics precisely at a time when universities have to cut back and government funding is slackening. It is likely that the discipline will shrink sharply in size over the next decade or so.

[39] I here draw on a paper I presented at the June 1992 SLAC conference on the "The Rise of the Standard Model" and on a draft for a workshop on the crisis in physics prepared by Tian Yu Cao and S.S. Schweber.

[40] Fredric Jameson, *Postmodernism, Or The Cultural Logic of Late Capitalism* (Durham: Duke University Press, 1991). A list of ends was given by Jacques Derrida in an 1984 essay entitled "Of an Apocalyptic Tone Recently Adopted in Philosophy."

[41] See the fine sets of essays in Saul Friedlander, ed., *Probing the Limits of Representation* (Cambridge, MA: Harvard University Press, 1992). Also Zygmunt Bauman, *Modernity and the Holocaust* (Ithaca: Cornell University Press, 1989).

[42] Everett Mendelsohn. Lechire, Harvard University, Fall 1989.

[43] A. Pais, *Bohr's Times, in Physics, Philosophy and Polity* (New York: Oxford University Press, 1991), p. 227.

[44] Perhaps monastery is an apter description of the Institute as long as Abraham Flexner was director. When Einstein first arrived at the Institute in Princeton in 1933, Flexner turned down an invitation to the White House that Einstein had received from Roosevelt, noting that Professor Einstein had come to Princeton "for the purpose of carrying out his scientific work in seclusion" and it was "absolutely impossible to make any exceptions." Abraham Flexner to F.D. Roosevelt, 3 November 1933. In *Science in America: A Documentary History, 1900–1939*, Nathan Reingold and Ida Reingold, eds. (Chicago: The University of Chicago Press, 1981), p. 451. The Institute did become a Temple during the thirties when Flexner gave over the reins to Frank Aydelotte.

[45] Recall the statement Einstein made in 1941. "Although it is true that it is the goal of science to discover rules which permit the association and foretelling of facts, this is not its only aim. It also seeks to reduce the connections to the smallest possible number of mutually independent conceptual elements. It is in this striving after the rational unification of the manifold that it encounters its greatest successes, even though it is precisely this attempt which causes it to run the greatest risk of falling prey to illusions . . ."; in *Science, Philosophy and Religion*. A symposium published by the Conference on Science, Philosophy and Religion in Their Relation to the Democratic Way of Life, Inc. New York, 1941. Reprinted in *Ideas an Opinions*.

[46] J. Bronowski, *Science and Human Values* (New York: Harper Torchbook, 1959). See also, Alan G. Wasserstein, Scientific Optimism: A Progress Report, *Raritan* **9/1** (1989), 114–133.

[47] "I want to know how God created the world. I am not interested in this or that phenomenon, in the spectrum of this or that element; I want to know his thoughts; the rest are details." Albert Einstein. From "A talk with Albert Einstein" in *The Listener*. September 1955. Quoted in M. Sachs, *Einstein versus Bohr* (La Salle, Ill.: Open Court, 1988).

[48] For Einstein's views on God and religion see *Ideas and Opinions*.

[49] "The real then, is that which, sooner or later, information and reasoning would finally result in, and which is therefore independent of the vagaries of me and you. Thus, the very origin of the conception of reality shows that this conception essentially involves the notion of COMMUNITY, without definite limits and capable of a definite increase of knowledge." Elsewhere Peirce asserts that "What anything really is, is what it may finally come to be known to be in the ideal state of complete information; so that reality depends on the ultimate decision of the community," and "But the reality of that which is real does depend on the real fact that investigation is destined to lead, at least, if continued long enough, to a belief in it."

There is therefore some notion to convergence toward truth – but it requires an ideal society which has complete information.

[50] In the final discourse at the centennial celebration of the NAS, Rabi pointed to this aspect of science as one of its greatest attraction. Rabi's depiction of the community was as follows: "Members of this community possess an inner solidity which comes from a sense of achievement and an inner conviction that the advance of science is important and worthy of their greatest effort. This solidity comes in a context of fierce competition, strongly held conviction, and differing assessments as to the value of one achievement or another. Over and above all this too human confusion is the assurance that with further study will come order and beauty and a deeper understanding." I.I. Rabi, "Science in the Satisfaction of Human Aspiration," in *The Scientific Endeavor. Centennial Celebration of the National Academy of Sciences* (New York: The Rockefeller Institute Press, 1963), pp. 303–309, p. 308.

JOSEPH AGASSI

CONTEMPORARY PHILOSOPHY OF SCIENCE AS A THINLY MASKED ANTIDEMOCRATIC APOLOGETICS

1. PHILOSOPHY SANS POLITICS

Let me begin by making explicit some background assumptions so as to prevent some false impressions. I wish to avoid the impression of an overall dismissal of philosophy or even of contemporary philosophy. This is too easy to do by the observation that for ever so many purposes philosophy is useless; this is true of every field of human activity. I remember full well how distasteful I found Hilary Putnam's declaration, during the turbulent days of the student revolt, when we sat on a panel together in Boston University: all the philosophy we teach, he declared, is just bourgeois crap. I do not agree. We teach the history of philosophy, and the history of theories about science, including the views of Bacon and Descartes and of Whewell and of Duhem and Poincaré, and these are neither bourgeois nor crap. I am speaking in the present essay of the political significance of the contemporary scene, not of philosophy in general, nor of the philosophy of science in general nor of significance in general.

Politically speaking, the situation within philosophy is rather simple: it is by no means restricted to the philosophy of science in the technical sense of the word, though it is deeply linked to attitudes towards rationality, and so to attitudes towards science in general. The two leading mainstream philosophical schools today, the Continental and the Anglo-American, so-called, or the phenomenological and the analytic, so-called, and both elitist – opting respectively for a cultural elitism (that is, a thinly masked anti-science irrationalism) or for a technological one (and so in favor of science-based technology, and thereby in favor of science as the tough-and-no-nonsense attitude that is but a thinly masked hostility to culture). Thus, inasmuch as they partake in politics, they are often elitist antidemocratic. The rest are mostly neo-Marxist or neo-Thomist and thus, likewise, more often antidemocratic than democratic when politically active. This situation is quite general today; it is very sad and it is reflected in the philosophy of science as much as in other sub-fields

K. Gavroglu et al. (eds.), Physics, Philosophy and the Scientific Community, 153–170.

of philosophy. This, to repeat, is only the political side of the situation, and only to the extent that the situation is political.

Nor is this the whole picture. Let me stress that not much philosophy is political and that philosophers may act quite differently in their capacity as citizens and as philosophers. And yet paradoxically, these two limitations raise many difficult issues. How much does the non-political aspect of philosophy impinge on political matters? That is to say, considering students of elitist philosophy who ignore matters political and then become politically involved, will the elitist philosophy have an impact on their political choice? If so, will they revise their philosophy, or will they be swayed by it to antidemocratic attitudes, or neither? Moreover, will the individuals who are democratically inclined as citizens but who in their politics are inclined towards elitism, as educators behave in their capacity more as citizens or more as elitists?

These are fascinating problems that merit sociological study. They are obviously significant, yet they are neglected. Regrettably they will stay neglected here. They are mentioned here because the following discussion is preliminary to their empirical study and borrows its significance from them.

To what extent is the situation in contemporary philosophy political? To what extent is philosophy, other than political philosophy proper, inevitably political? To what extent does it have to be political at all? Marxists say that every philosophical activity has a political aspect, intended or not: every intellectual activity, the Marxists say, has significant political implications. Marxists even go further and say that every human activity is inherently socio-economic and thus has an essential political component; willy-nilly, they say, each and all individual members of any community are politically involved, and if they do not actively undermine the regime they thereby support it. There is much to say in defense of this doctrine; it sounds very reasonable, but ultimately it is mistaken. Politics includes the choice of a lifestyle, and in the liberal doctrine of the choice of lifestyles an important item is the demand that individuals may choose to be politically neutral and that likewise individuals who are politically active may deem some of their activities politically neutral.

In this respect I support the attitude of Rudolf Carnap, in his famous disagreement with Otto Neurath, when he defended his right to study rigorous philosophy regardless of politics, against Neurath's Marxist position. There is, he assumed without debate, a vast area of philosophical activity that is neutral to all extant political conflict and controversy. And this assumption is sufficiently widespread that it may be taken here as true without any debate. Whether Carnap was right or not, it is well-known that he had his political convictions, that he was true to them rather bravely, and that he refused to discuss the possible relevance of his detailed philosophical activitics to his political ones. Possibly Carnap would have agreed that intellectual activity as such is not politically neutral, but he insisted, contra Neurath, that many philosophical researches are neutral to political conflict and controversy, so that he could do some worthy piece of research, without troubling himself about its political import and, moreover, regardless of its being politically rather neutral. He did not feel that he had to restrict his research to the (directly) politically significant. Neurath, in line with Marxist lore, demanded that one's actions be constantly politically relevant, that intellectual activities be heightened by one's awareness of their political significance and import. We need not go into that: surely, whether Carnap was correct in his judgement or not, he had the right to it. Nevertheless, clearly, there is no guarantee of success for one's intention to stay clear of politics, even of success for the intention to keep any single activity politically neutral. My discussion here will centre on this point, as I suggest that the political import of the current way of "doing" philosophy of science has some political implications that are too transparent to be ignored.

2. PRUDENCE AS POLITICS

Let me offer first an example of an innocent and unnoticed political implication of a particular piece of philosophical research. When Carnap tried to formulate a rule of conduct in his *magnum opus*, his *Logical Foundations of Probability* of 1950, he did not refer to politics, but he took no precaution to keep it out. He was careful in his wording in other respects; indeed, he worded his rule for conduct in accord with inductive logic very carefully: the final version of his book presents five versions of this rule, presented in increasing degrees of precision. Nor is it surprising, as this is the single rule of the book and it is central to its very thrust: probability is the guide for life, says the motto of the book, and this guidance is condensed and articulated in one rule, and so the rule is very carefully worded. In all its variants the rule is that of adoption of the calculus of probability in practical affairs, but the grounds for this advice is not a simple matter. In the final version of the rule Carnap offered the advice to business people to follow the rules of the calculus of probability because, he said, the rules promise *success in the long run*. Success in the long run unintentionally carries wide and powerful political implications. The proposal to consider the long run is even politically objectionable: in the long run we are all dead, said economist John Maynard Keynes against the economists who said that in the long run governmental nonintervention in the market is best.

Let me stress: Carnap did not intend in the least to contradict Keynes. As it happened, he was even more inclined to support government intervention in the market than Keynes, since he was a socialist more ardent than Keynes. Nevertheless, he advised business people, and in a manner that unintentionally conflicts with his own political convictions and with Keynes' economic theory. Whenever faced with such criticism, Carnap's technique was to respond with the claim that it is based on confusion, and he brought as evidence his own intentions, which, to repeat, are here not contested. Were he alive, he would perhaps say the long run in his work and in Keynes' works are different. (His response to Popper's critique of his theory of confirmation was of this ilk: he said his and Popper's concept of confirmation differed and so their views on confirmation only seemingly clashed but did not. See his "Replies" in *The Philosophy of Rudolf Carnap*.)

Assume this to be the case. The question then is, what advice would he give to economic reformers? Would he advise them to follow Keynes and to cater to immediate national economic needs since, if they do so repeatedly, then sooner or later, in the long run, they will be successful? Keynes wanted to attend to the immediate national economic needs since he wanted to see immediate relief. Carnap's theory does not offer immediate success ever, as it is based on mere probabilities, which cannot promise success except perhaps in the long run. Carnap's advice, then, seems to be inherently useful only for entrepreneurs and inherently useless for economic reformers. Will this be compatible with his intent to keep his research politically neutral? Perhaps. Perhaps he could say, his study (of 1950) is valid for entrepreneurs and the next study should follow in its wake and serve economic reformers. However this would justify Alan Musgrave's critique of inductive logic as a logic for commerce only, adding that if different logical systems are needed for reformers, for consumers, for parents, etc., then it scarcely deserves to be called 'logic' though it is what the "Whig" F. A. von Hayek called "the logic of the situation", or "situational logic"; this "logic" certainly is politically tainted, as was stressed by Hayek's mentor, the "Whig" Ludwig von Mises. Still, were this the only critique, it would at least be politically neutral. Is it the only critique?

What could Carnap's advice be for non-capitalists? To begin with, why did he address capitalists of all people? Or did he mean by business not the market activity which the expression usually denotes but any activity in the world of practice? I do not know. I do not know what his advice could possibly be to economic reformers. Perhaps really the logic of situations that require success in the short run differs from the logic of situations where success in the long run should suffice. If so, then the question will be, which problem-situations require solutions in the short run and which can wait? On this Keynes and Hayek disagreed. Will the disagreement be resolved by a short-run induction or a long-run one? This is but the first step in the infinite regress that bedevils what Carnap liked to call inductive logic, of course. Carnap said that the problem of induction is no problem. (Did he consider it a pseudo-problem?) So he may have had a solution to the difficulty posed here. If so, however, then Carnap's solution will settle a dispute in political economy! And Carnap surely did not intend his philosophy to solve cardinal political problems – certainly not incidentally!

There is ample evidence for this, indirect and even direct: towards the end of his life Carnap tried to grapple with the objections here voiced and moved away from the idea that prudence is the guide for life: he moved from probability to decision theory, where the choice between prudence and daring is often left to individuals to decide upon, and at times advice is offered to act boldly rather than prudently, when prudence is not expected to bring a good payoff but daring might. This is more in line with Keynes' sentiment, as he observed that most people handle their meagre financial resources quite prudently, yet for no good reasons.

This discussion goes too well with Neurath's critique of Carnap. Yet it is not meant to prepare for a *volte face*; I have no reason to withdraw my siding with Carnap's view that in principle one need not attend to politics unless one chooses to do so. Nevertheless, I acknowledge that Neurath's critique cannot be overlooked. Carnap did mix politics in his researches even though he was not always aware of it. Hence, the question arises, under what conditions is it safe to ignore politics without fear of making (poor) political judgements by default? I find this an intriguing and an unjustly neglected question, but I, too, will let it go for now. What I claim is that practically all of contemporary philosophy of science has political implications, and they are conservative and unsavory (at least for those who do not find conservatism so attractive as to take it as the default answer to all questions).

To conclude my discussion of Carnap, my concern is not with his legacy. I used his case only as an example of the difficulty of avoiding political implications of one's (philosophical) researches and views. My concern is with contemporary philosophy of science. In that field many of today's acknowledged leaders, such as Richard Jeffreys, Isaac Levi and Adolf Grübaum, follow Carnap's inductivism (more or less); they continue the search for the right rule for rational belief and for rational conduct, on the assumption that rational conduct is founded on rational belief and that rational belief is rationally binding. And though they have not found the right inductive rule, they know that it should justify established scientific opinion, and so they demand that everyone follow this still-not-known rule of induction and endorse received scientific opinion on the pain of being branded irrational. This is a political matter and these philosophers gain the powerful support of the scientific establishment which likewise does not possess any knowledge of the mysterious, all-powerful rule of induction. What concerns me in the present essay is the impact that this travesty on reason has on democracy.

How does the philosophy of science square with democracy? Is the obligation to follow the rules of inductive logic compatible with the democratically guaranteed right of conscience? Traditionally rationalist philosophy (to which the philosophy of science belonged as a matter of course) was linked not to any specific regime but to the idea of equality, of the siblinghood of humanity. This idea had a tremendous political import, as it deprived the aristocracy of its privileges. Though the scientific tradition was always liberal, it was not always democratic, and for the reason specified: liberalism may tolerate irrational views, democracy requires a view of rationality that permits dissent, and inductive logic forbids it.

It is important to the present discussion that it is easier to be liberal than to be democratic. To be liberal without democracy one needs to tolerate but not appreciate disagreement; to be democratic without much liberalism, one needs to tolerate some rational disagreements but not all, or at least not foolish ones.

The question raised here is not new: is there democracy without liberalism and vice versa? This question is still repeatedly heard. It seems to me clear that we can have democracy without liberty, if and when the people decide to send the aristocrats to the Guillotine, for example, or the Jews to the gas-chambers. Liberalism is endangered in a democratic society in which popularity can be found for the advocacy of gas-chambers, namely, to use a new (1991) Austrian expression, for the advocacy of the employment policy of the Third Reich. This is why we usually speak today not of democracy *tout court*, but of liberal democracy. Those who say a country which is not liberal cannot be democratic will not speak of liberal democracy any more than about wet water. The matter, thus, is largely a matter of definition, and so it is easily decidable to the satisfaction of all parties. The question, can there be a liberal regime without democracy is harder. On the face of it we have instances to that effect: we do have cases of regimes that are relatively liberal with no shade of democracy in them. We do not like them, however. We do not like them not only as democrats but also as liberals. The trouble is that the liberty they offer is not protected, as the rulers, whoever they are, can decide to become illiberal. In particular, the democracy that Karl Popper advocates is the one in which there are institutions for the safeguard of liberty. This is not so simple, since, as Popper stresses, there is no guarantee for the preservation of liberty even in a democracy. Indeed, the legitimate rise of the Third Reich is his example for that. Nevertheless, the fact remains that in modern western-type liberal democracy there are institutions for the safeguard of liberty and in other, possibly liberal regimes or countries, there are none. But in a liberal democracy there is only the safeguard of dissent and of free speech and free exchange of ideas; there can be no guarantee of the quality of dissent, and when dissent is always poor democracy may be in danger.

What can the philosophy of science tell us about politics in general and about democracy and about the safeguard of liberty in particular and of the safeguard of the quality of some dissent so vital for both democracy and liberty? This is the question of the present essay. The view it advocates is that though some philosophers of science are democrats and liberals and struggle to institute newer and better safeguards for liberty, most of contemporary philosophers of science, knowingly or not, do the opposite.

I shall now make two points and contrast them, thus bringing this essay to a close. First, let me observe what the philosophers of science recommend. Second, let me explain the current situation in political philosophy. Finally I will bring the two together. Let me sum up the rest of this essay now, however. First, the philosophers of science recommend that scientific ideas be admitted as obligatory objects of belief. More generally, they recommend prudence. Second, the current dispute is between democracy and technocracy: in the name of prudence the technocratically inclined advocate that technicians run society. Third, that the advocacy of prudence is the advocacy of the status quo and the advocacy of science as obligatory object of belief is technocratic and thus clearly antidemocratic. As a coda let me add that Paul Feyerabend is an ally of democracy who is more of a liability than an asset to its cause.

3. THE PHILOSOPHY OF SCIENCE AS NORMATIVE

The philosophy of science is a brand of rationalist ethics, not a theory of what constitutes science. This is puzzling to hear (I am reporting a repeated empirical observation here), presumably since the philosophy of science is generally taught as descriptive, not as prescriptive. But two incontestable, acknowledged observations will prove my point. First, the courses taught in universities in the philosophy of science never include the views of science held by irrationalists. Even the considerable body of writings of Martin Heidegger on science and on technology, regularly taught in other kinds of courses, even this is shunned by philosophers of science proper. I do not wish to blame them: shunning Heidegger is generally praiseworthy. But the fact proves my point: the philosophy of science is prescriptive. Second, take the critique that John Watkins and Paul Feyerabend launched against Thomas S. Kuhn in the famous bestseller, edited by Lakatos and Musgrave, *Criticism and the Growth of Knowledge*, of 1970. They say he describes science as oppressive and illiberal; moreover, they add, he conceals the fact that he advocates science as he describes it. Kuhn, in his celebrated "Replies" in the same volume, admits the first point and denies vehemently (and rightly) the second: he openly commends science though it is oppressive and illiberal. Kuhn has an excuse for this, of course: he considers the illiberal attitude of science consistent with political liberalism, since no one forces the individual students and researchers to stay in the university (namely Harvard University) that practices the illiberal manner of scientific instruction and research that he, Kuhn, describes: every one of them is free to leave.

I do not endorse this excuse. The same excuse is used against the attempts to improve the quality of working life and for industrial democracy in general: no workplace in any liberal democracy ever forces any worker to be there, and, every industrial concern's spokespeople say this exempts them from catering to demands for democratic control of their employment practices. Their error is in their presentation of the facts in detail and not in their overall characteristics. Workers are not obliged to go to any single workplace, and they are not forced to aspire to better jobs; yet workers have to find jobs somewhere, and it is in the national interest to create incentives for them to seek better jobs. Hence, some arrangements should be made to improve the quality of working life and for the worthwhile opportunities for involvement of workers – in their own attempts to contribute to the fate of their own workplaces.

Kuhn is not a sociologist, much less a sociologist of work. He relates that when Alexandre Koyré the great teacher of the historians of science of the post-war period (the paradigm maker, to use Kuhn's terminology), informed him, Kuhn, that he was advocating a specific sociology of science, he, Kuhn, was genuinely surprised. I am no paradigm maker (or so I hope) but I am now reporting that Kuhn has a sociology of work, and a distinctly politically conservative one, and one that has not gained the critical attention of any fellow philosopher or historian of science (except for myself; Popper and others have criticized his philosophy of science, but not his sociology of science). True, he was only generalizing the work habits of the physicists at Harvard in his own formative years, but as those were the days in which the Pentagon was the slave-driver of all the physicists it could bully to work for the defense industry, this was it. On this basis, Kuhn reports the sociology of the leadership of the scientific community, and so his peers dare not dissent from him on this.

Kuhn openly declares the task of the normal research scientist boring and he defends this situation; despite the tacit consent of his peers, it is no accident that this is not favorably looked at by the movement for the improvement of the quality of working life and for industrial democracy. The question this raised is, can the quality of life of the poor normal scientist be improved, and if so at what costs and under what conditions? Surely, Kuhn has not the foggiest idea; not enough attention was paid to this matter, and Kuhn has no scientific authority to back his sociological judgement – at most he has the force of tradition and of the scientific establishment to support him. Quite generally, Kuhn (follows Polanyi and) defends the tradition of the scientific community as that backed by inarticulate but decisive expert judgements on the authority of the experts as experts; and "experts" denotes those branded as experts (namely Harvard professors), namely the establishment. Now experts are traditionally known to oppose tradition now and then, and in the case at hand the experts in the natural sciences and in the philosophy of science side with Kuhn, but the experts in the sociology of work, the paradigm makers of the field in question, are active nowadays in the sub-fields of sociotechnics, of QWL (= the quality of working life), of *Mitbestimmung*, of OD (= organizational development) and others. Alas! they do not side with the experts in science. The same goes for scientific education: the expert scientists have strong views on scientific education, deeply rooted in tradition, and they practice them with a vengeance. The experts in scientific education dissent from them but have very little prestige and they are opposed by other experts in their field who gladly kow-tow to the scientific establishment.

The instruction and research as practised in leading academic institutions is gruesome and unnecessary, not to say an impediment to scientific progress, and should not be permitted in a public institution like an American university. I will not elaborate on this here, as here I discuss not scientific instruction or research but the scientific attitude and the way science is presented to the public at large, and I present the claim that the scientific attitude is democratic, yet the canonical views on philosophy of science present it as autocratic. Before leaving this point let me voice what is empirically found to be the strongest and most convincing criticism of the view I advocate; if the situation is as bad as I describe it, critics ask rhetorically, how come science flourishes? Let me demur: science does flourish, and if things go well someone has done the right thing. Does it not follow that we are doing the right thing? Does it not follow that at least the normal scientific researcher is doing the right thing?

No. Some researchers do the right thing, and as far as the general trend is concerned, mostly what is right is what our forefathers did. We are living off of the fruits of their labors. The very fact that normal, detailed and lifeless research is going on is hardly proof that what they do is right. Rather, it is proof enough that we are living off of old good ideas and traditions, which we are consuming and altering rapidly and unthinkingly. For example, we receive the new philosophy of science as advocated by Carnap, Kuhn, or others with no sensitivity to the imprudence involved, and even while preaching prudence to the lay people, while acting recklessly in our own back yards. We do not even notice that all that the views competing for the support of the scientific establishment share is their conservative political character.

The strongest argument against my view, then, is very poor: we still do not know what we are doing when instituting normal science and endorsing Kuhn's establishmentarian view of science and it advocacy of the close cooperation of research with the military-industrial complex, so-called. Nevertheless, for the purpose of my present discussion I need not insist that the autocracy of the scientist is detrimental even to science itself; all I argue is that there are here possible – repeat, *possible* – political implications to the most obvious disagreements in the field of the philosophy of science. This is still not to disregard, nor to disagree with, Carnap's view in his controversy with Neurath, as Neurath's view is that politics is utterly unavoidable, whereas the view advocated here is that it is at times avoidable and then with full right, and that when unavoidable, we better show awareness of this characteristic.

4. SCIENCE POLITICALLY VIEWED

My concern is now not with views that philosophers hold about science but the political proposals that they make and the political import that these proposals have – intentionally or not – which are consistently hyper-conservative and consistently hyper-establishmentarian.

The one thing that is quite intentional and quite damning is the defense of science, not to say the public-relations on behalf of science, that today goes under the name of the philosophy of science. First and foremost, this defense is quite cowardly even though when Galileo defended science it was a brave act on his part, of course. Even Galileo's case sounded strange to Einstein: what was it to Galileo, he asked, whether the Church of Rome was ready to admit Copernicans or not? The answer is, Galileo was a thoroughly religious Roman Catholic and it mattered to him profoundly that his Church should admit the doctrine he considered true and he bravely defended science before the religious authorities even though the defense was at the time punishable and punished by death at the stake. When Bacon and Descartes advocated science as preferable to academic intellectual activity, they were not in as much danger as Galileo and not in the least in an inner conflict, they were in the minority in the intellectual world. They soon became the leaders of the *avant garde*, though. The spokespeople of the Enlightenment movement were certainly expressing views accepted by most of the important and appreciated intellectuals of their day, yet they too were having an uphill struggle, or at least that is how things looked to them, as they were concerned with the education of the population at large. And they certainly were not favored by the political leadership of their day, except that to some extent the enlightened among the politically powerful tolerated them, but even then they did so at a price which often was found too high.

To defend science today is at best redundant, as nobody attacks science, not even the irrationalists among us, like Heidegger and his followers, who find science a threat, yet all they can do is declare that there are things in the intellectual world other than science, perhaps worthier pursuits for gentle souls than scientific research. When the philosophers of science are asked what is the point of recommending the faith in science, they have two answers. First, they say that they are busy attacking the irrationalists who allot to science too limited a role in society and in the life of our culture. Second, they say that they are engaged not in making recommendations but in the examination of existing recommendations. The second answer is satisfactory and to some measure it is even true, except that usually the examination is done within limits that do not allow them to question the first answer, which relates to the political aspect of the recommendation, to which I will now come, after a brief excursion into the cost of political neutrality.

It must be repeated here that the politically repulsive aspect of the established philosophy of science is no accident. The philosophers of sciences do not offer criteria by which to examine current science, but, on the contrary, they take it for granted that every satisfactory criterion will prove the choices made by the leaderships of the natural sciences correct. This is why there is an advantage to the view of Polyanyi and Kuhn, which takes the very endorsement of an idea by the establishment to be the sole criterion. Moreover, when this sole criterion is criticized by examples of poor choices made by the establishment, then the Polanyites will declare these to be examples of wrongs already rectified long ago, and thus irrelevant, or *sub-judice*; or, if they are as brave and honest as Polyanyi was, they will say, and even stress, that nobody is perfect, not even the establishment of the natural sciences. But they will not desist.

Everyone has the right to behave apolitically in one's own research. This is both a description of the law accepted in all democratic countries and a moral judgement, not a political description. As a legal description this statement needs some unfolding. In a democracy researchers are free to choose their own topics, methods, problems – as long as they operate within the law. It is a limitation of the law that it should not limit the choice of topics, methods, problems, only the practices that they incur. For example, the researcher is not allowed to torture humans not as a limitation on research but as a limitation on human conduct in general. On the contrary, the law specifically refers to permission for researchers to act somewhat more freely than others, within well-specified limitations and under well-specified controls. Even that, however, has a cost. As Kuhn rightly observes, researchers (as a community) often have a say on the degree of economic and social freedom allotted to a given piece of research, and at times they even decide such matters with no public interference or control, even though the allocation of research money always has some consequences for human lives. Feyerabend says we should not be biased against astrology and such, and, he adds, the amounts of time and money allotted to astrology should be democratically decided (he implies but does not say outright that he has in mind a referendum rather than the use of the normal instruments of control as practised in modern democracies, as he criticizes these as scientific dictatorships). I heartily disagree. But I disagree more heartily with Grünbaum, who goes to the other extreme and would have all philosophers of science confine their research to the study of his own output and of the output of his associates; he does all within his power to see to it. But all this is an aside: freedom costs something even in the best of circumstances, and as long as the cost is bearable, one need not complain. In any case, our present discourse is not about the level of academic freedom required (as this is a vast and difficult area). No matter how one views the case, the best response to my challenge to philosophers of science to show that they are useful is to describe some useful tasks that today's leading philosophers of science perform. And this takes us back to my discussion of my question, what tasks do philosophers of science perform today other than public relations for the scientific establishment? In particular, what problems do they concentrate on?

The central theories in the philosophy of science are the inductivist and the conventionalist. The standard problem of the inductivists is the problem of rationally justified belief or of empirically justified belief. It is the one problem that most studies in the field today are devoted to, and this includes the problems and paradoxes of confirmation, since the theory of confirmation gains its value from the claim that confirmed ideas are proper objects for rationally justified belief. We see clearly that the philosophy of science is today largely a part of ethics, and of a conservative, technocratic ethics at that. The conventionalist theory of rationally justified belief exists too, and it differs from the inductivist theory the way classical skepticism differs from classical rationalism. Classical rationalism recommends the acceptance of rationally justified belief only and the skeptics accept that and add that no belief is justified. Hence they recommend the suspension of all belief for ever and a day. The conventionalists likewise insist that there are no scientific beliefs to hold, that the different theories within science are not real alternative views but mere different modes of locution that should all be tolerated.

This is all plainly illiberal as it conflicts with the liberal doctrine of the freedom of opinion. The people who have frankly admitted this are Michael Polyanyi and Thomas S. Kuhn; but their testimony is redundant; though other philosophers of science do not say frankly that individuals who dissent from received opinions are at risk of losing their jobs, they clearly imply this when they say that received opinion is rational and that rational opinion is obligatory.

This illiberalism sounds paradoxical, and paradoxical it is: the scientific tradition was always liberal, since it took pride in the freedom of the intellect and the high value of rational argument. As Francis Crick stresses, the scientific establishment still leans backwards in its effort to be liberal, organizes hearings even for some pretty obviously lost causes. Why, then, does it sound as if the validation of science is illiberal? When is the challenge to validated ideas admitted and when is it dismissed as a bore? We do not know. This, indeed is the reason that the idea that the tradition of science is illiberal sounds paradoxical: not sufficient care is paid to the realm within which freedom of opinion is/should be tolerated and that in which not. For example, clearly, the individuals who deliberately distort the rules of debate so as to win at all costs are to be tolerated, but not as scientists. The same holds for the holders of the view that contradictions are good, for the advocates of old-fashioned unadorned Hegelianism and/or Marxism, and for the teachers of para-consistent logic so-called, not to mention other illiberal philosophers amongst us. That there is no need to argue against the dogmatic follows from the fact that they are, indeed, illiberal, and the reason one need not argue against the illiberal is that it is impossible to do so, and this impossibility is but the fact that the illiberal will not tolerate rational argument proper. An extreme example will make this clear, even though most examples are far from being extreme: when the National Socialists in the Weimar regime invited the public to their public meetings, promising the public the right to participate in the discussions after the speeches were over, they could easily get rid of those members of the public whose arguments they found annoying: these were the Jews amongst them, and one should be tolerant towards all but not towards Jews.

Can we ignore some ideas without being charged with dogmatism? The line between the illiberality of those who brand every unpleasant critic a Jew and those who refuse to argue with them is not always easy to demarcate. Some writers clearly declare this demarcation nothing but illiberality – for example the already mentioned Feyerabend. For, according to him only democratic vote can decide that magic should not be taught in schools, and then the vote cast this way proves the nation illiberal and that is all that there is to it, he says. What are we to do with the argument that all demarcation of dissent to good and bad is inherently illiberal? For one thing, I do refuse to argue with Feyerabend, as I think the debate between us has stagnated and become excessively boring. But he may say, as he does, that these facts are the outcome of my dogmatism. Where is the criterion to distinguish the rationality of one of our criteria from the other? Can it decide who is irrational, he or I? Here is the difference between the liberality and the illiberality of science and this is how the paradox is resolved: the scientific tradition allows and even encourages rational debates and shuns irrational debates. This is not politically illiberal as it allows people to stay outside the scientific tradition. This is why Feyerabend has the right to call the scientific tradition illiberal and terrorist if and to the extent that it demands of us all to join it.

Query: Does science tolerate the presence of individuals outside the scientific tradition? Answer: Yes. Science is not a proselytizing movement but a research activity and a system for the dissemination of information. Query: Do the philosophers of science allow for this or do they demand of everyone to uphold the beliefs advocated by scientists? Answer: No, they do not allow for belief not rationally justified and they demand that scientific opinion be endorsed by all and sundry. Hence, Feyerabend's complaint is valid against the inductivists philosophers of science who demand that our belief be empirically justified, but not against the scientists. Proof: when scientists are engaged in the act of proselytizing then we say that they act as philosophers or as politicians, not as scientists. When scientists as scientists encounter individuals who reject established beliefs, and the paradigm here is young Albert Einstein, then they reason with them or, if they find them tedious, they leave them or send them to the philosophers to argue with. End of proof.

5. DEMOCRACY AND SCIENCE

The difference between the need to tolerate opinions in a liberal democracy and the need to tolerate opinions within the scientific domain is of great importance but hard to describe. This is a very important question, since the rule of science and of scientific technology is technocracy, not democracy, as readers of Plato's *Republic* should find easy to comprehend. The first description of the difference between political and scientific tolerance that comes to mind is the view that liberal democracy is public and the scientific domain is private. This is true to a large extent but not quite, since the prestige and the social role of science in contemporary society makes it very costly to be a scientific outcast – or even a philosophical one – and so the concern of science is to a large extent well within the public domain. It is no accident that both Polanyi and Kuhn speak of the threat of loss of a job!

Should not jobs in science as elsewhere be restricted to the qualified, though? Is it wrong to prevent the unqualified from a prestigious position in a university? But can dissenters help it if they hold heretical views? Should they be penalized for their convictions? Polanyi himself was a dissenter, and moreover, his view became the received opinion half-a-century later. Yet, he said, he could and did teach the currently received opinion, though he dissented from it. Perhaps even that is too much; perhaps all that is reasonable to require of a university teacher is the ability and the readiness to teach the received opinion, not to assent to it. And this runs contrary to the received opinion in the philosophy of science, which is thus undemocratic and illiberal.

The first difference between scientific and political tolerance, then, is that regarding scientific jobs, some competence is required from candidates that has to do with familiarity with received opinion. The second difference between political and scientific tolerance that comes to mind is the view that politics must be much more tolerant than science since we have to tolerate politically the advocacy of "the employment policy of the Third Reich" though science can ignore it since evidently it is so stupid (this is not to say any stupidity is inherent: if someone makes a stupid view interesting, that step constitutes great intellectual progress). Moreover, even stupid ideas have to be examined closely from time to time: the need to scientifically contemplate possible atrocities is quite obvious. This is what Herman Kahn labelled with his unforgettable label: thinking the unthinkable.

This shows clearly that we need some new and important change of attitude before we can exit from this labyrinth without becoming hopelessly scholastic. First of all it should be made clear that the democratic attitude towards expertise is neither to belittle it, as Feyerabend suggests, nor to accept it blindly, as most philosophers of science today do. Experts are given to democratic control (K.R. Popper, *The Open Society and Its Enemies*). Yet this raises the problem of standards of expertise that even those in control need. I do not know if I have a proposal that is bold enough to help but I propose the following simple idea that might be good enough. I have not yet heard a criticism of it, though it is published in my *Science in Flux*. It is this: let us take as a datum the fact that we have all sorts of differences of opinion and all sorts of arguments about them. Let us take as a datum the fact that some arguments are rational and some not, and that every reasonable argument *eo ipso* renders rational the disagreement that it lends reasons to, no matter how many unreasonable arguments are involved in the same disagreement. Let us agree, then, to rephrase the problem of rationality as the following problem.

What differentiates reasonable from unreasonable arguments? I do not know if this will do the trick, and at times this looks to me suspiciously simple-minded. But at least it is clearly a recommendation for reasonable differences of opinions and the recommendation to try to make every difference of opinions reasonable, namely by the search for reasonable arguments for the minority view. Also, clearly, it is better to demarcate rationality than science since rationality is more egalitarian and less exclusive than science, especially today, when the problem of demarcation of science is becoming increasingly the unpleasant problem of who deserves praise. With this I can stop for a while.

ACKNOWLEDGEMENT

This is a slightly expanded version of a paper delivered in the Department of Philosophy of Karl Franzens Universität, Graz, while I was visiting professor there.

Tel-Aviv University and York University, Toronto

TIAN YU CAO

A PHILOSOPHER LOOKS AT SCIENCE

The philosophical understanding of science that I shall present in this essay is a product of the confrontation of some of my prejudices, or preconceptions, with contemporary developments in science and in reflective thinking. The preconceptions can be roughly summarized by a label 'scientific humanism', which was handed down to me through the works of Descartes, Leibniz, Kant, Hegel, Marx, and other thinkers in the same tradition. According to this tradition, our species as a rational subject is able to discover objective truth and universal norms, upon which systems of thought and action can be built, so that nature can be conquered, social life rationally restructured, and the human being emancipated from ignorance, superstitions and prejudices, from poverty and scarcity, and from all kinds of domination by irrational traditions and arbitrary authorities. In other words, human history is a progressive process toward emancipation, of which reason – its most sophisticated and effective achievement is scientific knowledge – is the major source and vehicle.

The conceptual scheme associated with scientific humanism has successfully rendered understandable and meaningful many developments in science and its applications, such as some of those in physics, biology and other natural sciences, also in logic, mathematics, computer science and cognitive sciences, and in sociology, economics and other social sciences as well. The success, however, has never been a total one. Conceptual difficulties have already appeared in its encounters with quantum physics, metamathematics and, in particular, with anthropology. The tension created thereby between this metascheme and 'empirical facts' provided by the sciences is further exacerbated by an ever-stronger attack from some currents in theoretical discourse; and this has threatened its rupture. These opposing currents, the post-empiricist history and philosophy of science and neo-pragmatism in Anglo-American philosophy; phenomenology, hermeneutics, existentialism and post-structuralism in Continental philosophy, are by no means brand-new. They have a long tradition, a tradition of criticising modernist discourse, of which scientific humanism is a paradigmatic case. In a sense,

K. Gavroglu et al. (eds.), Physics, Philosophy and the Scientific Community, 171–187.

contemporary cultural life in the west can be characterized by the conflict between modernist and post-modernist discourses. The latter has successfully drawn its strength in the conflict from the deepening intellectual crises (scepticism, relativism, moral collapse and cynicism), and in return has further deepened the crises. Not surprisingly, therefore, there are attempts to find a way out, and the following analysis is one of them.

I.

About half a century ago, Ernst Cassirer claimed that science was the last step in man's mental development, the highest and most characteristic attainment of human culture, and the summit and consummation of all our human activities. One may disagree with Cassirer's neo-Kantian position, yet it seems hardly deniable that science, having provided modern man with both a critical frame of mind and a unified outlook on the world, is a vital part of modern thought. In contrast with the irreversible decline of classical metaphysics and medieval religion and theology since the fifteenth century, science enjoys high prestige in the modern mind, and the scientist is taken to be a cultural hero in contemporary society. The source of this prestige is obvious: We are living in a new world that has been successfully constructed by science and science-based technology; science is one of the foundations – or even the foundation, as many have claimed – of the western industrial civilization, despite the fact that it has also, from time to time, brought about efficient barbarism.

Why is science successful? Is this because science reveals the secrets of nature? Or because of its specific method? The question is a subject for persistent debate and touches upon the very foundation of science. Before turning to this topic, which lies at the center of the philosophy of science, some preliminary observations are needed.

(i) *Characteristics.* Science is a public enterprise. Among the factors that directly contributed to the birth of modern science, we find craftsmanship of the later medieval time, the principles of experimental method prescribed by Francis Bacon and Robert Boyle, the neo-Platonic conception of mathematics and of the world cherished by Kepler, Galileo and others. All of these factors require public demonstrability. Neither in the empirical nor mathematico-conceptual dimensions of science, is there anything private. In addition, in the transformation of man's permanent curiosity about the cosmos into modern science, the economic needs of the early bourgeoisie in the later Middle Ages played a decisive role. More generally, science as a human project for mastery over nature (human nature included) is essentially a communal activity, and its social dimension cannot be underestimated. For, it is society that gives science problems, motivations and rewards, material and technical resources, and, in particular, ideas, metaphors and conceptual schemes for solving its problems and interpreting its results.

The public character of science has many ramifications. It entails that science can only exist and develop in a certain social context and a certain socially given conceptual context. With the change of its contexts over time, science changes historically both in terms of its problematics and of its interpretive framework. It also explains science's striving for objectivity, or at least intersubjectivity. This in turn explains its dogmatic tendency: Science, as systematically organized experience that has to be acceptable to the public, requires first a stablization and consolidation of flowing perception and thoughts, and then a standardization and fixation, in terms of its established results, of its logical (inductive, abductive or hypothetico-deductive) procedures – from crude experience to more and more sophisticated theories – and of various methods adopted in the process of reasoning.

While the dogmatism of science is rightly attacked by, among others, romantic critics (often with exaggeration), we should not forget that science is after all a critical enterprise. First, modern science from its genesis served the enlightenment of the human mind, fighting against superstitions, prejudices, traditions and established authorities, metaphysical principles and religious dogmas. This critical impulse of early modern science, if understood in a wider cultural and historical context, had its social root in the conflict between the rising bourgeoisie and the declining feudal classes of the time. Second, the scientific spirit requires that all its results, methods and approaches be tested against ever-renewing experience, and subject to communal interpretation and reinterpretation. Such a critical examination involves dialogue, questioning, negotiation and argumentation, and thus serves to counterbalance its dogmatic tendency and prevent it from falling into rigid dogmatism. Third, intrinsically, there is in science an interpretive dimension, and also a pragmatic dimension. They indicate a great flexibility enjoyed by science. I shall return to this point later.

(ii) *Functions.* For internal functions, we can list description, explanation and prediction. Thus, gratifying man's curiosity, science serves, intrinsically, one of the most humane desires of man; and it is also an efficient instrument for controlling various courses in the world (nature, society and the human mind) and for creatively manipulating and transforming, or, in short, for conquering the world. In external terms, however, as one of its social functions science is always translated into an efficient aid to the ruling power. As a form of consciousness coexisting with art, religion and philosophy, science can also function as an ideology (scientism), providing society with world views, such as the mechanical view, the evolutionary view, and the probabilistic view. These views themselves are often just a projection upon nature of social norms of the time. Yet the norms are reinforced and quasi-legitimated through the mediation of science. Moreover, under different circumstances, depending on whether the dogmatic tendency or the critical spirit takes the upper hand, science can be either stifling or liberating in the spiritual life of the time.

(iii) *Social and cultural impact.* On the positive side, science has greatly helped to increase the efficiency in producing material wealth and in social organization and administration, thus having laid down a foundation for the industrial civilization. A long-term impact of this efficiency, within the context of capitalism, is that a wider space is created for allowing mass participation in economic, political and cultural experience (trade unions, democracy and mass culture). Aesthetically, in contrast with the traditional arts, which express the experiences of the man-sized world, science, with the help of mathematical imagination and manipulation and of conceptual analysis, has opened a new world of experiences, experiences of objects of cosmic or subatomic size, with unexpected complexity and harmonious structures, widened the horizon of man's creativity and imagination, and created new types of beauty and enjoyment. Besides, the positive aspects of the internal political ethics of the normal scientific community – self-assured individuation, voluntary cooperation and solidarity – provide a model of democratic life, a living model for a Jeffersonian Republic.

On the negative side, science worship or scientism is a paradigmatic case of instrumental rationality. This is a calculating spirit without moral judgement or practical wisdom, focusing solely on the efficiency of the means without any critical evaluation of the end. The worst result of this is efficient barbarism. Another result is authoritarian rule by the elites who monopolize scientific knowledge and hence have a manipulating power over nature and over other people. Aesthetically, the penetration into various life processes of standardization and fragmentation, initially issued from the dogmatic tendency of science, dehumanizes the processes and spoils our tastes.

A central point of these observations is the ambiguous character of science: science is both dogmatic and critical; it can be a servant of efficient barbarism as well as of human emancipation; it is sometimes boring or frustrating but sometimes fascinating and encouraging. In more than four decades, Professor Robert S. Cohen, to whom this essay is dedicated, has paid much attention to the ambiguity of science. His insightful observations, regretfully not assimilated by other scholars in their entirety, have provided the major inspiration for the present reformulation, in the light of current philosophical and cultural debates, of that ambiguity. A proper articulation of the ambiguity, in my view, is crucial for any further philosophical understanding of science.

II.

The widening of the human horizon in the period of the fourteenth to sixteenth centuries, which was brought about by the Renaissance, the religious upheavals, the geographical discoveries and, in particular, by the Copernican Revolution, has accelerated the decline of the authority of the church, of faith in an anthropocentric cosmic order, and in ethical principles, thus causing widespread doubt, fear and anxiety, and making agnosticism and scepticism quite acceptable. Montaigne, for example, was popular and influential in his criticism of the validity claims of scientific knowledge and in his scepticism concerning the power of human reason. In response to this challenge, Bacon appealed to experience and Galileo to mathematics. Yet the first philosophical attempt to articulate and answer the anxiety for a firm foundation to ground scientific knowledge was made by Descartes.

Descartes separates the knowing subject from the external world and takes it to be the foundation of knowledge. A difficult question for Descartes, and for all those who pursue the subject-centered epistemology, is how the subject can grasp the object once the two are separated. The internal dynamics of the subsequent development of epistemology is determined by this problem. Descartes' own solution is to take knowledge as the inner representation of an outer world. What is crucial for Descartes is that knowledge as an inner representation can be analysed in abstraction from what it represents. Thus, our knowledge would be indubitable if we acquired it by a rational method of reflexive clarifying; for the existence of the subject is self-evident in terms of self-consciousness, and his self-consciousness is indubitable and can secure an indubitable rational method. The final certitude of our knowledge, according to Descartes, lies in our reflexive consciousness or reason, and the validity claim of knowledge and the power of reason are grounded on human reason itself. Obviously, such a self-grounding is to take the old metaphysics (mind) to ground new science, and thus has been exposed to severe criticisms.

The Cartesian anxiety for an Archimedean point initiated a new style in philosophizing: foundationalism. Traditionally, the discourse of foundationalism is almost always congruent to the discourse of rationality. At the present time, however, few philosophers would defend foundationalism, although rationality remains a topic for hot debates. More recently, the focus of attention, among some critics, has been shifted onto the representational construal of knowledge, which is said to underlie foundationalism and has wide ramifications in the modern age. The idea is to articulate a new notion of human agency so that the nature of science can be brought into new light. But how did the reflection of knowledge come to this point, both in terms of its internal dynamics and of the impact on it of the development of science?

Classical rationalism is contaminated with a priori metaphysics, such as the assumption of preestablished harmony. Classical (sense data) empiricism worships the passivity of the mind. Only in Kant's critique is the active and constituting role of the mind in the genesis of science appreciated without falling back into the old metaphysics. Yet Kant is inspired by the most successful sciences available to him, Enclidean geometry and Newtonian mechanics. He takes some features of them, as necessary, universal and timeless, the characteristics of transcendental categories which make experience possible. Hegel replaces Kant's transcendentalism with dialectics, according to which knowledge loses its timeless character and becomes generated, mediated and evolved in a dialectical process.

Hegel's speculation upon an Absolute Knowledge of the spirit, which is nothing more than the self-consciousness of the mind, despite its significant influence, was soon abandoned by his successors. In consequence of the decline of Hegel's system, two orientations in philosophical movement developed. One of them tries to return to the old tradition of epistemology and to find a firm foundation for the positive sciences. Different mixtures and combinations of neo-Humean and non-Kantian doctrines result in different schools: Positivism, neo-Kantianism, conventionalism, logical atomism and Husserl's phenomenology. The vicissitudes of these schools are largely impelled by the subsequent developments in mathematics and sciences, such as the discovery of non-Euclidean geometries, Frege's formalization, relativity theories, quantum theory and Gödel's theorems. One of the achievements of this movement was logical empiricism, which became the received view in the philosophy of science in the Anglo-Saxon world from the 1930s to the 1960s.

The other orientation, however, moves away from the positive sciences, drawing its inspiration mainly from psychology, history, anthropology and linguistics, and is very critical of rationality. For the existentialists and other life-philosophers, philosophy is an expression of life. Since our life experience is essentially irrational (vague, uncertain, and full of contradictions), philosophy has nothing to do with rational knowledge. For Nietzscheans, the rejection of logocentrism and the taking of power relations as the determinant factor in our interpretations of 'facts' undercut any claim to knowledge, truth or rationality. For Peirce, objective knowledge is still definable, even if only as a limit, as a result of consensus reached in a long-enough period within a community. Yet for later pragmatists or instrumentalists, Dewey in particular, knowledge is defined only in terms of its function as a means. In all these movements, with the exception of Peirce, there is no place for epistemology, let alone foundationalism.

In a sense, Hegel's notion of mediation, which is absent in the first of these orientations, is inherited by the second orientation, although only in a distorted way. As a descendant of Nietzsche, Heidegger breaks with Husserl's fascination with scientism. He argues resolutely against the tyranny of 'presence' in understanding science. For him, knowledge, scientific knowledge included, is mediated by a historically shaped world-view structure (a pre-understanding of the lifeworld), which is embodied in a language system. That is, argumentative thought and propositional truth make sense only within a given linguistic and preconceptual framework. Thus from the world-disclosing aspect of language and its historicity, Heidegger argues that we cannot escape from linguistic historicism and meaning indeterminacy. Similar arguments are presented by the late Wittgenstein and other linguistic philosophers, and are adopted, later on, by Derrida and Rorty. Derrida and Rorty, in their respective ways, deploy these arguments in terms of an endlessly-disturbing intertextuality as the mediation of knowledge, against the notion of knowledge as a representation of an independent world. The idea of mediation is fully exploited by the hermeneuticists, Gadamer in particular, in order to introduce the notions of tradition, authority and prejudice into our understanding of knowledge. While Gadamer's view seems to have at least left enough room for further elaboration, Foucault's is more biased. For Foucault, also a Nietzschean, the only mediation for knowledge-production is power relations, which become the start and the end of our knowing activities, if not of all human activities.

One cannot ignore the radical implications of this way of exploring Hegel's notion of mediation, namely, historicism, relativism, and ultimately, scepticism. As a matter of historical fact, the dominance of the second orientation on the continent since the 1920s, after the dominance of neo-Kantianism, is largely due to widespread scepticism and pessimism over rationality, the Enlightenment and progress, first in defeated Germany after World War I, and then after World War II in France and Germany, the intellectually leading countries on the continent, due to the frustration with the occurrence of efficient barbarisms: the Nazi movement, fascism, and Stalinism.

Interestingly enough, the first orientation, through its internal developments, has recently evolved and converged upon the second orientation with an accelerated pace. The turning point was Quine's critique of logical empiricism and his introduction of holism and conceptual relativity. The critique was reinforced by the later Wittgenstein's self-criticism and his reflection on language, and by Popper's devastating criticism of inductive logic, and thereby of confirmation theory, which had effectively killed logical empiricism, and his idea of falsification, which provided him a mechanism necessary for his evolutionary view of scientific progress. A further consequential step within the evolutionary framework was taken by Kuhn with his revolutionary view of the history of science. According to this, the history of science involves a series of radical paradigm shifts. Knowledge claims in different paradigms are incommensurable, yet the evolutionary progress can be measured in terms of ability at puzzle-solving. Central to Kuhn's view is the notion of paradigm, which, as a result of historical development and as a precondition for scientific enquiries, has brought historical and conceptual relativity into the understanding of science. More radical than Kuhn are the social constructivists, who have brought social, economic, political and cultural dimensions, as the determinants, into the understanding of science, and thus rejected the very idea of scientific rationality. Another off-spring of Quine's critique are the neo-pragmatist arguments developed by Putnam, Davidson and Rorty, which are interwoven with the arguments presented in the debates provoked by Kuhn's work. Rorty takes up historicist and pragmatist arguments in the post-empiricist history and philosophy of science and interprets them in a relativist and sceptical way. He assimilates the deconstructive, post-structuralist and hermeneutic arguments from Continental philosophy, brings all these together and deploys them against the modernist discourse of reason, truth and progress, thus becoming a central figure in the contemporary cultural debates.

There are many disagreements and debates among these post-modern philosophers, particularly between Rorty and some more moderate people. Yet common to them all are the beliefs that (i) science has no firm and fixed foundation; (ii) science is not a representation of a metaphysical world, which presupposes a Cartesian split of the knowing subject from his environment, for (iii) this subject and the world in which he or she lives intrinsically penetrate into each other because of the very nature of human beings as active agents. While it seems reasonable to accept this set of beliefs, they are by no means destined to be incompatible with conclusions other than relativism and scepticism. Such different conclusions can be drawn from analysing science and arguing philosophically in a different way.

III.

It is a cliché among historians of modern philosophy that the style of philosophizing is very sensitive to the change of climate in science. For example, many commentators notice that the rise of epistemology (taking knowledge as representation and seeking for a firm foundation) was at least correlated with the rise of mechanics, and that the rise of conventionalism was almost a direct result of the discovery of non-Euclidean geometries, and so on. It is interesting to notice, for my present purpose, that there has been a consequential shift in basic metaphors in philosophizing since the late nineteen century, a shift from a mechanical model to a biological model. Here what I am referring to is more than a shift from metaphors of a unilinear causal chain (knowing as direct, passive mirroring or representing) and of reduction of a whole into its parts (foundation) to metaphors of organic connections and of reciprocal interactions (knowing as participating and dealing with), and functional and holistic analysis. All these can be found in Hegel's work already and even can be traced back to Aristotle. What I want to stress here is a shift to the model of Darwin's evolutionary biology. There is a shift to the metaphor of the organism/environment relation: the knowing subject is always dealing with and thus internalizing what is to be known, and this internalization as an essential part of adaptation for survival is a precondition for the subject's existence, and thus is largely appropriate; also the object to be known is not immune to change and transformation caused by the action of the subject. In addition, the metaphors of heredity (conceptual, social and historical preconditions for knowing) and variation, competition and selection, speciation and evolution, etc. are also adopted by philosophers. Even a cursory inspection of work on knowledge (outside of technical elaborations strictly along the line of logical empiricism) would show that the fertility and suggestive power of these metaphors in philosophizing are beyond question.

The impact of science on philosophizing did not stop at the stage of evolutionary biology. In the twentieth century, with the emergence of quantum theory, some new metaphors have entered into philosophizing. Most important among them are chance and emergence. Of course, chance is a very old notion and historically has many applications in various areas. Yet in the mechanical model, according to the ignorance interpretation of chance, the notion of chance is, ultimately, a derived category and can be removed in principle. This assertion applies to biology as well. According to Oparin, life is the product of chance; according to genetic biology, a new species of life is the result of accumulated random mutations. Yet a successful molecular biology would remove the notion of chance from all these areas. However, in quantum theory the notion of chance has acquired a metaphysical significance and becomes an ultimate category. Here, not only is a deterministic law replaced by a statistical one and no individual event can be predicted with certainty, but (and this touches upon our understanding of a material world even deeper) even the ultimate object the theory is dealing with is not an actual fact any more. It is a tendency to become something, a propensity, or a potential. Heisenberg revives Aristotle's notion of *potentia* and gives it an endorsement from the perspective of the most advanced physics. Then the actualization of a *potentia* in quantum theory is taken to be a result of our measuring.

The mystery of the measurement problem may be somewhat mitigated if we remember that the object in the quantum world is small enough to be hypersensitive to the disturbance brought by the measuring operation. There is no room for idealism, nor for subjective arbitrariness. The interaction between a quantum subject and the measuring equipment is physical and shows some regularity (even if only statistically). Yet the lesson drawn from the measurement problem leaves no doubt that our knowledge of the quantum world is the result of our participation or intervention (with measuring equipment) into the quantum world. What is even more important is that there is no metaphysical quantum world available to us free from our participation. The observation of the plasticity of the world becomes inescapable only in quantum theory because in classical theory, including relativity theory, the disturbance brought by measuring can always be ignored, and a rigid physical world free from our participation, waiting there to be measured, seems to be a good approximation.

Some philosophers waste no time in taking the lessons learned from the measurement problem as a basis for our new conception of the physical world, and even taking them as a model of all sciences. And this has brought a fatal blow to foundationalism and can become (even if not necessarily) a major source of scepticism.

Similar to the notion of chance, the notion of emergence is also very old and had various applications. For reductionists, however, emergence is only an appearance, has only epistemological meaning, without ontological significance. With causation of emergent properties, the reductionist program has many difficulties. The problem is, are these difficulties only practical ones? Or, are they intrinsic flaws of the program? People disagree on this issue, but only on the basis of intuition, without convincing argument. However, beyond the *epistemological* emergence recent developments in quantum field theory have indicated that the idea of *ontological* emergence can be convincingly argued for, or even demonstrated, in the domain of elementary particles, (usually regarded as the very foundation of reductionism) on the basis of such discoveries as 'symmetry breaking', 'renormalization group', and 'decoupling of low energy phenomena from high energy processes'. The concept of a 'quasi-autonomous domain' emerging from these arguments has provided a basis for a pluralist view of the physical world. (Here pluralism holds not in the sense that different domains have no connection to each other, but in the sense that the dynamics of one domain cannot be reduced to or derived from that of another domain. For details, cf. "The conceptual foundations and philosophical aspects of renormalization theory" by Tian Yu Cao and Sylvan S. Schweber, *Synthese* **97**: 33–108, 1993.) And this has removed any heuristic support for foundationalism.

In addition to these, in the big sciences of the present day, no direct interpretation of facts is available to us because of the conceptual and social complexities involved in them, and thus our knowledge of the object appears as a result of negotiation. Thereby, the dialogical, intepretive, pragmatic and social dimensions of science become no longer ignorable, and a firm foundation seems more and more intangible and elusive.

IV.

The metaphors and prototypes provided by the contemporary sciences mentioned in the last section, combined with some insights offered by the contemporary philosophers, seem to suggest that:

(i) Science is not a subjective representation of an independently existing object; rather, it is a result of participation by the knowing subject in the object to be known, of interaction between a plastic, changeable and transformable object and an also plastic and adaptable subject. In addition to intervention (experiment), as the pragmatists, Marxists and hermeneuticists would suggest, our understanding of an object also involves interpretation (theory) and application (industry, technology, or social project).

(ii) The mediative character of science has its root in the nature of the knowing subject as an active agency. The knowing subject is neither a *tabula rasa*, nor a disengaged rational being. Rather, biologically he inherited a set of cognitive faculties which, as a result of adaptation, enable him to internalize and interpret the outer world in an essentially appropriate way. Socially, the root of his identity is in a community which is delimited by a shared language, a shared tradition, a shared network of economic and political structures, and a shared culture, including shared preconceptions of the world and some psychological tendencies toward the world. Thus knowledge is always mediated by these pre-given specificities, which are in some sense detached from individual knowers. Both the dogmatic tendency and the hermeneutic dimensions have their roots in this mediation mechanism.

(iii) The critical impulses are intrinsic to science. The cumulative effects of experiencing and acting on the object will affect both the object and the subject, due to their plasticity, and sooner or later will lead to a conflict between the newly-acquired knowledge and the pre-given structures if these structures are institutionalized. The conflict often results in a transformation of the pre-given structures so that the newly-acquired knowledge can be properly accommodated. Two points are worth stressing: First, the pre-given structures, as preconditions of knowing, form no closed texture. Otherwise no progress can be made which would lead to a bursting of the shell of this closed system. They could not possibly be so because of their genesis from the on-going interaction between the subject and the object, and their very nature as a means for continuing adaptation. Second, accordingly, science itself is also an open texture without a closure.

(iv) Science is knowledge of an outer world. Here Husserl's analysis of intentionality is compatible with our biological metaphor of internalization. Moreover, different from absolute fluidity, plasticity (which is defined in terms of a relationship between a subject and an object) at any given moment has its limit. The limit of the plasticity of an object at any given moment is historically set jointly by its pre-given state and by the subject's ability of penetrating and transforming this state. Thus the intervention of the subject will meet resistance if it reaches the limit of the object's plasticity. This ontological sense of the plasticity of the object has its epistemological counterpart. That is, our knowledge is inevitably constrained by a concrete outside world. In some cases, the constraints can be very rigid. In general the limitations of the plasticity of the outside world explain the continuity of the scientific enterprise: Even at the most dramatic revolutionary moment in science, most of our knowledge always remains unchanged, although nothing is unchangeable in principle or from a long-run point of view. Thus both social constructivism, which is incompatible with the idea of nature's resistance to human action, and foundationalism are defied.

(v) The positive sciences cannot be used to naturalize philosophy, not even to naturalize the theory of science, because the naturalization is incompatible with the ontological pluralism derived from the idea of emergence. Granted that neurophysiology and neurpsychology may give us a lot of knowledge concerning the cognitive power of the knowing subject (sensation, memory, imagination, abstraction, the ability for analysis, inference and synthesis, etc.,) and also granted that the study of the machinery of cognition may be conducted not only ontogenetically but also phylogenetically, yet a naturalized theory of science is still impossible. The mind operating in scientific activities operates not only at the cognitive level, but also at the critical and normative levels, not as an isolated, disengaged individual, but as a member of a community with a certain cultural and historical tradition. These normative, critical, cultural and historical dimensions of science can neither be exhausted by the positive sciences, nor be eliminated from science. Thus the impossibility of replacing philosophy of science, let alone philosophy in general, with the positive sciences.

A natural question is: what is the relation of philosophy to science? Moreover, up to this point, I have not addressed the relativism and scepticism rampant at the present time in the wake of the retreat of foundationalism, a situation not unlike the one that appeared in the sixteenth century after the collapse of the church's authority. Actually, the two questions are related and I shall take them together in the next section.

V.

A different way of expressing the impossibility of naturalizing the theory of science is to state that science alone is unable to discredit relativism, which renders scientific rationality (objectivity, truth, progress, etc.) quite suspicious, if not totally discredited. The impotence of science in fighting against relativism is rooted deeply in its very nature. A science is not a collection of facts, but involves interpretation and understanding. Scientific interpretation and understanding are possible only within certain conceptual frameworks and, due to the continuity of science with everyday life, within certain cultural traditions, and within certain social and historical contexts. Even the way of perceiving and conceiving facts itself is heavily constrained by these frameworks, traditions and contexts. These determinants of scientific knowledge vary spectacularly and show a great diversity in space and time. Thus some form of relativism is an inevitable conclusion unless we assume that the diversity can be reduced to a unity. The trouble is that the assumptions of reduction and of unification are metaphysical rather than scientific ones.

Moreover, these metaphysical assumptions were undermined by scientific discoveries since the nineteenth century. Both anthropological discovery of alien civilizations and various comparative studies by other social sciences have led to a definite anti-ethnocentric conclusion: social norms, artistic tastes and even criteria for truth are radically different and would not be understandable if one adheres solely to the western ones.

The belief in non-reducible pluralism suggested by social sciences was reinforced since the 1960s, both from a cultural climate of anti-establishment (scientific reductionism included) and from a deep doubt concerning foundationalism in philosophical reflections. The latter was marked by the decline of logical empiricism and the wide spread of the incommensurability thesis. The works of Gödel, Popper and Quine had already led to a sense of crisis in methodology in the logical empiricist camp, and required a historically situated non-algorithmic and more flexible understanding of method. Around 1960, when Peter Winch published his work on social sciences and Gadamer published his work on philosophical hermeneutics, the verificationist view of meaning and the covering-law view of explanation were challenged by social scientists and humanities scholars. They argued that, different from natural sciences, understanding in social sciences and humanity subjects has an indispensable hermeneutical dimension. The argument was soon taken up by historians and philosophers of science and applied to natural sciences too. The slogan was incommensurability: Among different paradigms (cultures, traditions, language games, forms of life, historical epochs, etc.) there is no higher standard than respective paradigms which can be appealed to for justifying claims to objective truth and progress.

Granted that there are no fixed, absolute and universal criteria which can apply to different paradigms. Granted also that statements in different paradigms are untranslatable. Nonetheless, one can argue that understanding the statements in a different paradigm is still possible, and comparing and judging statements in different paradigms are still possible. The possibility of understanding a different paradigm is most forcibly argued by Gadamer in terms of the openness of language, and in terms of the fusion of horizons, each of which is delimited by a paradigm (or tradition). The Gadamerian understanding is a special way of realizing Habermas's more general communicative (intersubjective) rationality by continuous dialogue, communicating, questioning and testing.

The ground for comparing and judging different paradigms (when a logical algorithm is impotent as noticed by Kuhn) is more difficult to clarify. It involves critical reasoning and practical rationality. Notice that a scientific enquiry deals with facts (the potential in the quantum world has to be actualized so that it can be grasped by the quantum scientists), and aims at understanding and controlling the object underlying the facts, without questioning the aim underlying the enquiry and the indirect, implicit and potential relevance of the enquiry to other areas than its direct object. Thus scientific rationality cannot go beyond the scope of instrumental rationality. Different from science, critical reasoning examines an object in all its potentials and tendencies, with a perspective on the expectations of the wisdom of the time. Its concern with potentials and tendencies requires imagination, vision, insight and wisdom, which are atypical in routine scientific enquiries. These peculiarities of critical reasoning enable it to judge an object properly on the basis of practical rationality, which concerns, ultimately, the right way to live. More specifically, acknowledging that the ground for judging is shifting and historicized, relativism can be rejected not by appealing to teleology, but by taking all possible tendencies into consideration and making a value judgement concerning different directions or tendencies. Kuhn's idea of value judgement in theory choice during scientific revolutions is a good example of practical rationality. It clearly illustrates its informal character. Since values are heavily context- (culture-, tradition-, prejudice-) dependent, the value judgements cannot be freed from incoherence and contradictions. And this requires further critical examination. Thus endless confrontations in the enterprise of practical reason, directed by the human desire to find a right way to live.

It should be clear now that philosophy is the furthering of the critical component of science, and concerns its aim and direction. The critical examination of science by philosophy is beneficial because first, science, as a continuation of everyday life, always involves tradition, prejudices and illusions, which should be brought into a confrontation with new experience and the new perspectives acquired and accumulated over time. Second, only a critical examination can overcome the dogmatic tendency of science and the positivist view of science, and help to reach a proper understanding that science is more than sense data plus logic, but involves human participation and penetration into the object under investigation, involves metaphysical presuppositions, cultural traditions, and social and historical contexts; and that scientific knowledge is the result of negotiation within a community, constrained by a plastic object. Finally, only philosophy in general as self-understanding, as a concern with practical control of human destiny, can put science in proper perspective, as a part of the human project. Since self-understanding cannot be obtained solely by scientific reasoning (even with the help of cognitive sciences and social sciences), but requires a dialectical process of critical reasoning and hermeneutic confrontation, a proper general framework can no longer be scientific humanism but had better be dialectical humanism.

Yet as I have noticed in Section III, the revival of the dialectical view itself is a result of elaborating the metaphors suggested by the recent developments in the sciences (biology and quantum physics). Thus, we seem to have closed a hermeneutic circle in terms of the relation between science and philosophy.

Center for the Philosophy and History of Science at Boston University

MARJORIE GRENE

ANIMAL MECHANISM AND THE CARTESIAN VISION OF NATURE

In the fifth part of the *Discourse*, a few pages are devoted to the thesis that was to become notorious as the doctrine of the *bête-machine*, the doctrine that all animate bodies are machines, and, in particular, that all non-human animate bodies are *only* machines (A.T. *VI*, pp. 40–60, esp. pp. 57–59).[1] In looking back to consider the impact on modern thought of the Cartesian revolution, this seems a minor and peripheral theme, no more weighty (except to animal lovers) than the pages it occupies in proportion to the whole volume published in 1637 (A.T. *VI*, pp. 1–485). Yet if we start from this document and put it, along with other texts allied to it, into the framework of the Cartesian enterprise as a whole, it turns out, I think, to guide us to some fresh insights – and also, I confess, to some fresh problems. I want to consider the implications of animal automatism, then, in four contexts: first, as one of the two allegedly objectionable features of Cartesian natural philosophy, which Descartes's disciples felt called on to defend; second, as an aspect of the Cartesian effort to reduce all natural change to local motion; third, in the context of Descartes's interest in automata; and fourth, closely related to the third, in the context of Descartes's defense of his new program as related to what he calls (with a capital M) *Mechanica*.

1. Although I have taken the thesis of the *bête-machine*, in view of the space it occupies in the volume of 1637, to be a relatively minor theme in, or by-product of, Descartes's major program, it appeared more central to some at least of his critics. The celebrated Father Daniel, in the sequel to his Cartesian *Voyage*, called it "the very spirit and sap of pure Cartesianism" (in a Latin version: *spiritus est et succus, si ita loqui diceat, puri Cartesianismi*, Daniel, 1694, p. 3). It gave support, of course, to the anti-vivisectionist rhetoric of Daniel and other anti-Cartesians – although, repugnant though the doctrine is, it does not seem a necessary condition for the practice of vivisection in these, or any, centuries. Harvey, who was no Cartesian, was a happy vivisectionist. Reading the history of biology is an uncomfortable occupation, to say the least. But that is by the way. What I want to stress here is the first,

K. Gavroglu et al. (eds.), Physics, Philosophy and the Scientific Community, 189–204.

and narrowest, context in which the doctrine assumes significance in the history of Cartesianism: that is, in the defense of Descartes by his disciples against the charge that some of his doctrines were dangerous, if not directly impious. Cordemoy, for example, in 1668 published expressly for this purpose a letter addressed to one Father Cossart, in which he argued that "all that M. Descartes has written of the system of the world and of the soul of beasts can be drawn from the first chapter of Genesis" (Cordemoy, 1968, p. 257). In the course of his exposition he cited the same passage from Leviticus that Descartes had relied on in attesting his own orthodoxy in reply to the criticism of Froimont – one of the earliest exchanges concerning the *Discourse* and its accompanying essays (see A.T. *I*, p. 414). It seems to me a questionable source, since the Jews are there forbidden to consume blood because the blood contains the soul of beasts. So, say Descartes and Cordemoy, animal souls are but blood and not souls. But if blood were just plain stuff like any other, what did Moses, or God, or whoever, think was so special about it, and why might good Jews eat flesh but not blood? Never mind – it was a scriptural source, so there. In his *Discernement*, two years earlier, Cordemoy had also discussed "natural machines," although his example there is chiefly the circulation, following Descartes following and "revising" Harvey, rather than those pure machines, the beasts (Cordemoy, 1968). In the same year La Forge had published his *Treatise on the Mind of Man*, in which he defended the thesis of the beast-machine (La Forge, 1974, e.g., pp. 120–124, 156, 295–6), following and to some extent elaborating Descartes's own arguments; here, however, as in the Master's own developed work, the automatist theory seems to be a relatively minor by-product of a firmly dualist metaphysic. On the other hand, Jacques Rohault also, one of the most popular advocates of the Cartesian system (and Clerselier's son-in-law to boot), in his *Entretiens sur la Philosophie*, published in the same year as his persistently popular *Traité de Physique* (Rohault, 1671; pagination of modern photocopy), took up the two doctrines to which devout (or official?) people had chiefly taken exception: in the first dialogue, the explanation, so to speak, of transubstantiation, and in the second, the thesis of animal automatism. The dialogues have a rather charming pre-Berkeleyan ring and it is perhaps worth reporting the gist of them, or rather of the second, before we go on to place our thesis in its broader Cartesian contexts.

One Monsieur N., a friend who has had to abandon the cassock for the sword, and so has lost touch with developments in the "new philosophy," has come to inquire of the author about the alleged dangers associated with this novel doctrine, especially with respect to the miracle of the Eucharist. Satisfied with the author's explanation of this (in the first dialogue), he returns the very next day to report that he has heard tell of another strange development: there are some people, he has been told, who believe that animals do not act from knowledge, but are pure machines (p. 138). (It is the *knowledge* of animals that seems to be the chief point at issue in many of these debates.) This, he says, seems absurd. Yet it is a probable view, the author persuades him, if we consider probability as attested by the phenomena, rather than by the majority opinion. For what we have before us, in appearance, is simply a variety of movements.

N. agrees, but raises several objections, all easily answered. First, there are so many movements. Answer: a clock with only ten chief parts can tell the time beautifully; think how much better God could do with the parts at His disposal. N.: clocks must be rewound; R.: beasts need food. N.: animals behave "purposively"; R.: so do magnets. N.: beasts flee from danger; R.: magnets also move to the opposite pole. N.: what of their cries of pain? R.: that's like organ music. There is always some mechanical analogy. N.: complex behavior, like nest building or insect metamorphosis, demands knowledge; R.: stones move toward the center of the earth and flowers, too, which certainly don't *know* anything, exhibit complex development. So this must all be a question of the movements of the parts of matter. Besides, says the author, "this argument, drawn from the exactitude with which beasts perform their various actions, proves too much: for one would have to conclude that beasts have more perfect knowledge than men" (p. 143) – and that would indeed be a scandal! N. now concedes that it is probable that beasts are machines; but he still asks, "will it be necessary to say they have a soul?" (p. 143). The author explains how confused the scholastic view of soul has been; so that's all right. But N. has heard, he says, some arguments to the effect that this opinion about animals may lead to atheism. No, says his host, it is the view that beasts think that is more dangerous, since it threatens the principal proof of immortality, and he proceeds to recapitulate, via question and answer, the proof implicit (though of course not explicit) in the *Meditations*. (Rohault notes, by the way, that, strictly speaking, body as such is not destructible, only the complex arrangement that is a particular body; but of course soul, being indivisible, is not destructible in any sense (p. 147). This remark seems to support the view that in Cartesian terms there is only body, not, in the last analysis, bodies.) N. is of course as convinced by the author as Hylas will be by Philonous; but he still worries that almost everybody does think beasts act with knowledge, and so libertines will argue that since their souls are mortal, ours are, too. "You see then," says the author, "how important it is, not to attribute to beasts something of the certainty of which one has not the least proof, and of which the contrary is very probable." (p. 147).

Fine, says N., but surely those who hold to the existence of animal souls also have sure-fire proofs of our souls' immortality. The author has of course answers to these alleged proofs: first, that we have knowledge of universals; R.: so do animals, if they have knowledge at all. Second, perhaps beasts just have less knowledge than we do – it's a question of degree. R.: in that case, we are a kind of beast! Irrefutable answer! Third try: we can reason; answer: so do beasts, if we permit them any knowledge. Fourth: we can reflect on our thoughts. Answer: reflection is but a return of thought to itself; no harder than a return to the past, which any being capable of thought can manage. All this shows, finally, says the author, that you can't get good consequences from a bad principle. And one last fling, anticipating a later century: some may say, that if beasts are machines, men are machines as well. N. replies: to say so, they must speak, and so contradict their assertion of their own mere-machine nature. All is well. The thesis of the *bête-machine*, far from endangering orthodoxy, secures the mind, in its indestructibility, against the challenge of our lower, bodily nature.

That is the most direct and obvious message of animal mechanism for the Cartesian system, and it is of course in connection with the contrast between our rationality and the mindlessness of other animate beings that the thesis is first presented in the *Discourse*. And in this context, of course, it forms part and parcel of Cartesian ontology: there is God, there are finite (human) minds and finite bodies (or body) and the modes of both of these – and that is all there is. Yet Cartesian ontology was invented (or discovered?), I hope we can agree, at least on one level for the purpose of reforming physics: even before he wrote *Le Monde*, apparently, Descartes had discovered that he needed a metaphysical foundation for his scientific enterprise (see, e.g. Rodis-Lewis, 1971, vol. *1*, chs. 3 and 4; Garber, 1992). But the doctrine of the beast-machine forms an integral part of that enterprise; that is the second context in which I want to look at it. Let me attempt, first, to defend my statement by means of a very general contrast between Cartesian and scholastic explanations, and then to back up this (over) generalization by looking at one piece of evidence: the criticisms of Froimont and Descartes's reply in the fall of 1637.

Descartes wanted to alter the method and therewith the doctrines of scholastic physics. The School, he held, had never solved a single problem, while with his method one could go on, in the right order, to answer all the questions that would ever confront the human mind. Now both parties to this dispute agree, of course, that God made the world as He made it. The question is: how to describe the regular, yet subtle, changes that occur in God's creation. Both parties agree, also, that what nature consists of is substances and their modes or accidents. The problem is to explain satisfactorily the changes that take place in these naturally existing entities. Now the position of the School, according to Descartes, is roughly that of common sense, which has elaborated with pompous verbiage and so solidified the prejudices of infancy and childhood. On this view, which the scholastics share with Father Aristotle, the most conspicuous, and paradigmatic, example of substance (on this earth) is living things. And what is most conspicuous about living things is that they come into being and perish: that is the first and most fundamental kind of change for Aristotle – and, as Gilson has persuasively argued, it is in connection with this phenomenon that the concept of "substantial form" was needed (Gilson, 1975, ch. I, esp. pp. 143–176). Every time a new living entity comes into being, a new "substantial form" comes into being with it: for this is a substance, uniquely characterized by its form, this instance of the form of its kind. Further, there are changes of quality: an ignorant man becomes musical, a hairy man goes bald, an unripe apple ripens from green to red, and so on. And there is growth: the child grows taller and becomes a man. Finally, also, substances move about: hares and even tortoises go from place to place, as do less clearly substantial entities like water or fire. In four of the ten categories, substance, quality, quantity and place, there is change, and it is the task of the physicist to study all of these. Most of this, Descartes believes, is empty talk, both childish (we have all thought in our pre-Cartesian youth the world was so full of a number of things, or kinds or qualities of things) and pretentious (in Descartes's time one was taught, even in the best schools, such as the one he had attended, that substantial forms, real qualities, intentional species and the like could support in our philosophical maturity those infantile delusions). In fact, Descartes has discovered, there is only one kind of physical change: local motion. Birth and death, qualitative alteration, growth are all reducible to this one, measurable, manageable type: the change in position of bodies relative to one another. Granted, if growth is change in quantity, that should be all right, from the Cartesian perspective, since it is measurable and hence quantifiable – but if we think of growth as one of the characteristics of animate existence, as what happens typically to little plants and animals, then it too must go. The very phenomenon of life, archetypical for ordinary people as well as for ordinary schoolmen, is and must be overlooked and indeed abolished. The reduction of animal existence to a congeries of movements is a necessary consequence of this fundamental shift in the concept of natural change. Thus the thesis of the *bête-machine* forms a coherent part of the style of explanation initiated by Descartes, first, of course in *Le Monde* and in the *Traité sur l'Homme* (A.T. *II*), but, publicly, first in the essays of 1637. (For Descartes's own idiosyncratic use of "substantial forms," see Grene, 1991.)

We can find a small instance, or confirmation, of this general theme in the exchange, via Plemp, with Froimont in September and October of 1637. Froimont raised eighteen objections; I certainly shan't go through all of them, but mention several that illustrate the general contrast that seems to be involved. To begin with, Froimont confuses Cartesian explanation with Democritean or Epicurean – a common enough mistake. Descartes, he fears, may be, like Ixion, embracing a mere cloud when he thinks he has Juno in his arms. Often, Froimont writes, "he falls unconsciously into the physics of Epicurus, a rudimentary and rather coarse (*pinguiusculam*) physics, insufficiently polished, as most believe, for the texture of exact truth" (A.T. *I*, pp. 402–3). He then appends, as he puts it, "a few hasty remarks" as examples of his difficulties with the text (*loc. cit.*). First, "the author seems to be saying that heat like that produced in the liver can perform all the operations of the animal in the human body, except for actions proper to the rational soul. Thus the heat of the liver, without any sensitive soul, can see, hear, and so on. Such noble operations do not seem to be able to proceed from so ignoble and brutal a source (*ex tam ignobili & bruta causa*) (p. 403). Second, Froimont remarks: "It is the same when he says that if an automaton had the internal organs and external shape of a brute, there would be no way to distinguish between a real animal and such a machine. What need would there be then to suppose substantial souls in brutes, if the heat of the liver were sufficient for the operations of all the internal and external senses and of the appetites? This is perhaps to open the way for atheists, so that they may attribute the operations of the rational soul (*animae*) to a similar cause, and exclude it from the human body, or rather cram into us a material instead of an immaterial soul. It is not fitting to attribute such lofty operations to such humble causes" (p. 403). Three is a relatively minor point about the blood's rarefaction in the heart (which Descartes answers easily). Fourth, Froimont is sceptical of the analogy of the blind man and his stick: we don't touch the sun's rays when we see it, he says, "as the hand touches a stone via a stick" (p. 404). And fifth, how can Descartes deny the *intentional species* of the colors, since it is necessary to vision that these be depicted on the back of the eye? (*loc. cit.*). After voicing some worry about the tennis ball analogy for the explanation of refraction and doubts about Descartes's claim that air resists the passage of light more than does water, he returns, in his eighth objection, to the necessity of *intentional species* for the explanation of vision. How could one believe, he asks, "that the faculty of sight will distinguish, for example, between the colors purple and yellow, if not through the quality of an 'intentional species,' but that it is only a wholly local impact (*impulsum*) that determines vision?" (pp. 405–6). Ninth, he wonders how all sensation can be in the brain, since touch seems so clearly to be located elsewhere. Tenth, coming to the *Meteors*, he is not convinced by the account of the composition of bodies in terms of the shapes of their parts. This seems again "too coarse and mechanical an explanation" (pp. 405–6). "Many particles of water, for example," he says, "have the same form; and one can no longer distinguish these sorts of hooks and clasps in the parts of parts without end. Thus we must necessarily admit another, integral union of adjacent parts, without any heterogeneity or inequality of shapes" (p. 406). Similar difficulties arise, in the eleventh place, from the explication of the composition of water out of elongated parts "like eels." "As if one could imagine, in the calm water of a lake, parts that extend from east to west instead of from north to south; or as if such bodies were nothing but a fabric of fibers or filaments, like the flesh of animals!" (p. 406). And 12: "what a paradox too to say that the same small corpuscles, if they strike the sense of touch lightly, produce the sensation of cold, and that of heat if they strike more vigorously! As if there were so much difference in such and such a local impulse, not in the qualities themselves that affect the organ of touch in different ways" (pp. 406–7). Omitting various physical questions in points 13–16 (where, in 14 and 15, both Froimont and Descartes in his reply invoke the authority of Archimedes), and also omitting 18, which deals with the explanation of winds, let us notice finally point 17, about the difference between salt and fresh water. Froimont complains: "He says that the salty taste of sea water comes from this: that the larger parts of this water strike the pores of the tongue with their points rather than transversely. Would it then have a different flavor, if the parts happened to lie across the organ of taste? He hopes through position alone or local motion to explain too many things that cannot exist without other *real qualities*, – or I understand nothing!" (p. 408).

Let us run quickly through Descartes's replies (A.T. *I*, pp. 413–431). The first objection holds only if beasts see as we do, but of course they don't; the same reply goes for point two, and besides, Descartes, far from opening the way for atheists, is reaffirming the statement of Leviticus about the identity of animal soul with blood. In point four, Froimont has confused his explanation with the atomists' – but far from resembling Ixion in his situation, Descartes avers, *he* is indeed embracing the true Juno. About the denial of intentional species in vision, he admits, ironically, that if the blind man needs them, then so do the sighted! But he has (in reply to the eighth objection) in fact explained the differences between colors without any need to resort to scholastic terminology. That all sensation is in the brain, of course, is proved by the case of an amputee (a particularly horrid history in this instance). About the composition of bodies (point ten), Froimont is complaining of just what Descartes is proudest of: I shall return to that response in connection with my fourth and final context. On the problem about the parts of water, he refers the unhappy Froimont to a passage in the *Discourse* in which the relation between proofs and explanations in these essays is elaborated, and illustrates how by his method he might explain these and similar problems. Of course, on hot and cold, the example of tickling (yielding pleasure) and hard rubbing (which produces pain) is enough to defeat Froimont's over-qualitative thinking. To the objection about salt water, Descartes replies that one does not prick with a needle except with the point nor wound with a sword except with the cutting edge. And finally, in reply to the objection that he wants to explain too much through position and local motion, he boasts that if Froimont "wants to enumerate the problems I have explained in the Treatise on *Meteors* alone, and compare them with those that have been handed down till now on the same material, in which he himself is so well versed, I am confident that he will not find so much occasion to despise my somewhat coarse and wholly mechanical philosophy" (p. 430).

The contrast is obvious; I need not stop to summarize it again. But I want to move to another context of the *bête-machine* doctrine, which is explicitly touched on in the *Discourse*: that is Descartes's interest in automata. A perfectly constructed replica of an animal would exhibit no difference from a natural one: animals are God-made, not man-made, automata, but automata all the same. Men, on the other hand, can be distinguished from automata on the basis of the two criteria for the possession of reason: speech and the multiplicity of our skills. Descartes's fascination with automata, which perfectly matches that of A.I. enthusiasts in our time, is exhibited in more detail in the science fiction example of another response to criticism of the animal-automaton doctrine that I have dealt with at length elsewhere (Descartes to Reneri for Pollot, 1638, A.T. *II*, p. 514; Grene, 1985, pp. 35–38). And of course in the *Treatise on Man* (A.T. *II*, pp. 130–2) he dwells lovingly on the tricks effected in the Royal Gardens, where visitors approaching a bathing Diana cause her to hide in the reeds, and where if they try to follow her a Neptune menaces them with his trident. Just so, he suggests, do external objects affect the movement of the spirits in the brain – and they in turn are like the force of the water that makes these delightful machines work. When the Danes threatened to think of politics in the last century, their king built them Tivoli and they lived happily ever after – so the story goes. God, it appears, has built us a huge Tivoli with all sorts of enchanting self-running, and even self-reproducing machines. If art imitates nature, nature, the original of art, copies, does, what art does, only infinitely more ingeniously (Gilson, 1967, p. 422). Or, to put it a little differently, nature also imitates, or parallels, art: but now it is God's art that is in question, how much exceeding our poor efforts. Yet our arts, imitating God's, enable us to understand nature better than would the direct (obscure) perceptions of nature herself. The latter have been given us to guide us in this life, while we are mind-bodies rather than pure minds. They have – with some exceptions, as we shall see – no direct cognitive import. Sensations and feelings, one supposes, would have been of use in the disciplines of medicine and morality, had Descartes lived to develop them as branches of the tree of physics with its root in metaphysics. But in the foundations of Cartesian science as such they have no place: it is precisely the turning away of mind from the senses that needs to be effected through the discipline of the *Meditations*. The stress on automata assists in this endeavor, helping us to see nature itself as an aggregate of machines like the ones royal landscape architects have devised, and so to free us as minds from the ensnaring, confusing quality-rich things we used to find around us in this many-faceted world.

There is yet another context for the interest in automata in general, however, and hence for the thesis of the beast-machine in particular. It is true that, when it comes to the science of physics and its explanations of natural events, Descartes insists on restricting change to local motion, and especially to the changes of position of small parts of bodies, thus eliminating substantial forms and real qualities from his ontology. When Froimont labels Descartes's philosophy "mechanical," however, and when Descartes replies proudly to that seeming insult, it is not just "mechanistic explanation" in the theoretical sense that is involved, but the everyday practice of Mechanics. I have never searched Adam and Tannery (or consulted the Descartes lexicon) for references to (capital M) mechanics, but the term is conspicuous both in the exchange with Froimont and in a letter from and reply to Huygens in the same period (Huygens' letter of September, 1637 and Descartes's reply two days after the letter to Plemp for Froimont, A.T. *I*, pp. 431–488). Huygens asks for "three leaves on the subject of the foundations of mechanics, and of the four or five engines one can exhibit there" (A.T. *I*, p. 395). Descartes replies with sketches of half-a-dozen sorts of engines for lifting weights. "Mechanics" here, as in the Galilean *Mechanics* Mersenne had translated, and which Huygens mentions, refers to the craft of engine design, the kind of practice involved also in the design of fountains, automata and the like. This corresponds to Boyle's account of the "stricter and more proper sense of the term" in 1671 (OED). Mechanics in this sense is certainly involved also in Descartes's defense (in reply to Froimont) of his approach in the *Dioptrics* and the *Meteors* as well as in the *Discourse*. Here is what he said about his account of the composition of bodies:

I don't understand what he is objecting to. . . . For if my philosophy seems to him too crude, because it considers only sizes, shapes and motions, as Mechanics does, he is condemning in it what I consider one ought to praise above all things, that I philosophize in such a manner, that I use no argument that is not mathematical and evident, and such that its conclusions are confirmed by true experiments; so that whatever I have concluded can be done on the basis of my principles, really can be done (*fiat revera*), as long as actives are appropriately applied to passives (*quoties activa passivis, ut par est, applicantur*). I am surprised he himself has not noticed that what has been practised till now, Mechanics, is nothing but a part of the true Physics, which, because it could find no place with the vulgar Philosophy, withdrew to the Mathematicians. Thus this part of Philosophy remains truer and less corrupt than others, since, because it relates to use and practice, those who make mistakes in it usually suffer by the loss of their expenses: so that if he condemns my style of philosophizing for its similarity to Mechanics, this seems to me the same thing, as if he were to condemn it because it is truc (A.T. *I*, pp. 420–1).

The reference here to Mechanics in the sense of the practical design of workable machines casts what is for me a new light on the Cartesian project for a mathematical physics. Descartes boasts often enough that his physics is purely geometrical, and it is plain that the cleavage of mind from body is meant to enable the uncluttered pure and attentive mind to read aright the geometrical nature of extended substance. At the same time, it has often been objected that Cartesian physics is still qualitative, that it fails to achieve any sophisticated mathematization (e.g. Clarke, 1982). But there is a humbler and more practical kind of "mathematics" involved here. Builders, whether of automata or pumps or cannon or what not, do actually have to make measurements. They cannot sit there and play with fancy terminology; they have to measure and cut and fit pieces precisely – real pieces of stuff in the real world. The piece of wax will turn out to be but something flexible and extended, and those are not "real qualities" in some confusedly sensed or imagined entity that mirrors our inner sensings; they are concepts intelligible to a mind. But all the same it is a real piece of wax, a changing part of extended substance, as is the bee that produced it. The senses mislead, and the Aristotelian senses mislead systematically. Yet in the application of the concept of extension and of change in size and shape, we do find a place where thought and external reality meet. The art of Mechanics knows that, crudely and practically, if you like – but Descartes will prove it clearly a few years later in Meditation VI. Body *as geometrical* really exists and we as mind-bodies (creatures embraced by the third common notion) can manipulate it. This is the point, the only point (at this stage in the development of Cartesian science) at which the austerity of the real distinction permits us to manipulate a real nature in a way that the more "refined" scholastic approach, mired in the diversity of "sensible" things, forever prevents us from doing. The significance of this connection is apparent, indeed, from the preface to the French version of the *Principles*: for it is not only morals and medicine that will branch out from physics, the trunk of knowledge, but also *Mechanics*, the very art exemplified, and praised, in the correspondence with Huygens and Froimont (A.T. *II*, p. 14). Thus as late as 1647, it still has its place as one of the major applications of Cartesian *scientia*. But it is even more than that: it is the crucial juncture at which Cartesian geometry can become physics. Like Descartes's seventeenth century disciples, we late interpreters are inclined to stress the real distinction, and rightly so, I believe. The third common notion seems an anomaly – as, indeed, it appears in the correspondence with Elizabeth. It seems, I have thought until now, to have nothing to do with Cartesian science, only with the everyday needs of practical life. Yet Cartesian science is the knowledge by mathematicizing intellects of a *real* extended, geometrical world. And the humble practice of engine or automaton building seems to be the very craft to exercise this meeting of mind and matter. The measurements a craftsman performs are, indeed, safely, obscured by no scholastic verbiage, both mental (mathematical) and bodily at once. No wonder Descartes's faithful disciple Rohault used to go and visit artisans to study their practices. (Indeed, it is worth noting that all three branches of the tree of knowledge, medicine, morality and mechanics, entail the unified mind-body rather than the intellect of the real distinction.)

Even Cartesian geometry itself, as Emily Grosholz emphasizes, depends at some junctures on the actual use of certain simple machines (Grosholz, 1990). The rather cumbersome-looking apparatus built of a number of rulers depicted in Part Two of the *Geometry* was used to implement the author's argument. The proofs needed to support the new geometry had in part to be physically constructed. In the light of the real distinction, of the claim that it is pure mind that understands the geometrical essence of bodies, this seems an anomaly. But is it? This is just where mind meets body in Cartesian science – never in everyday life, where only crudely practical maxims prevail. At least that will be the case until Cartesian medicine and morals receive their full development as consequences of his system – as, of course, they were never to do. In everyday life, meantime, and presumably forever, although God never deceives, He may even misguide us for our good: see the excuses for God put forward in the Sixth Meditation. But one place we really can safely apply the scientific intellection of pure mind to matter is in gadgetry. Animals too are gadgets, if God-made gadgets, and we understand them best, as we do everything else in this vast spread-out desert called nature, by analogy with the gadgets we ourselves invent. Both the argument on the essence of body in the Fifth Meditation and the proof of the real existence of bodies *as geometrical* in the Sixth support this thesis. It is the practice of Mechanics in this humble, traditional sense that exhibits the plainest meeting point of Cartesian mind and Cartesian body and that even gives to the mind-body, the object of the third common notion, a place in the initiation of that firm and stable structure that will be Cartesian physics, short of the development of morality or medicine.

Perhaps I have been laboring a thesis that has been obvious all along to more "mechanically minded" readers of the Cartesian corpus; but it does seem to me to shed a new and rather different light on Descartes's project. At the same time, it also raises some problems, especially for the reading of the *Meditations* and the *Passions* (and of course the correspondence with Elizabeth associated with the latter). For example, does the stress on the practical craft of Mechanics weaken the real distinction? Perhaps it need not, if we see that Mechanics embodies the first, and seemingly unique, place at which Cartesian mind can safely apply itself to Cartesian matter.

Further, one might wonder how the stress on Mechanics as providing the meeting point of mind and matter relates to other discussions of the third common notion and the experiences dependent on it, or rather on its denotatum, that is, the intimately coexisting substantial unity of mind and body. When Descartes discusses this unified entity, it is usually in connection with sensation, or feeling, or the passions. I can't recall any discussion of this peculiarly craftlike mind-body cooperation. This may be, of course, simply because the discussions of *sentiment* arise chiefly, first in the correspondence with Elizabeth and then in the *Passions*, in the context in which Elizabeth's enquiries placed them: in the everyday life of the emotions. The humdrum concerns of artisans and engine builders were presumably beneath the notice of the Palatine Princess, even in her exile; the life of the emotions, on the contrary, was certainly something she had to wrestle with. And that is where Descartes's treatment of the passions began, at her behest, rather out of the order of his overall program.

There are, however, two small connecting links between our present thesis and Descartes's exposition of the passions. I can only point them out, but have no well worked out interpretation to offer. In the correspondence with Elizabeth, Descartes refers to his little treatise on animals, which she has seen, and which is presupposed in his account of the passions. It does not come in directly in the eventual treatise on this subject; one can only guess that a proper understanding of the bodily machine (most easily exemplified in animals, where there is no soul to interfere with the operation of the mechanisms in question) is a necessary condition for the understanding of mind-body interactions as in the passions (and hence, indirectly, in morality).

The other link is even slighter, but perhaps potentially even more revealing. You may have noticed that, in the defense of Mechanics against Froimont's attack, Descartes claimed that everything deducible as possible from his principles can in fact be made (*fiat revera*), "so long as actives are properly applied to passives" (A.T. *I*, p. 421). This is an odd remark, but in a way it parallels (with differences I must overlook here) the discussion of activity and passivity in general at the beginning of the *Passions*. The relation between action and passion in Descartes remains a difficult topic, although Paul Hoffman has dealt ingeniously with this question and some of those allied to it in a recent essay (Hoffman, 1990). The little I shall say about it here, in conclusion, is indebted to Hoffman's work, not only in the essay just mentioned, but also in his paper "On the Unity of Descartes's Man," some of the implications of which I have touched on in a recent lecture (Hoffman, 1986; Grene, 1991). Each human mind is a complete substance, according to Descartes, body as a whole is substance, and each particular body, while it moves as a unity, may be taken to be a substance, too. The unity of mind and body is also substantial, but this is a case of a compound substance, such that the mind is in this life a mode of the substance body, or I should think, the body a mode of the substance mind. Either, being substantial, can exist separately; but for Descartes it is only the possibility, not the actuality, of separate existence that marks a finite substance; every finite substance need not always in fact exist on its own. And in this life, mind and body are in fact interdependent. Each therefore can affect the other as its accident: the pain of a pinprick is the effect of a bodily movement on my soul, and my attempts to turn the wheels of an exercise bicycle impose the effect of my will (such as it is, poor thing) on the muscles of my bodily machine. Each is action to the other's passion. Indeed, the passions might be said to be consequences in the soul of the actions of the body. And the actions of the mind in controlling the body could equally well be called the passions of the body. Only Descartes (sounding a bit like Froimont) says we prefer to talk of passions in the case of the soul only, because of course it is "nobler" than the body! But, to go back to our craftsman, he is actively measuring, applying his ability to act, with respect to the passive thing, a piece of wood or steel or wire, that he is using in his construction. Something like that, I presume, is involved in the "appropriate application of actives to passives." It is the active manipulation by minds of passive bodies that makes Mechanics – and Cartesian science – work. And the same will hold of the other two branches of science, medicine, which prescribes rules for the regulation of our bodily machine while we dwell in this vale of tears, and morality, which teaches the mind to control the unruly bodily movements impinging on the pineal gland under the same temporal, and temporary, circumstances. But to return to mechanics: rulers as well as the spaces they measure are passive, of course, but those who use them to measure act to make out of passive tools and materials what their ingenuity has contrived, what, on the basis of correct principles, they understand to be possible. And of course even though we often speak of the "action" of one body on another, it is, strictly speaking, only minds that are active. When rightly used, they can control the passive nature that surrounds them, confronts them, and submits to them. Animals too are included in the "passives" to which "actives" may be appropriately applied. Since, manifestly, they are not rational, that is all they could be.

Department of Philosophy at
Virginia Polytechnic Institute and State University

NOTE

[1] Reference to Descartes 1971–1976 are abbreviated A.T. plus vol. no. throughout the text.

REFERENCES

Clarke, D. 1982. *Descartes' Philosophy of Science*. Manchester: Manchester University Press.

Cordemoy, G. 1968. *Oeuvres Philosophiques*, ed. Clair, P. and F. Gribal. Paris: Presses Universitaires de France.

Curley, E. M. 1986. Analysis in the *Meditations*: The Quest for Clear and Distinct Ideas, in Rorty, A., ed., *Essays on Descartes's Meditations*. Berkeley: University of California Press, pp. 153–176.

Daniel, G. 1694. *Novae Difficultates*, bound with *Iter per mundum Cartesii*. Amsterdam: Wolfgang.

Descartes, R. 1971–6. *Oeuvres.*, ed. Adam, C. and P. Tannery. Paris: Vrin, 2d ed.

Garber, D. 1986. *Semel in Vita*: The Scientific Background to Descartes's *Meditations*, in Rorty, A., ed., *Essays on Descartes's Meditations*. Berkeley: University of California Press, pp. 81–116.

Garber, Daniel. 1992. *Descartes' Metaphysical Physics*. Chicago: University of Chicago Press.

Gilson, E., ed. 1967. *Descartes, Discours de la Méthode, Texte et Commentaire*.

Gilson, E. 1975. *Études sur le rôle de la Pensée médiévale dans la formation du système cartésien*. Paris: Vrin.

Grene, M. 1985. *Descartes*. Minneapolis: University of Minnesota Press.

Grene, M. 1991. *Descartes among the Scholastics*. Milwaukee: Marquette University Press.

Grosholz, E. 1990. *Cartesian Method and the Problem of Reduction*. Oxford University Press.

Hatfield, G. 1986. The senses and the Fleshless Eye: The Meditations as Cognitive Exercises, in Rorty, A., ed., *Essays on Descartes's Meditations*. Berkeley: University of California Press, pp. 45–80.

Hoffman, P. 1986. The Unity of Descartes's Man. *Phil. Rev.* **95**, 339–370.

Hoffman, P. 1990. Cartesian Passions and Cartesian Dualism. *Pac. Phil. Quart.* **71**, 310–332.

La Forge, L. de. 1974. *Traité de l'Esprit et de l'Homme*, ed. Clair, P. Paris: Presses Universitaires de France.

Rodis-Lewis, G. 1971. *L'Oeuvre de Descartes*, 2 vols. Paris: Vrin.

Rodis-Lewis, G. 1986. On the complementarity of Meditations III and V: From the "General Rule" of Evidence to "Certain Science," in Rorty, A., ed., *Essays on Descartes's Meditations*. Berkeley: University of California Press, pp 271–296.

Rohault, J. 1671. *Entretiens sur la Philosophie*. Paris: Le Petit.

GERALD HOLTON

MICHAEL POLANYI AND THE HISTORY OF SCIENCE

BIOGRAPHICAL BACKGROUND

I shall be discussing Michael Polanyi's career as a scientist and philosopher from the point of view of history of science as well as his contribution to the history of science. But in order to arrive at my conclusions, I shall have to make some excursions into other fields, such as the branches of politics and epistemology that were fashionable during his career. For it is fair to say that Polanyi (1891–1976) was a unique person in the history of science, not least in the spectrum of his interests. The *Dictionary of Scientific Biography* (Vol. 18) lists his professional fields as chemistry, philosophy, and sociology.

Polanyi's early family upbringing in Budapest must have counted a great deal in the eventual choice of his ideological direction. His parents' home was a place for "regularly held literary gatherings that attracted the leftist intelligentsia, some of them Marxists."[1] Moreover, the three children of the family joined left-wing youth movements at an early age. But as in the case of so many Central-European intellectuals, the subsequent course of history radicalized Michael against authoritarian political movements.

He went on to study medicine at the University of Budapest, graduating a year before World War I. Even before obtaining his diploma he was publishing in chemistry, and although qualifying as physician, he also earned a Ph. D. in chemistry with a thesis on thermodynamics. In the Albert Einstein Archives is a considerable amount of Einstein–Polanyi correspondence. It begins with a letter of January 30, 1913, from Einstein in Zurich to professor Bredig in Karlsruhe, who had been unable to judge Polanyi's thesis on entropy at high pressures and who had sought advice on it from Einstein. The latter responded, "The publications of your Herr Polanyi please me very much." Einstein had checked them and found them altogether useful and full of fortunate thoughts. (He only wished that it had been at less length.) Then follows a series of handwritten letters to Polanyi, in which Einstein does find some points of disagreement on matters of thermodynamics. But they clearly speak as

K. Gavroglu et al. (eds.), Physics, Philosophy and the Scientific Community, 205–223.

equals. Michael's son, John C. Polanyi, later noted that Polanyi's first scientific paper was published on Einstein's recommendation, adding, "it really marked the first step in my father's academic career."[2]

Polanyi's interest in the history of science was perhaps triggered by his personal experience of historic episodes. When he presented his theory at a meeting in Berlin, at which Einstein was another participant, his theory was rejected and it took more than a decade until his views began to gain acceptance. Eventually, Polanyi became a researcher at the Kaiser Wilhelm Institute for Physical Chemistry in Berlin, where he developed his best-known work on dislocation theory. Leaving Germany when the Nazi party seized power in 1933, he became Professor of Physical Chemistry at the University of Manchester, where he continued a period of high achievements in physical chemistry. But perhaps under the pressure of contemporary history, his interest turned to the social sciences, and he resigned his professorship in 1948 to move to a chair in social studies. After his retirement in 1958 he continued his researches in the sociology and philosophy of science at Merton College, Oxford.

TWO SOURCES OF POLANYI'S APPROACH

To understand the animus of his philosophical-sociological views, one must remember that throughout most of his career as a scientist, a main model for philosophy of science for the scientific community came from the writings of logical positivists or logical empiricists, who in their most extreme writings seemed to say that science is based only on sense data and that all questions that cannot be subjected to tests in the laboratory are meaningless. But Polanyi's own work had convinced him that, while objective knowledge is the basis of science as transmitted within and by the scientific community as a whole, during the nascent period of individual scientists, they draw on resources quite different from those that follow the logic of justification of proven achievements. Some prominent philosophers thought otherwise. Thus in a passage quoted by Polanyi, Hans Reichenbach, in the essay "The Philosophical Significance of the Theory of Relativity," had written that "the philosopher of science is not much interested in the thought processes which lead to scientific discovery; he looks for a logical analysis of the completed theory, including the relationships establishing its validity. That is, he is not interested in the context of discovery, but in the context of justification."[3] To be sure, Reichenbach at once modified this severe judgment by adding: "But the critical attitude may make a man incapable of discovery; and as long as he is successful the creative physicist may very well prefer his creed to the logic of the analytic philosopher." Yet, one should also note that no sooner had he made this generous gesture he followed it up by writing: "The philosopher has no objections to a physicist's beliefs so long as they are not advanced in the form of a philosophy."[4]

In opposition to all this, Polanyi proposed the value of what he called "Tacit Knowledge," the "personal notions and concepts" that are essential for progress and motivating, but that may be neither sanctioned by the scientific theories of the moment nor even always consciously known to the creating scientist. I shall have much to say about this later, including a surprising test of the concept. But Polanyi went further by also opposing reductionism, the widely-held conviction that ultimately all biological and social phenomena will yield to the primacy of explanation in terms of mathematics, physics and chemistry – and one should add that since the triumph of quantum mechanics most physicists tend to think that chemistry itself is just that part of physics which really works.

A radicalizing event in Polanyi's life appears to have been his visit to the Soviet Union in the mid-1930s, where, he later wrote, he first encountered questions of philosophy. During a discussion with Bukharin in Moscow in 1935, Polanyi was especially appalled by the concept of "planning and guidance of scientific research,"[5] and returned from the trip eager to devote himself to work on behalf of the freedom of scientists to choose the content, subject and means of their work. Thereby he ran head-on into the opposition from a strong movement in the United Kingdom, led in opposite direction by such well-known scientists as J. D. Bernal, J. S. Haldane, and P. M. Blackett. The Society for the Freedom of Science, which Polanyi founded and which attracted many adherents in this country also (for example, P. W. Bridgman at Harvard), was no match.

As it turned out, with the explosion of research opportunities in the post-World War II era, the support system in most countries in the West showed itself capable of providing scientists with the necessary elbow room for the pursuit of their own ideas. But during the pre-World War II period in Britain, the prevalence of ideas expressed by scientists with Marxist leanings was reinforced by a famous conference on the history of science held in the early 1930s, in which a number of Soviet scholars made presentations. The most impressive and memorable among them was Boris Mikhailovich Hessen, who published a work entitled "The Social and Economic Roots of Newton's *Principia.*" In it, he followed the demand of Friedrich Engels, whose views on the history of science were the most commanding element in his whole approach. I should inject here that when I visited China a few years ago to give some lectures on the history of science under the auspices of the Chinese Academy of Sciences, I found that for most Chinese scholars the basic work in the history and philosophy of science was still thought to be the set of notes Engels had prepared for himself in the period between 1873 and 1882, published in 1927 under the title *Dialectics of Nature*. To Engels, in his words, "A single achievement of science like James Watt's steam engine has brought in more for the world in the first 50 years of its existence than the world has spent on the promotion of science since the beginning of time."[6] Science itself should thus be examined in the light of Marx's theory of historical materialism. As Engels had put it, "It is not the consciousness of men that determines their existence, but their social existence that determines their consciousness."[7] And Engels added, "From the very beginning the origin and development of the sciences has been determined by production."[8]

In this spirit, Hessen, on British soil, attacked the very icon of abstract science, Sir Isaac Newton, and proclaimed, as he put it, "the complete coincidence of the physical thematics of the period, which arose out of the needs of economics and technique, with the main content of the *Principia*."[9]

TOWARD THE TACIT DIMENSION

Opposition to such currents of thought, accepted in the 1930s more widely in the West than we now may believe, surely determined Polanyi's motivation in large part. His visit to the Soviet Union, of which I spoke earlier, provided a force in the same direction. To document this, I can do no better than quote a part of Polanyi's introduction to his book, *The Tacit Dimension*, which he completed in April 1966. There he bared to us his soul in these words:[10]

> I was struck [during the discussion with Bukharin] by the fact that this denial of the very existence of independent scientific thought came from a socialist theory which derived its tremendous persuasive power from its claim to scientific certainty. The scientific outlook

appeared to have produced a mechanical conception of man and history in which there was no place for science itself. This conception denied altogether any intrinsic power to thought and thus denied also any grounds for claiming freedom of thought. . . . My search has led me to a novel idea of human knowledge from which a harmonious view of thought and existence, rooted in the universe, seems to emerge.

I shall reconsider human knowledge by starting from the fact that *we can know more than we can tell.*

One sees in the italicized phrase the head-on attack against the instrumentalism of the positivists who, since the time when Galileo made the division between primary and secondary qualities, declared that we can know for certain only what can be rendered in quantifiable, shareable terms. And it is also a confrontation with the Wittgensteinian positivism which, in the last sentence of the *Tractatus*, declares that whereof we cannot speak, thereof we must be silent – the implication being, of course, that such things are not worth thinking about.

Perhaps I should be permitted here a brief digression of a personal nature. I came to know and like Michael Polanyi during the last two decades of his life, when he often visited the United States. He was of course always gracious to younger colleagues; but we also shared a reaction against the more extreme form of positivism – although in those post-World-War-II decades, the Vienna Circle type of positivism had become greatly ameliorated by taking interest in sociological, psychological, and historical components, and had moved away from the strict form that characterized its early phase. I had discovered through the study of the history of science that the model of science in terms of observable phenomena and the logic of analysis and mathematics alone by no means accounts for creation, discussion, acceptance, rejection, and ultimate fate of any scientific advance. A clear distinction had to be made between two meanings of science: the personal stage of science, and the corpus that gets into textbooks as public science.

This distinction became particularly clear to me while assembling and studying the archives of Einstein's correspondence and manuscripts at Princeton, from the mid-1960s on. Therefore in my experience I found myself in need of a way of thinking about the growth of science that would add a dimension to the logical-analytic and the phenomenomic dimensions. In my case I opted for including a third dimension in the study of the origins of scientific thought, the thematic one. That is not the subject of this chapter; but it indicates why I had sympathy for Polanyi's wish to escape from a two-dimensional view of science.

But to return to Polanyi's presentation of the *Tacit Dimension*. Immediately after the passages which I quoted, ending with the point that "we can know more than we can tell," he adds, "This fact seems obvious enough; but it is not easy to say exactly what it means." And he gives the example of pattern recognition – for example, the recognition of somebody's face. Polanyi tries to make it plausible that one can know and act on what one cannot tell by referring to Gestalt psychology. He refuses to think that Gestalt psychology can be reduced to the disposition of impressions on the retina or the brain. Rather, to him Gestalt is "the outcome of an active shaping of experience performed in the pursuit of knowledge. The shaping or integrating I hold to be the great and indispensable tacit power by which all knowledge is discovered, and once discovered is held to be true."[11] And it is not only perception which is an instance of tacit knowing. More generally, "our bodily processes participate in our perceptions";[12] and once we understand that this is the case we can "throw light on the bodily roots of all thought, including man's highest creative powers."[13]

This is analogous to Dilthey's and Lipp's teaching that there is an empathy or in-dwelling required for a proper knowledge of man and the humanities, an aesthetic appreciation that may not be possible to render directly in language. Polanyi believes the same to be the case for the natural sciences also. "To rely on a theory for understanding nature is to interiorize it. For we are attending from the theory to things seen in its light and are aware of the theory, while thus using it, in terms of the spectacle that it serves to explain. This is why mathematical theory can be learned only by practicing its application: its true knowledge lies in our ability to use it."[14] Formalizing all knowledge, "to the exclusion of any tacit knowing," which is the aim of those who hold to the model of "strictly detached objective knowledge . . .", is "self-defeating, for in order that we may formalize the relations that constitute a comprehensive entity, for example the relations that constitute a frog, this entity, i.e., the frog, must be first identified informally by tacit knowing."[15] And Polanyi adds that the most "striking concrete example of an experience that cannot possibly be represented by any exact theory" is simply "the experience of seeing a problem, as a scientist sees it in his pursuit of discovery."[16] What you need most at that stage is "the intimation of something hidden, which we may yet discover." It is this "tacit foreknowledge of yet undiscovered things"[17] which provides the sometimes passionate motivation to uphold a direction of work or a theory against heavy pressures from the outside.

ELABORATION AND RESPONSES

This approach has two results. One is that he is in head-on conflict with Karl Popper, by severely disagreeing with Popper's doctrine that "The scientist is not only indifferent to the outcome of his surmises, but actually seeks their refutation."[18] Polanyi responds, "This is not only contrary to experience but logically inconceivable. The surmises of a working scientist are born of the imagination-seeking discovery. Such effort risks defeat, but never seeks it. It is in fact his craving for success that makes the scientist take the risk of failure. There is no other way."[19] Obviously, a working scientist is speaking here.

The second result is that Polanyi touches here, without saying so, on the very old problem of how major discovery is possible in the first place, given the obvious, severe limitations of the human mind, faced with the infinitude of natural phenomena and their connections. Einstein himself tried to answer it in 1918 with the daring suggestion that our minds are guided by "what Leibniz termed happily 'the preestablished harmony'."[20] You will recall that Gottfried Wilhelm Leibniz had postulated that our ability to discover the laws concerning material bodies is one aspect of the unity from which God created the two apparently separate entities of the universe, the spiritual and the material. Each of these obeys its own laws, but they can interact in sympathetic unison, somewhat in the way one string instrument goes into resonance and picks up the sounds made by a second one which is tuned to it. Or to use Leibniz's own words to explain this possibility of harmonious interaction, in which he uses an image that must have delighted Einstein, "the souls follow their laws and the bodies follow theirs, but nevertheless these two beings of entirely different kind meet together and correspond to each other like two clocks perfectly regulated at the same time. It is this that I call the theory of preestablished harmony."

In the early 19th century, the Danish physicist Hans Christian Oersted also struggled with this problem; in his way of reading Immanuel Kant, which was typical for the *Naturphilosophen*, Kant's insistence that phenomenal facts are not things in themselves but mere appearances, culminated in the warning that the study of these appearances and the connections between them are an interaction not with nature but with one's own mind. As Kant had put it in the *Critique of Pure Reason*,

> That nature should direct itself according to our objective ground of apperception, and should indeed depend upon it in respect of its conformity to law, sounds very strange and absurd. But when we consider that this nature is not a thing in itself but is merely an aggregate of appearances, so many representations of the mind, we shall not be surprised that we can discover it only in the radical faculty of all our knowledge, namely in transcendental apperception, in that unity on account of which alone it can be entitled the object of all possible experience – that is, nature.[21]

Oersted, and I think Polanyi also, found this idealism uncongenial to a working scientist's mind, and Oersted therefore invented a modification, in his splendid conception of an "anticipating consonance"[22] existing between the mind of the scientist and the workings of nature. Polanyi comes very close to this notion. For example, he writes, "When a discovery solves a problem it is itself fraught with further intimations of an indeterminate range, and . . . when we accept the discovery as true, we commit ourselves to a belief in all these as yet undisclosed, perhaps as yet unthinkable, consequences. This is of course not explicit knowledge, and [he acknowledges] there is no explicit justification for the perception of a dawning truth."[23]

Still, Polanyi never quite admitted that these elements of tacit knowledge and of intimations of undisclosed consequences are, more often than not, simply wrong, although as a working scientist he must have observed this to be the case. But he would also have known that there is a certain pattern to these intimations, or anticipations. Almost by definition, a major scientist is one for whom this mechanism somehow works, at least often enough. At any rate, whether these anticipations are correct or turn out to be "a delusion"[24] (his words), Polanyi holds it futile to search for strictly impersonal criteria of their validity, "as positivistic philosophies of science have been trying to do for the past 80 years or so."[25]

In the climate of the decay of logical positivism after the mid-1950s, Polanyi's conception of tacit knowledge, or personal knowledge, did not remain a prominent target of attack. In any case it was never really a completely thought-through theory of scientific creativity. On the other hand, a concept that Polanyi held to be directly related to the tacit dimension, namely "emergence," seems to me to have been then, and remains now, a focus of debate and opposition, particularly among biologists. To rescue biology from reductionism, from being dissolved into mere physics and chemistry, Polanyi announced "the principle that the operations of a higher level can never be derived from the laws governing its isolated particulars,"[26] hence that "none of the biotic operations can be accounted for by the laws of physics and chemistry."[27]

Here we encounter a newer version of the old debate that so agitated scientists and philosophers in the 19th century, of mechanism versus vitalism. Of course Polanyi did not deny that there is "a great deal of truth in the mechanical explanation of life";[28] but he wanted to insist that living functions are "determined at all stages by a combination of a mechanism with organismic regulation."[29] At the very least, he said, "a principle not present in the inanimate must come into operation when it gives birth to living things."[30] Such views, coming from a prominent physical chemist, found probably a much more willing audience outside the laboratory than in it.

Also, to Polanyi, the principal interest of evolution was the rise of man from "lower" beings. To him, the problem of evolution seems to boil down to understanding how we reached "our position as the highest form of life on earth, and our own advent by a process of evolution."[31] But we are now in the age of anti-speciesism, in which even the spotted owl still has some political clout; so one cannot expect much resonance nowadays with Polanyi's call for a reshaping of "the problem of evolution deformed by the current theory of evolution."[32] The origin of species was a preoccuption that he thought can only make us "lose sight" of that more fundamental question. Properly understood, evolution is an expression of the concept of the "stratified universe of living things,"[33] in which progress from one level to the other cannot be done via reduction, or even by the continuation of the logic of one level with respect to the logic of the second above it, but rather by emergence – "the first emergence by which life comes into existence being the prototype of all subsequent stages of evolution."[34] Polanyi is quite frank that such ideas connect with earlier versions encountered in the history of science, for example, that of Teilhard de Chardin.[35]

With such tools, Polanyi struggled with what he called the concept of the "potentiality for obedience to higher demands,"[36] and "the capacity to feel reverence for men greater than oneself,"[37] both of which he regarded as aspects of the process of evolution. One must acknowledge that within a very different system of concepts than Polanyi's, sociobiologists such as E. O. Wilson are in fact struggling with very similar problems, summarized under the heading, altruism. I mention this only to indicate what to an historian of science is again and again so impressive: the continuity of preoccupations of the same sort within very different frameworks and worldviews.

CONSEQUENCES OF POLANYI'S DOCTRINES

As if by simple extrapolation, we can guess where Polanyi's thought might land next. It is the modern base for moral belief. How, he asked, can "intellectual powers, grounded in tacit knowing and descended from evolutionary emergence . . . exercise the kind of responsible judgment which we must claim if we are to attribute a moral sense to man. In a world where, it is widely held, scientific rationalism has impaired moral beliefs . . . by shattering their religious connections, where the Enlightenment weakened ecclesiastical authority, and modern positivism denied justification to all transcendent values,"[38] where can one find a theory for reestablishing the justification of moral standards? Control through established ecclesiastical authority appealed to Michael Polanyi as little as the control of science itself. Thus he wrote, "It was only when the philosophy of Enlightenment had weakened the intellectual authority of the Christian churches that Christian aspirations spilled over into man's secular thoughts and vastly intensified our moral demands on society. The shattering of ecclesiastical control may have been morally damaging in the long run, but its early effect was to raise the standards of social morality."[39]

What he feared most of all was the fusion of scientific skepticism and moral perfectionism. He saw that hybrid represented by modern existentialism and by what he called "an angry absolute individualism."[40] And the same hybrid also, by demanding a total transformation of society as a utopian project, expressed itself in Marxism as a political doctrine. In fact, it would be difficult to associate Polanyi's ideas with allegiance to any "ism"; for by putting centrally the concept of tacit thought as an indispensable element of all knowing, "The transmission of knowledge from one generation to the other must be predominantly tacit,"[41] and therefore cannot become concretized in a uniquely shareable ideology at a given time or through history.

And yet, apparently paradoxically, Polanyi sees a way of attempting total individual understanding with one's own mental faculties. That alternative is "entrusting oneself . . . to a teacher or leader."[42] St. Augustine observed this when he taught, "Unless you believe, you shall not understand." However, Polanyi does not hold out great hope that religion as now understood could fill this place for the need for tradition. "Modern man's critical incisiveness must be reconciled with his unlimited moral demands first of all on secular grounds. The enfeebled authority of revealed religion [as he called it] cannot achieve this reconciliation; it may rather hope to be *revived* by its achievement."[43]

Polanyi's final paragraph in *The Tacit Dimension* indicates his puzzled frame of mind on this point: "Perhaps this problem cannot be resolved on secular grounds alone. But its religious solution should become more feasible once religious faith is released from pressure by an absurd vision of the universe, and so there will open up instead a meaningful world which could resound to religion."[44] This view is connected, I believe, with Einstein's much better developed ideas on Cosmic Religion, with which Polanyi was undoubtedly acquainted.

I don't see it as my task to provide a rebuttal to, or even a general assessment of, Polanyi's thoughts. This has been done many times, for example, in the volume *Intellect and Hope*, edited by Langford and Poteat. But perhaps a few words are appropriate about the way comments on *The Tacit Dimension* generally run. Let me refer here to Robert S. Cohen's essay in the volume edited by Marjorie Grene, *Interpretations of Life and Mind*, an essay entitled "Tacit, Social and Hopeful." Cohen, both a physicist and a philosopher of science, acknowledges right away that the tacit dimension of knowing appears to him "acceptable and well-established."[45] Polanyi did not discover tacit knowledge, but he discovered at least how important it was in his own epistemology, and made more of it than many others. For Polanyi, "knowledge is situated within a background of clues, or a tacit background." That "means that there is a reality hidden behind the discovered objects. And so objects as we know them become clues to an as yet undiscovered and deeper level of reality."[46] For philosophy, this poses an immense challenge owing to the whole series of progressively more hidden realities it implies.

But, Cohen asserts, there is in all this a hint that Polanyi has a novel and interesting though undeveloped view of how the history of science progresses. "Polanyi asserts that different epochs of science offer different cue-maps, different forms of in-dwelling."[47] Any working scientist who has passed through the development of his or her own field over a couple of decades (and nowadays that can be an immense distance) – with the possibility of vast changes of mind and attitudes – is likely to assent to this picture.

THE "BIG BOOK" AND "THE STORY OF RELATIVITY"

The work that most closely connects Polanyi with the field of the history of science as scholarship is of course chiefly what he called his "big book," *Personal Knowledge: Toward a Post-Critical Philosophy*, published in 1958. As he says at the beginning of his preface,

> This is primarily an inquiry into the nature and justification of scientific knowledge. But my reconsideration of scientific knowledge leads on to a wide range of questions outside science. I start by rejecting the ideal of scientific detachment. In the exact sciences this false ideal is perhaps harmless, for it is in fact disregarded there by scientists. But we shall see that it exercises a destructive influence in biology, psychology and sociology, and falsifies our whole outlook far beyond the domain of science. I want to establish an alternative ideal of knowledge, quite generally. Hence the wide scope of this book and hence also the coining of the new term I have used for my title: personal knowledge. The two words may seem to contradict each other: for true knowledge is deemed impersonal, universally established, objective. But the seeming contradiction is resolved by modifying the conception of knowing.[48]

That new conception is based on the view that the personal participation of the knower in acts of understanding does not make such understanding *subjective*. The act of comprehension is "a responsible act claiming universal validity. Such knowing is indeed *objective* in the sense of establishing contact with a hidden reality."[49]

The book was based on his Gifford Lectures, delivered in 1951–52 at the University of Aberdeen. But he confesses that he spent "nine years almost exclusively on the preparation of this book."[50] Nevertheless, as even his friends and followers admit, it is by no means an easy book to read or accept. In their introduction to the collection, *Intellect and Hope. Essays in the Thought of Michael Polanyi*, the editors, Thomas A. Langford and William H. Poteat, begin with the sentence: "*Personal Knowledge* is an exasperating book." They add,

> If one does not find it exasperating, one has not *really* read it . . . There can be no doubt that *Personal Knowledge* comes to us with its rhetoric all out of focus. It is a mixed bag . . . Philosophers by and large, at least English-speaking philosophers on both sides of the Atlantic, find *Personal Knowledge* annoying because it is dangerously loose, innocuous because it says what has been said elsewhere and better, or irrelevant because its preoccupations are no legitimate concern of philosophy or of philosophers . . . One never 'gets going.'[51]

There is no doubt that the book is maddening in spots. For example, Polanyi writes, "The principal purpose of this book is to achieve a frame of mind in which I may hold firmly to what I believe to be true, even though I know that it might conceivably be false."[52] It has been suggested that it may be best to consider it an example of the confession literature, with such distinguished antecedents as Augustine and Rousseau.

Polanyi begins by going over a version of the development of the ideas of the solar system during the scientific revolution of the 17th century as a consequence of the Copernican model, and he draws on various well-known anecdotes to show how foolish it would be to hold that these early scientists forbade themselves to "go beyond experience by affirming anything that cannot be tested by experience."[53] Quotations of ecstatic passages from Kepler easily show the opposite to be the case. But Polanyi devotes only a few pages to these matters, for his main proof depends on what he calls "the story of Relativity." That theory was indeed taken by the positivists to show that through instrumentalist thinking Einstein had freed 19th-century physics from its metaphysical underpinnings, and thereby made the breakthrough to modern science. Polanyi correctly points out that every textbook of physics tried to present the rise of relativity as the necessary response to an experimental situation, namely the supposed null result of the Michelson–Morley experiment searching for an ether drift in 1887 – fully in accord with the sensationist or positivist view of how theories must proceed. (As well, we should add, as being the easiest *pedagogic* method of convincing students that they must take seriously what otherwise would be so counter-intuitive.) But, Polanyi declares, "the historical facts are different."[54] He noted that Einstein, in his publication, had not mentioned the Michelson–Morley experiment at all, and concludes from it that this theory was proposed "on the basis of pure speculation, rationally intuited by Einstein before he had ever heard about it."[55]

AN EXPERIMENTAL PROOF OF TACIT KNOWLEDGE

Let us stop at this important point in Polanyi's book and consider what an historian of science would now do on the basis of such a personal hunch or presupposition. It is an interesting enough case to give it serious treatment. You would begin by searching the literature of the period around the publication of the theory, encompassing perhaps a decade to either side of it, and not only of Einstein but of his contemporaries, to see who says what, if anything, about the Michelson–Morley experiment but also about the others of the same sort which were available by 1905. Then you would try to consult available documents in the archives of the main persons involved in the genesis and debates, pro and con, of the special theory of relativity, in the hope of finding contemporaneous exchanges or unpublished drafts and manuscripts from Einstein, but also from H. A. Lorentz, H. Poincaré, etc. You would also try to consult oral history interviews, autobiographical writings at a later stage, and so forth. If a promising Ph.D. candidate had come to me with such a project, I would have estimated it would take a year or two of research and quite a bit of travel to archives. In fact, I published a long article on this case in *Isis* in 1969, entitled "Einstein, Michelson, and the 'Crucial' Experiment", and it did take me the better part of a year.

But Michael Polanyi did none of that. As he tells us in his book *Personal Knowledge*, he availed himself of a remarkable shortcut. After all, he had been in touch with Einstein since 1913. Taking advantage of his entré, and in order "to make sure" of his hunch that Einstein's theory was based on "pure speculation, rationally intuited by Einstein before he had ever heard" of the Michelson–Morley experiment, Polanyi got in touch with the physicist N. Balazs who was working with Einstein in Princeton in the summer of 1953, and asked his fellow former-countryman to submit this speculation to Einstein himself.

As a consequence, Balazs had an interview with Einstein on that subject, describing it to Polanyi in a letter of July 8, 1953. He reported that Einstein concurred that (as Balazs wrote) "The Michelson–Morley experiment had no role in the foundation of the theory. He got acquainted with it while reading Lorentz's paper about the theory of this experiment (he of course does not remember exactly when, though prior to his papers), but it had no further influence on Einstein's considerations, and the theory of relativity was not founded to explain its outcome at all."[56] What did matter during the genesis, Einstein had told Balazs, was his concern with a series of more fundamental problems, such as the impression an observer, moving with the velocity of light, would have while viewing a light wave, and the lack of symmetry of action between coils and magnets when they are moved with respect to each other in producing the induction of currents in the coil.

Needless to say, this second-hand report of what Einstein may have said to Balazs, which Polanyi strangely chose to quote only in a footnote in his book, was not found convincing either by philosophers of science or by historians of science, the more so as the book as a whole was using this report as a tool in an otherwise quite idiosyncratic attempt at a new epistemology. Not until years later, when all the supporting work that I have mentioned above as necessary was actually done, would there be the kind of impact on the scholarly community that Polanyi had hoped to make by his shortcut.

And yet, and yet. . . . Polanyi *was* right. His hunch, of which he was so convinced that he tested it only in the most perfunctory way, through a third party rather than even taking the trouble to put the question to Einstein himself, was borne out later by a great deal of more laborious work. To be sure, Polanyi overreached when he declared that Einstein's theory was framed "on the basis of pure speculation, rationally intuited"; but the theory was evidently experiment based chiefly both on the speculation about a thought (that of travelling with the speed of light along a light beam) and some old, well-established 19th-century experiments long before Michelson's, those of Faraday, Fresnel, and of stellar aberration.[57]

How could that be? Was it merely an accident that Polanyi's presupposition was borne out on the whole? Perhaps. But I prefer to think of it in Polanyi's own terms. After all, for decades he had been a very prominent and successful scientist himself, engaged both in experiment and theory. He had *internalized* how scientists think, and had observed how others do their work, in finished publications as well as in conversations and in debates, for example during his time in Berlin, when Einstein was also there and Polanyi saw much of him. In short, if there is such a thing as apperception, personal or tacit knowledge, and indwelling, we must allow Polanyi to have had those capabilities as a scientist himself. Or to put it in Hans Christian Oersted's terms: Polanyi's prediction of how serious research in the entirely different profession of the history of science would illuminate the genesis of relativity was an act of anticipating consonance with the real state of affairs, one that would be made fully clear only later.

In short, I would like to hold open the possibility that it is precisely Polanyi's *lack* of having made a serious study and yet having reached the right sort of conclusion that constitutes, as it were, an experimental verification of his concept of personal and tacit knowledge. We all know that this sort of mechanism has worked in science, from the days of Kepler and Galileo who made advances to which their purely scientific knowledge of the time did not really entitle them. Polanyi is the first example I know of where the same sort of thing happened in the pursuit of the history of science itself.[58]

NOTES

[1] Ferenc Szabadváary, *The Dictionary of Scientific Biography (DSB)* (New York: Scribner's 1981). vol. XVIII, p. 718.
[2] Letter of 5 June 1964 from J. C. Polanyi to Helen Dukas (who had been Einstein's secretary).
[3] H. Reichenbach, in *Einstein: Philosopher-Scientist*, ed. Paul A. Schilpp (Evanston, Il.: Library of Living Philosophers, 1949), p. 289. This reference is given by Polanyi in *Personal Knowledge* (Chicago: U. of Chicago Press, 1958), p. 14, footnote 1 (along with a quote from H. Mehlberg expressing a similar point of view.)
[4] *Ibid.* Similar passages can be found in many places, for example in Karl Popper's *The Logic of Scientific Discovery* (New York: Harper and Row, 1959), p. 31.
[5] See Ferenc Szabadváary's summary of Polanyi's views in *DSB*, p. 719: "In the mid 1930's, Polanyi visited the Soviet Union. After his return, he wrote several articles and a book criticizing Soviet economic notions, and on planning and guidance of scientific research. . . . He was a supporter of complete freedom in scientific research, not only concerning its content but also in the choice of the subject."
[6] F. Engels, "Outlines of A Critique of Political Economy", in *Collected Works* (New York: International Publishing Co., 1975), vol. II, p. 428. Quoted and referenced by Robert S. Cohen in *DSB*, vol. XV, p. 136.
[7] F. Engels, *Preface to A Contribution to the Critique of Political Economy* (New York: 1970), pp. 20–21. Quoted by R. S. Cohen in *DSB*, vol. XV, p. 136.
[8] F. Engels, *Dialectics of Nature* (Moscow: 1954), p. 247. Quoted by R. S. Cohen in *DSB*, vol. XV, p. 136.
[9] B. M. Hessen, *The Social and Economic Roots of Newton's Principia*, introduction by Robert S. Cohen (New York: H. Fertig, 1971), p. 26.
[10] M. Polanyi, *The Tacit Dimension* (New York: Anchor Book A540, Doubleday, 1967) pp. 3–4. The concept of tacit knowledge in science has also been treated by others, though not necessarily with the same terminology. For a recent example, see Gerhard Grössing, *Das Unbewusste in der Physik* (Vienna: Verlag Turia & Kant, 1993).
[11] *Ibid.*, p. 6.
[12] *Ibid.*, p. 7: ". . . bodily processes are prominent in the operations of perception." Similarly, Einstein reports that what he called "the elements of thought" were "in any case of visual and some muscular type." [Jacques Hadamard, *The Psychology of Invention in the Mathematical Field* (Princeton, NJ: Princeton University Press, 1945), pp. 142–143] See also Edmund Husserl, who writes: "Moving freely within the moment of experience which brings what is present into my intuitional grasp, I can follow up these connections of the reality which immediately surrounds me. I can shift my standpoint in space and time, look this way and that, turn myself forwards and backwards; I can provide for myself constantly new and more or less clear and meaningful perceptions and representations, and images also more or less clear, in which I make intuitable to myself whatever can possibly exist." [Edmund Husserl, *Ideas*, trans. W. R. Boyce Gibson (New York: Collier Books, 1962), chapter 3, section 27, p. 92.] See, further, Maurice Merleau-Ponty: "We grasp external space through our bodily situation. A 'corporeal or postural schema' gives us at every moment a global, practical, and implicit notion of the relation between our body and things . . . A system of possible movements, or 'motor

projects', radiates from us to our environment. Our body is not in space like things; it inhabits or haunts space. It applies itself to space like a hand to an instrument . . ." ["An Unpublished Text by Maurice Merleau-Ponty", trans. Arleen B. Dallery, in Maurice Merleau-Ponty, *The Primacy of Perception*, James M. Edie ed. (Evanston, Il.: Northwestern University Press, 1964), p. 5; first published as "Un inedit de Maurice Merleau-Ponty", in *Revue de metaphysique et de morale*, no. 4 (1962), 401–409.]

[13] M. Polanyi, *The Tacit Dimension*, p. 15.

[14] *Ibid.*, p. 17.

[15] *Ibid.*, p. 20.

[16] *Ibid.*, p. 21.

[17] *Ibid.*, p. 21: "For to see a problem is to see something that is hidden. It is to have an intimation of the coherence of hitherto not comprehended particulars." Again, compare with Husserl: "What is actually perceived . . . is partly pervaded, partly girt about with a *dimly apprehended depth or fringe of indeterminate reality* . . . Moreover, the zone of indeterminacy is infinite. The misty horizon that can never be outlined remains necessarily there." [E. Husserl, *Ideas*, p. 92.] See also Merleau-Ponty: "The characteristic property of the Visible is to have a layer of invisibility in the strict sense, which it makes present as a certain absence." ["Eye and Mind", trans. Carleton Dallery, in M. Merleau Ponty, *The Primacy of Perception*, p. 187; first published as "L'Oeil et l'Esprit" in *Art de France*, vol. 1, no. 1 (January, 1961).] For a discussion, see E. Wigner and R. A. Hodgkin, *Bibliographical Memoirs of Fellows of the Royal Society*, vol. 23 (1977), p. 430, footnote.

[18] Polanyi, *The Tacit Dimension*, p. 78. Polanyi quotes Popper's words and gives the reference in footnote 10, p. 98: "This view has been persuasively expressed by K. R. Popper, e.g., in *The Logic of Scientific Discovery*, New York, 1959, p. 279."

[19] *Ibid.*, pp. 78–79.

[20] A. Einstein, in an address given at a celebration of Max Planck's sixtieth birthday before the Physical Society in Berlin. In *Ideas and Opinions* (New York: Bonanza Books, 1954), p. 226.

[21] I. Kant, *Critique of Pure Reason*, A 114.

[22] H. C. Oersted, "On the Spirit and Study of Universal Natural Philosophy," in *The Soul of Nature*, trans. Leonora and Joanna B. Horner (London, 1852; reprint London, 1966), p. 450.

[23] Polanyi, *The Tacit Dimension*, p. 23.

[24] *Ibid.*, p. 25: "The anticipation of discovery, like discovery itself, may turn out to be a delusion."

[25] *Ibid.*, p. 25.

[26] *Ibid.*, p. 36.

[27] *Ibid.*, p. 42: "Moreover, the conclusion that machines are defined by the fact that boundary conditions expressly left open by physics and chemistry are controlled by principles foreign to physics and chemistry, makes it clear that it is in respect of its characteristic boundary conditions that a mechanically functioning part of life is not explicable in terms of physics and chemistry."

[28] *Ibid.*, p. 42, immediately following the preceding quote.

[29] *Ibid.*, p. 43.

[30] *Ibid.*, p. 44. (The very definition of emergence.)

[31] *Ibid.*, p. 47.

[32] *Ibid.*, p. 46.
[33] *Ibid.*, p. 50.
[34] *Ibid.*, p. 49.
[35] See *Ibid.*, p. 46. See also *Personal Knowledge*, p. 388, where he associates his views with "noogenesis" and cites Teilhard in a footnote.
[36] *Ibid.*, p. 52.
[37] *Ibid.*, p. 52.
[38] *Ibid.*, p. 56.
[39] *Ibid.*, p. 57.
[40] *Ibid.*, p. 59.
[41] *Ibid.*, p. 61.
[42] *Ibid.*, p. 61.
[43] *Ibid.*, p. 62.
[44] *Ibid.*, p. 92.
[45] R. S. Cohen, "Tacit, Social and Helpful", in M. Grene, ed., *Interpretations of Life and Mind* (London: Routledge and Kegan Paul, 1971), p. 137.
[46] *Ibid.*, p. 138.
[47] *Ibid.*, p. 140.
[48] M. Polanyi, *Personal Knowledge*, preface, p. vii. I thank Mary Jo Nye for the observation that Polanyi's "tacit knowledge" and "personal knowledge" are two notions that only partially overlap.
[49] *Ibid.*, p. vii.
[50] *Ibid.*, p. ix.
[51] T. Langford and W. Poteat, eds., *Intellect and Hope. Essays in the Thought of Michael Polanyi* (Durham, NC: Duke University Press, 1968), pp. 3–7.
[52] M. Polanyi, *Personal Knowledge*, p. 214. T. Langford and W. Poteat quote this passage in their introduction to *Intellect and Hope*, p. 14, in the context of comparing *Personal Knowledge* to the *Confessions* of St. Augustine.
[53] The import of such passages is likely to be Polanyi's opposition to the teaching of scientist-philosophers such as Ernst Mach, of whom Polanyi says (p. 9) that "his book, *Die Mechanik*, published in 1883, founded the Vienna school of positivism."
[54] *Ibid.*, p. 10.
[55] *Ibid.*, p. 10.
[56] *Ibid.*, pp. 10–11, footnote 2.
[57] I have discussed the relation between the Michelson–Morley experiment and Einstein's work in chapter 8 and on pp. 477–480 of G. Holton, *Thematic Origins of Scientific Thought: Kepler to Einstein* (Cambridge, MA: Harvard University Press, 1988).
[58] I wish to record my thanks to Anne Davenport for helping me convert the manuscript for a lecture into an article.

SELECTED BIBLIOGRAPHY

Cohen, Robert S. "Tacit, Social and Hopeful," in *Interpretations of Life and Mind. Essays around the Problem of Reduction*, Marjorie Grene, ed. (London: Routledge & Kegan Paul, 1971), 137–148.

Gelwick, Richard. *The Way of Discovery. An Introduction to the Thought of Michael Polanyi* (New York: Oxford University Press, 1977).

Holton, Gerald. "Einstein, Michelson and the 'Crucial' Experiment," in *Thematic Origins of Scientific Thought: Kepler to Einstein* (Cambridge, MA: Harvard University Press, 1988), chapter 8 and pp. 477–480.

Langford, Thomas A. and William H. Poteat, eds. *Intellect and Hope. Essays in the Thought of Michael Polanyi* (Durham, NC: Duke University Press, 1968).

Polanyi, Michael. "The Growth of Science in Society," *Minerva*, vol. 5, no. 4 (Summer 1967), 533–546.

Polanyi, Michael. "Life's Irreducible Structure," *Science*, vol. 160 (21 June 1968), 1308–1312.

Polanyi, Michael. *Personal Knowledge. Towards a Post-Critical Philosophy* (Chicago, IL: University of Chicago Press, 1958).

Polanyi, Michael. "The Potential Theory of Adsorption," *Science*, vol. 141 (13 September 1963), 1010–1013.

Polanyi, Michael. *Science, Faith and Society* (London: Oxford University Press, 1946).

Polanyi, Michael. *The Tacit Dimension* (Garden City, NY: Doubleday & Co., Inc., 1967).

Szabadváary, Ferenc. "Mihaly (Michael) Polanyi," *Dictionary of Scientific Biography*, vol. 18 (New York, NY: Charles Scribner's Sons, 1981), 718–719.

Wigner, Eugene and R. A. Hodgkin. Biographical entry for Michael Polanyi, in *Biographical Memoirs of Fellows of the Royal Society*, vol. 23 (1977).

SHIGERU NAKAYAMA

COSMOLOGICAL OUTLOOKS AND TECHNOLOGICAL TRANSFERS: A COMPARATIVE VIEW FROM EASTERN PERIPHERY

In the following article, we shall use the dichotomy of East-West in the course of the history of science as defined in the following fashion. Namely, the shift of the center of the history of Western science took place from Babylonia, to classical Greece, the Hellenistic World (India), the Arabic world, the medieval Latin West, Renaissance Europe, seventeenth century England, eighteenth century Paris, nineteenth German universities, twentieth American laboratories and so on, while in the East the center remained in China until European expansion, Korea, Japan and Vietnam remaining satellites. Thailand, Indonesia and Malaysia were interesting meeting places of East and West in this historical mapping, to be explored further with profit in future.

When we name India as one of the seats in the main current of Western science, many Indians who try to explain history in terms of the dichotomy of materialistic West and spiritual East (India), disagree or at least feel uncomfortable. A seventeenth century Japanese Confucian scholar, Mukai Gensho, in his preface to a book on Western cosmology by the apostate Jesuit missionary Christovao Ferreira stated, "Those who write vertically and eat with chopsticks take the basic doctrines of *li-qi* (organization-matter) and the five elements for scholarly investigation. Their countries are Japan and China. . . . Those who write horizontally and eat with their bare hands instead of chopsticks do not comprehend the doctrine of *li-qi* and believe in four elements. Their countries are India and South Barbary [Portugal and Spain]." This is evidence that the Japanese in the premodern era regarded India culturally as foreign as the West, while maintaining affinity with Chinese culture. I would like to point out further that the major difference between scholarship and culture in the East and the West is the preference for the writing and recording tradition in the East and that for the logico-rhetorical debating tradition in the West.

K. Gavroglu et al. (eds.), Physics, Philosophy and the Scientific Community, 225–233.

COSMOLOGICAL OUTLOOK

Morphological Aspect of Cosmology

Cosmology is an equivocal term. It may mean a discipline in modern technical astronomy. More generally, it refers to the interpretation of the outer world and its relations with human nature not only from the scientific but also from the religious, philosophical and social viewpoints. Recently it has also been used to signify the world-view of contemporary ecologists. We think it appropriate to start with the most common element between the East and the West at the time of the beginning of modern East-West contact; namely the morphological aspect of cosmos, particularly the spherical earth theory.

Sphericity of the Earth

One may be surprised to find that such a civilized culture as China totally lacked the notion of a spherical earth until the arrival of the Jesuits. It can be hardly imaginable that Chinese astronomers, whose largest duty was the prediction of eclipses, possessed no knowledge that the lunar eclipse is the phenomena of the round earth shadowing the moon, but professional astronomers never mentioned it, perhaps because they simply were not in the habit of writing about physical causes. Although the curvature of the earth was early recognized, the sphericity of the earth was never mentioned. Neo-Confucian scholars during the Sung dynasty still adhered to the time-honored notion of 'circular heavens and square earth,' which envisaged an earthly flat plate floating in the midst of the spherical celestial globe.

While the scientific center in the West moved from one area to another, thus ensuring that the physical center of the earth was not limited to one particular place on its surface, the center of activity in the East remained in China almost throughout its history, thereby fostering an ethnocentricism in that country. One may conjecture that Sino-centric cartography, which put the traditional origin of land-surveying in central China and claimed it as the 'center of the earth', was incompatible with the sphericity of the earth, whose center was, as Matteo Ricci portrayed it, reserved for the hell of the damned.

With the advent of the Jesuits, however, the reception of the sphericity of the earth in the East did not seem to have encountered stubborn ideological resistance. Chinese objections were only piecemeal, argued from the viewpoint that local irregularities prove that the earth has no definite form.

At the time of reception, the afore-mentioned Mukai commented that unlike Westerners the lack of aggressiveness of Eastern gentlemen had left them with little interest in pursuing conquests outsidc their own borders. This he said gave the Japanese no need to attempt to measure the circumference of the earth.

In the eighteenth and nineteenth centuries, a Japanese Buddhist group campaigned against the introduction of Western cosmology on the basis of the flat-earth meru cosmology that originated in India, but they remained only a minority even among Buddhists. More importantly, they refuted the Aristotelian–Jesuit Western version of a finite cosmos, preferring instead the idea of a vast, infinite and plural 'three thousand worlds' notion of Buddho-Indian cosmology.

Dynamic and Physical Cosmos

After all, the morphological aspect of cosmology was not as important for Eastern scholars as the cosmos for their Western counterparts; the matter of 'form and appearance' remained superficial, in contrast to the fundamental *qi* (matter-energy in Needham's translation) and *Yin-Yang* dynamism that permeates the phenomenal world. However, Easterners began to recognize Western superiority at the level of figure and appearance. Then what ensued was how to cope with the imported cosmology. One way to maintain peaceful coexistence between them was to classify them into two types of cosmology, metaphysical and physical. Some of the features of Western physical cosmology, such as the sphericity of the earth, were inrefutable and hence taken for granted. On the other hand, Confucian scholars did not want to lose their own identity at the metaphysical level. Thus, they juxtaposed the traditional metaphysical principles of *Yin-Yang* and Neo-Confucian *Taiji* (fundamental principle) theory, which give quasi-causal explanations for phenomena in the natural world. This dichotomy later developed into other dichotomies in the same vein, such as '*wakon yosai* (Japanese soul and Western competence) and '*Zhongti, xiyong* (Chinese learning as the basis and Western learning as the instrument).

The metaphysical principles of the West, the Aristotelian trivium or the Thomist theology, did not interest even the Chinese converts of the

Jesuits, as they were quite culture-bound. When Newtonianism was first brought to East Asia, it did not satisfy the metaphysical mind of the Japanese translator, Shizuki Tadao; thus, he used the metaphysical principle of *I* or 'Change' (as explicated in the ancient Chinese classic, the Book of Changes) as the basis of Newtonian laws.

Newtonianism did not satisfy the Western theological mind either; thus it was followed by the development of deistic theology. Such trends remained totally unknown to Shizuki, who thought something necessary for filling the fundamental lacuna of the Newtonian physical cosmology.

If we look more closely, another dichotomy that existed in the Western history of science becomes apparent; a more concrete dichotomy between morphology and dynamics rather than physical and metaphysical. Expressed in a different way, the dichotomy may be identified with Ptolemaic–Copernican geometrical cosmos vs. Aristotelian–Newtonian physical causality.

Geometrical schemes are visible and hence more easily understandable than causal explanations. The latter are more culture-bound, and thus the Aristotelian four-elements theory failed to attract the interest and understanding of Confucian scholars. Newtonian causally rigorous physics, the quest for why an apple falls, was quite novel for the Eastern tradition, but Confucian vocabularies were not rich enough to render Newtonian natural philosophy adequately into classical Chinese.

Geometrical cosmology was closely associated with the Western medieval cosmos. Thus, Copernican heliocentricism played quite a revolutionary role in the Western mind. This was never so in the East. Shizuki translated it into 'earth-moving theory', which is the accepted translation of Copernicanism in East Asia today. For Eastern cosmology, the matter of mere figure and position is not meaningful; the principal question in the natural philosophy of *I* (Change) and the *Yin-Yang* dichotomy is the agency lying behind the force permeating the universe.

The World of Natural History

Heaven, earth and mankind; those who are well-versed in these three aspects of the cosmos are qualified to be Confucian scholars. This phrase represents the naturalistic stance of Confucian scholars. A long tradition of recording celestial portents, local products and human affairs can be found in Chinese dynastic histories. These were quite legitimate disciplines, which were missing in Western scholasticism.

With the early availability of paper and printing in East Asia, scholars accumulated voluminous knowledge and devised systems for classifying it. Under such circumstances, miscellaneous knowledge from Europe was easily accepted and classified into the existing system. Moreover, the importation of *materia medica* from the West must have encouraged scholars to identify them with those of the East, and then articulate their differences.

This activity was bilateral. East is East, West is West; there were certainly horizontal differences but not vertical ones. Something curious imported from far away was enthusiastically welcomed; no resistance was offered. While physiological and pathological theories differ fundamentally and no interchange or synergism was envisaged between them, foreign *materia medica* were easily incorporated into the different classificatory setting and played subsidiary roles at a syncretic clinical level.

Presumably, the Western endeavor of finding and bringing home something mysterious and interesting from the East was founded on a stronger motivation than that of Eastern countries, as they did trade more actively than the Chinese, whose official foreign commerce was conducted within the framework of their tributary relationship with peripheral countries. Evidence may be found in the fashions of *chinoiserie* and *japonesque* in Europe.

Ecology and Diet

The late Lynn White, Jr. once told me privately that when he presented a now-celebrated thesis that Christianity was the source of the exploitation of Nature, he was not serious but just joking. His thesis was combined with Schumacher's idea of 'Buddhist economy' and led to a rather naive view among the Western ecology movement activists that the Buddhist outlook on nature was much more ecological, and in Buddhist countries nature was preserved much better than in the Christian world. We have an impression that generally Buddhists are more naturalistic than Christians, as shown in the holy drawing of the Buddha on a lotus plant. Traditional Eastern paintings are more naturalistic than the highly inorganic style of medieval paintings found hanging in Catholic cathedrals.

We cannot say anything definite, however, until we find out how influential religion is in the modern Eastern world. Buddhism is not regarded as having played any significant role, either pro or con, in the modernization and the industrialization of Japan.

Still, it is worth closely examining the elements of Buddhist influence, if any, from the viewpoint of environmental history. I believe, at least in dietary habits – macrobiotic and the like – that Buddhist culture must have exerted a significant effect.

At the time when the Buddha was born, there was not much food prohibition going on in India. In China, pig and poultry were consumed, and Buddhist style vegetarian foods have been recorded in Chinese cook books from the sixth century onward. We notice lots of specialist vegetarian restaurants in Chinese cities now, but China does not seem to have as large a vegetarian population as India. We doubt how young people could survive without animal protein. Archaeological studies, in fact, proved that in prehistoric Japan people ate great quantities of animal meats since lots of bones are excavated from tumuli.

The introduction of Buddhism into Japan must have redirected Japanese dietary habits in the direction of vegetarianism. An Imperial Edict was issued to prohibit eating meat and fish in 676 A.D. In the earlier phase of Buddhism in Japan, only aristocrats came into contact with incoming Buddhist culture, thus coming under its heavy influence; it later gradually spread to the general populace. The eating of mammals was often officially banned during the Edo period in Japan due to the governmental agronomic policy of protecting animal labor. Miscellaneous evidence shows, however, that the prohibition was not at all strictly observed. Unlike the religiously strict vegetarian cultures elsewhere, the Japanese had not been thoroughly anti-carnivorous. Hence, upon contact with the West in the late nineteenth century, the average Japanese started to eat beef without too much hesitation.

In China and Japan too, all families refrain from taking meat on parents' anniversary days even today. Buddhist monks stay vegetarian at least for a certain length of time during their training period. Such Buddhist habits must have contributed in history, and even now, to the fact that the Japanese remain much smaller consumers of meat than Europeans.

TECHNOLOGY TRANSFER

In spite of the deep cosmological gap that existed between the East and the West, technology was one of the most easily transferable cultural artifacts, to be examined in the following as a liquidator of the gap.

East to West Transfer

The flow of technology transfer was usually from center to periphery. If we accept Joseph Needham's second Puzzle, "why Chinese civilization created technology superior to the West until the fourteenth century?" as a valid question, we must ask why and how the West overtook China in technology after the fifteenth century. Apparently the center of technology shifted from China to the West. What happened in the East as well as in the West at the time of center shift is also a legitimate question to ask. Let me present a still untested model of center-shift in technology transfer. Technology at the center is transferred to the periphery either through migration or information gathering. Technology, unlike science, is quite closely linked with local social conditions (including cosmology). Unless social conditions are favorable, the transfer will turn out to be a failure. On the contrary, if technology is transferred into more favorable conditions, it will flourish better than at the center. This causes the center to move from one place to another. In such cases, the conditions at the periphery are also different from those at the center; and then technology develops in a somewhat different direction. Such are the cases of technology transfer of paper-making, printing, magnetic needle and gunpowder from China to the West. Under different social conditions and cosmological outlooks, paper and printing were widely employed for journalism, the magnetic needle for ocean navigation, and gunpowder for warfare in the Renaissance West. Technology transfer is better studied at the periphery than at the center, in this particular case in the West.

East-East Transfer

Throughout history, the ultimate source of learning in East Asia continually remained in China. The transfer of knowledge, technology and cosmology was always one-way traffic. Chinese Confucian scholars worked without reference to the scholarly activity going on at the periphery, as they thought that the criteria of judging scholarship remained permanently in China. With the advent of European expansion, such a unilateral pattern of the transfer of information and technology was relaxed. People at the periphery started to question the possibility of another sort of paradigm and another source of information. For Chinese satellite countries, the new opening to the West in the age of European expansion undoubtedly created intellectual tension.

This was not the process of a center shift from one place to another, as often happened in the history of Western science and technology. A peripheral satellite like Japan never displayed ambition to replace the Chinese central position. If no other paradigm had appeared, the Chinese paradigm would have lived longer. It is merely a matter of paradigm choice between China and the West. We must recognize that before the advent of the Industrial Revolution, there was nothing particularly influential to prove Western superiority.

Technology Transfer Through Migration and Through Information Gathering

There have been two modes of transmitting technology from one culture to another. The first was through the migration of technical experts, and the second involved sending students to find and bring back information.

The former case of migration mostly remained within the realm of individual activities with a natural personal motivation, though sometimes skilled experts were invited under a state policy. In any case, many of the descendants of immigrants continued to engage in specially assigned jobs for their family, thus carrying on the tradition. Without human migration, such knowledge as technical expertise could hardly have been transmitted and disseminated. In technology transfer, the role played by written sources was very limited.

The latter case involving the dispatch of students was an intellectually higher means of and an expressly conscious effort at transmitting mainly written knowledge. Student envoys were usually sponsored by a state or religious order, which guaranteed a social position for them after returning home. Such students usually spent considerably long periods at the center of learning and took books and other equipment home with them. The transfer was planned, targeted and pointed and hence often failed due to changes in the political, economic and ideological environment, while the natural infiltration of migration was constantly assisted by the state of the market, a factor that encouraged them, and hence their technological knowledge to gain a strong foothold in society. State dispatches of students usually resulted in the importation of technological knowledge into the upper strata of society, thereby limiting its applicability and circulation to those who did not engage in manual technology.

In the case of East-East or West-West technology transfer, private technology was well transmitted with minimum government intervention. In the case of West to non-West technology transfer in modern time, technological immigration was rare because of cultural and language barriers. The Japanese have experienced both modes of transfer, the former in ancient times and the latter in modern, so that we may hopefully compare the above two modes with meaningful findings.

Kanagawa University

AZARYA POLIKAROV

SOME QUESTIONS CONCERNING LIMITATIONS OF THE RANGE OF VALIDITY OF KUHN'S MODEL OF THE HISTORY OF SCIENCE

The main question posed by Kuhn in his monograph,[1] as is obvious by its title, is about *the structure of scientific revolutions* (R). This structure consists in a *transition from one paradigm* (P_1) *to another* (P_2).

The *paradigm* (P) may be defined as the dominating scientific conception (theory), which means that it is shared and applied by (the basic part of) the *scientific community* and characterizes so-called *normal* science.[2] During this stage of the history of (a given branch of) science, teaching and research go on under the sign of a P. This is the criterion for evaluation of the activity and the promotion of the specialists.

Notwithstanding the circumstance that the empirical material is interpreted along the lines of the P, the possibilities for such an adaptation are not unlimited and the P may come up against anomalies that develop into a *crisis*. This shakes its dominating position and presages the transition to a new P, i.e. to *extraordinary science* or revolution.

In connection with this account several questions arise that need clarification.

1. What is meant by *science*, or which sciences are envisaged?

In the spirit of the Anglo-Saxon tradition science comprises natural (physical and biological) sciences. Accordingly, Kuhn considers mainly revolutions in astronomy (Copernicus), physics (Newton), chemistry (Lavoisier), and biology (Darwin).[3] There are many case studies applying Kuhn's conception to (or testing it in) some other areas (geology, ecology, psychology), including branches beyond the realm of the natural sciences (e.g., mathematics). The results of these studies are ambiguous: some of them support, and others reject the view in question.

This allows us to conclude that Kuhn's conception has a *restricted validity* even in the domain of (history of) the natural sciences.

2. What is meant by P?

We shall not dwell on the ambiguity of Kuhn's definitions or uses of P, a point emphasized long ago, especially by Masterman.[4] This forced Kuhn to introduce the concept of 'disciplinary matrix' denoting the belonging to a given discipline, and various kinds of ordered elements.[5]

K. Gavroglu et al. (eds.), Physics, Philosophy and the Scientific Community, 235–240.

What we have specifically in mind is the ambiguity stemming from the *degree of generality* of the P or the hierarchy of Ps (within a P). Thus, for instance, within the framework of mechanicism as a general (philosophical) P there are more concrete mechanistic (sub-) Ps in physics, chemistry, biology, etc. (physicalism, chemism, preformism, etc.). Analogously, there is the general atomistic idea involves chemical and physical atomistics. The latter is represented by the molecular-kinetic theory, atomic physics, and particle physics.

The general Ps remind one somewhat of Holton's general themata,[6] and their more concrete varieties correspond to the integral constituents (stages) of Lakatos' scientific research programme.[7]

In our opinion the inclusion of Ps of a lower level into a more general P means the elimination their qualitative difference, and correspondingly definite transitions between them may lose their revolutionary character. Thus, e.g., whereas according to some authors Darwin's doctrine and the results of Mendel's investigations are two revolutions, other specialists consider them as one P and hence as one revolution. There is a similar divergence of view with respect to the conceptions (theories) and the changes in contemporary physics. Along with a widely accepted view about two – relativist and quantum – revolutions, there is also the view that this is a single (quantum-relativist) revolution.[8]

Hence, the possibility of a different approach to the Ps *relativizes Kuhn's formula* about revolutionary transitions.

There should also be taken into account the possibility of a retrospective correction (in the sense of both negation and recognition) of the paradigmatic character of a definite conception, and accordingly in the characterization of a change as revolutionary.

3. Are revolutions always preceded by a crisis?

Yielding to criticism, Kuhn admitted – as an exception! – the possibility of revolutions without a state of crisis in the P.

We believe that the crisis may be essentially a fictitious notion due to an implicit (methodologically erroneous) presumption about the unlimited validity of one theory or another. Such a crisis indicates the boundaries of applicability of a given system and the necessity of framing a new theory that will make precise (restrict) the validity of the foregoing one. Characteristic in this respect is the construction of general relativity theory which – carrying out the program of field physics – overcomes the limitations of special relativity. This occurs without any symptoms of crisis and without a conflict between the new theory and the old one.

In general then, *crisis is not an indispensable transition stage* of revolutionary changes.

4. Is every revolution a paradigm-shift?

Along the lines of the adduced scheme, Kuhn makes a correction in his conception in the sense that it needs not be a large change nor needs it seem revolutionary to those outside a single community (p. 181).

There are several scientific achievements, especially discoveries that are considered as revolutionary, but do not consist in a substitution for a P. Such, e.g., is the case with the discoveries in modern astronomy or of the psi-particles (psions) in high energy physics.

Those revolutions associated with the framing of an initial P in a branch of science, i.e., of a P that has no predecessor, deserve special attention.

Kuhn also speaks of a pre-paradigmatic period and first Ps. Such are the Ps in ancient mathematics and astronomy, physical optics (Newton), as well the P of biochemistry which arouse by division and recombination of specialties already matured (pp. 13, 15).

Unlike Kuhn, we argue that conceptions like Aristotle's physics, or caloric, or phlogiston, etc., cannot be classified as Ps. Therefore, classical mechanics, molecular-kinetic theory, Lavoisier's chemistry, which Kuhn considers as P-shifts, are initial Ps. This in our opinion is a widespread case, to which belong the framing of electrodynamics, quantum theory, chemical atomistics (Dalton), Darwin's doctrine, molecular biology, cybernetics.

Such cases may be referred to as 'revolutions within a revolution' (J. Illy), i.e. changes with a revolutionary character within the framework of a more general P.

Here also belongs the case in which the appearance of a new P does not invalidate the available one, but they continue to coexist. This is characteristic for revolutionary achievements in mathematics such as the creation of non-standard (Non-Euclidean) geometry which does not mean a rejection of the standard (Euclidean) one.

Frequently the collision between Ps is due to (excessive) claims for universality and it may be surmounted by establishing a limit to the range of their validity.

The paradigmatic change associated with the framing of the theory of relativity (special and general) seems close to Kuhn's scheme. Yet here affairs turn out to be more sophisticated. Special relativity is considered by Einstein as an evolutionary continuation of the electromagnetic theory.[10] At the same time it leads to a radical reconstruction of mechanics (relativistic mechanics). Something like this is valid for the general theory, which is a next step in respect to the special theory, but in principle represents a new gravitation theory (compared with Newton's).

Besides, it is possible that the new P is but the vindication of an existing conception. Such seems to be the case in the recognition of plate-tectonics in geology.

Several revolutionary achievements have a *complex character*. The P may emerge as a result of unification of two (or more) conceptions or, conversely, two conceptions may be framed out of a single P. Thus molecular-kinetic theory and electromagnetic theory dissociate from mechanics, and molecular-kinetic theory and thermodynamics unify into one P.

One may speak also about combined revolutions, e.g., of revolutionary discoveries and framing of a new theory.

This differentiated examination allows for a *classification of the kinds of revolutions*, viz. 10 in number, joined in 4 classes (one of them being the Kuhnian kind).[11] Thereby the conception of a standard character of revolutionary changes is surmounted.

Hence, in contrast to Kuhn's thesis, it turns out that *not every revolution is a transition between Ps*.

On the other hand, a negative answer also raises the question of whether every transition (from P_1 to P_2) is a revolution, in the sense that in the descending line of Ps we arrive at *transitions of the sort that are not conceived as revolutionary*.

5. Further, the question is posed: Are supplanting Ps incommensurable? Is this a 'Gestalt switch' in which there is no continuity between the Ps (and thus, no possibility of communication between their proponents)?

Against this thesis have been adduced several arguments such as the correspondence principle (in physics) or the possibility of managing conceptual change.[12] Without entering into this discussion, we shall note that in our opinion the propounded *incommensurability* applies only to certain particular kinds of revolution and therefore *it is not valid in general*.[13]

6. Lastly, we could also wonder whether the succession of scientific revolutions follows a linear pattern (as with Kuhn). Our answer is

negative here too, at least as far as it concerns sophisticated (natural) sciences. Thus in physics, as a result of various kinds of synthesis, there appear to some extent as independent paradigmatic systems: mechanics (M), electrodynamics (E), thermodynamics (T), special (SR) and general relativity (GR), quantum theory (QT). Between these so-called completed theories (Heisenberg) various links and interactions (including retroactive ones) are established which constitute something like a set. We have tried to represent this in the scheme (Figure 1). As we see, in this respect also *the real course of revolutions deviates from the model* suggested by Kuhn.

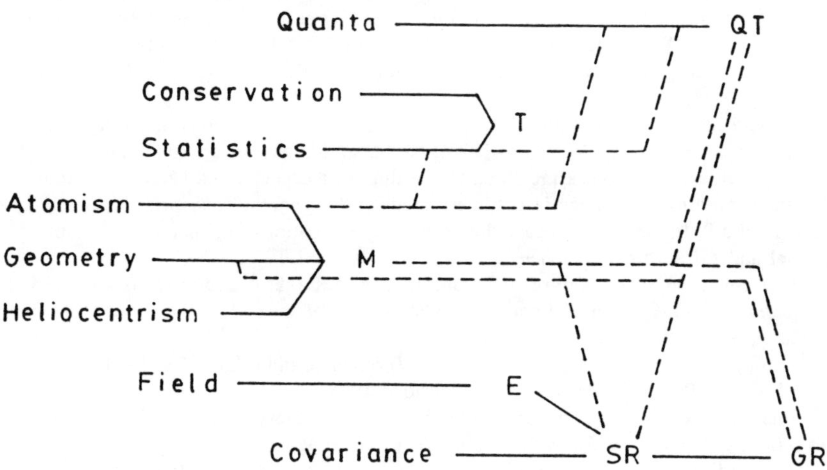

Fig. 1.

This issue also has another aspect. Let us admit that after the first revolution in a branch (= transition from P_1 to P_2) a second one (= $P_2 \rightarrow P_3$) occurs. Because of the noncommutative character of these changes (according to Kuhn) this case is reduced to a standard P-shift. Along with such a development, however, transitions are also conceivable with some continuity between P_1 and P_3 in the sense of Hegel's triadic scheme 'thesis-antithesis-synthesis'. Curiously enough, in order to substantiate a lack of directedness of scientific growth (in the ontological aspect) Kuhn argues that in some important respects Einstein's general theory of relativity is closer to Aristotle's teaching (!) than the views of either of them are to Newton's theory (p. 207).

The general conclusion is clear: Kuhn's model of history of science can only claim a restricted validity.[14] This conclusion differs from the overwhelmingly polarized standpoints of either an unqualified acceptance of the model under consideration or its rejection.

NOTES

1 Kuhn, T. S. *The Structure of Scientific Revolutions*. Chicago 1970.
2 Kuhn distinguishes two meanings of the concept P: as a set of beliefs, values, techniques etc. shared by the members of a given community (sociological); and as models or examples for concrete solutions of puzzles of annual science (Kuhn, T. *loc. cit.*, p. 175). In this way he attempts to overcome the logical circle in the definition of P as what the members of the scientific community share, and the latter as consisting of people sharing a P (*Ibid.*, 176). With this purpose in view, he introduces the community as the practitioners of a scientific specialty (discipline) with a similar educations and professional initiations (177).
3 I should like to note that this was the first of a series of questions that I put to Kuhn during our conversation (Princeton, 1973). He began his answer approximately with the words: "You are right to ask me about that. I did not answer it, and I have to". Actually the understanding of science is predetermined by the general conception, i.e., it plays the role of a P. In the concrete case this leads to qualifications like mature and immature, normal and extraordinary science.
4 Masterman, M. "The Nature of Paradigm", in: I. Lakatos and A. Musgrave (eds.), *Criticism and the Growth of Knowledge*. Cambridge 1970.
5 Kuhn, T. *loc. cit.*, p. 229 ff.
6 Holton, G. *Thematic Origins of Scientific Thought*. Cambridge, MA. 1974.
7 Lakatos, I. *Philosophical Papers*. Cambridge 1978.
8 Ginsburg, V. L. "Notes on Methodology and Development of Physics and Astrophysics", *Voprossi Filosofii*, 1980, No. 12 (Russ.).
9 This kind of relativity of the subsuming of objects and processes under definite (pairs of) categories is examined in Polikarov, A. *Matter and Knowledge*. Sofia 1961, chap. 5 (Bulg.).
10 Einstein, A. *Ideas and Opinions*. N. Y. 1981, p. 225, 240–241; Einstein, A. "Autobiographical Notes", in: P. Schilpp (ed.), *Albert Einstein: Philosopher-Scientist*. Evanston, Il. 1957, p. 37.
11 Polikarov, A. "On Various Kinds of Scientific Revolutions in Physics", *Epistemologia* **XVI** (1993), 213–234.
12 Krajewski, W. *Correspondence Principle and the Growth of Science*. Dordrecht 1977; Pearce, D. *Roads to Commensurability*. Dordrecht 1987.
13 Polikarov, A. "Is There an Incommensurability between Superseding Theories", *J. Gen. Phil. Sci.* **24** (1993), 127.
14 Or, if we use an expression of Kuhn, his conception "seems to have mistaken a few selected elements for the whole and then forced them to serve functions for which they do not perform in practice". Kuhn, T. *The Essential Tension*. Chicago 1977, p. 14.

JÜRGEN RENN

HISTORICAL EPISTEMOLOGY AND INTERDISCIPLINARITY

> The basic ideology of science is mastery of nature, a basic component of modern society. But that mastery must be mastered, by philosophy and politics, which is to say, by insight, activity and change.
>
> Robert S. Cohen[1]

The subject of interdisciplinarity concerns the philosophy and the history of science in a twofold way: on the one hand the emergence, the development, and the mutual relationships between disciplines are, of course, one of the central subjects of research in the history of science; on the other hand philosophy and history of science are fields that are themselves of an interdisciplinary nature. In fact, these two aspects are also closely related to each other, as viewing philosophy and history of science as forerunners of an emerging interdisciplinary field may also serve to emphasize the as yet immature character of our understanding of the phenomenon of interdisciplinarity. A *Festschrift* dedicated to Robert S. Cohen and his interests in the inseparable bonds between the philosophy and history of science provides a natural occasion to deal with both of these aspects: part one of the following essay therefore discusses some features of the development of historical epistemology as an interdisciplinary field, while part two addresses the new technological conditions under which this development takes place today, including some sidelights on the impact of these conditions on the future of interdisciplinary research in general.[2]

1.

Since the emergence of scientific disciplines is a process involving both social and cognitive factors, only an historical theory of scientific cognition which comprises both the social and the cognitive structures of science will be able to cope with the challenge to our understanding of science that is created by its growing interdisciplinary character. Such a theory, which I would like to call "historical epistemology" –

K. Gavroglu et al. (eds.), Physics, Philosophy and the Scientific Community, 241–251.

following a suggestion by Marx Wartofsky[3] – exists, however, if at all, only in very rudimentary forms. In the following I will first indicate some of the difficulties associated with the emergence of such a theory and then attempt to outline some of its general features, both on the basis of an analogy with evolutionary biology.

Although it has by now become customary to speak of the "philosophy and history of science" as if they were already *one* field, the ordinary practice of research and teaching is still far from corresponding to this nominal unification. In spite of the efforts of Cohen and others (in the *Boston Colloquium* and in the *Boston Studies*, to give just two examples), and in spite of pathbreaking examples for combined historical and philosophical studies (among the earliest those of Feyerabend, Holton, Kuhn, and Lakatos), the dichotomy between analysis and description, between a philosophy of science that emphasizes ahistoric methodology and metaphysics, and a history of science that is focused on an account of particulars rather than on the explanation of general structures, continues to dominate the academic reality of these fields. It is hence by no means obvious that philosophy and history of science are ready to give rise to an historical epistemology that would need not only to add the models and instruments of the social and cognitive sciences to the traditional methodological arsenal of the philosophy and history of science, but also seek a theoretical coherency that goes beyond exploiting historical case studies in order to flesh out preconceived philosophical opinions. This account of the present situation may serve to illustrate how urgent it is to seriously acknowledge the interdisciplinary character of the task to build an historical theory of scientific knowledge.

Obviously, however, a well-posed question does not guarantee a satisfactory answer, although the realization of the interdisciplinary character of a question may well be a necessary precondition for searching for answers beyond long-standing disciplinary entrenchments. But are there any indicators that such a search may be successful in the case of historical epistemology? In the following, I will attempt to answer this question by an historical analogy. The present situation of science studies seems to be comparable to that of biology before the advent of Darwin's theory of evolution. In fact, just as science studies today, biology in the first half of the nineteenth century was split into a variety of mutually unrelated subdisciplines, from botany to zoology, from morphology to paleontology. Darwin's theory of evolution radically changed this situation by making it possible to establish systematic conceptual links between "contextual" factors such as the geographical distribution of the species and "internalistic" factors such as the structures described by morphology.[4]

From an analogy between pre-Darwinian biology and present-day science studies we can first of all learn that the opposition between contextualist and internalist approaches in the philosophy and history of science may possibly be a consequence only of our lack of understanding of the mutual relation between external and internal factors. But the analogy may not only help to provide a diagnosis but perhaps also suggest a therapy for improving our understanding of this complex relation. This hope has its basis in the fact that the analogy does not just refer in general to a process of conceptual integration starting from a variety of heterogeneous subdisciplines but that it compares quite specifically theories of development, biological on the one hand, cognitive and social on the other. In fact, at least this much seems to be certain: if it should be possible at all to formulate a theory of scientific thinking in its interdependence with other areas of human culture, then it must be a developmental theory which does not separate scientific from non-scientific thinking in any absolute way, but which deals with the emergence of scientific thinking within its cultural and social contexts. It is only in this perspective that the tensions between the philosophy and history of science, between internalism and contextualism, between the essence and the appearance of science, between its rational and its irrational moments can be conceivably addressed within a single theoretical framework.

Let me now turn to a closer examination of the analogy between a theory of the development of scientific thinking and Darwin's account of the evolution of biological species: what were the conditions that enabled Darwin to forge a conceptual unity out of a bundle of mutually unrelated pursuits? The historization of nature that is the hall-mark of Darwin's achievement was in part based on an experiential base that did not have an obvious bearing on the question of the historical development of species, namely the practice of breeding. Nevertheless, this practice did supply Darwin with controllable experiments on the change of biological life forms that he could exploit for a theory in which not man but nature herself was to take on the role of the breeder. There is indeed a comparable experiential base that can be made fruitful for a developmental theory of scientific thinking in a way similar to Darwin's use of breeding for an understanding of biological development: studies of cognition in the widest possible sense, including psychology, educational research, and cognitive science.

This large field of study offers both a wealth of empirical knowledge as well as a pool of theoretical models which can be brought to fruition in a historical theory of scientific thinking. That the full potential of this field for the creation of an historical epistemology has so far not been realized may be due to the fact that it is not sufficient to simply transfer the results of laboratory studies into the field of history. Just as was the case during the emergence of the biological theory of evolution, it is rather necessary to elaborate a genuinely historical theory that will necessarily challenge also many of the theoretical presuppositions implicit in the laboratory studies.

Although the analogy with evolutionary biology certainly does not provide a blue-print for an historical epistemology, it may nevertheless help to formulate some of the theoretical questions to be answered by such an approach. For instance, in both cases mechanisms ensuring the continuity of the development as well as its innovative powers are required. An explanation of continuity in the development of scientific thinking is suggested by noticing that what hereditary transmission is to biology, tradition, and in particular the transmission of artifacts from one generation to the next, is to cultural evolution, comprising the history of thinking.

If the transmission of the material artifacts involved in scientific thinking processes – understood in a wide sense so as to include language, scientific formalisms and other means of external representation – forms indeed the back-bone of the social accumulation and tradition of knowledge, then this development is necessarily a self-referential and thus irrevocably historical process in the same sense that biological development is. In other words, in both cases we are not dealing with the autonomous development of structures (life forms, forms of thinking) within an invariable external context (the physical environment, nature) but within a context that is itself defined by the development of living or cognitive structures, as the case might be (natural resources are shaped by the biological development, our experience of nature as it is perceived by science is mediated by material artifacts already endowed with cognitive content).

In biology, the explanation of structural innovation and hence the explanation of the discontinuous aspects of development refer to a steady and by itself continuous process of the exploration of new life resources by the living organisms. In the history of scientific thinking, a similar explanation is required which replaces the still widely accepted but essentially empty idea of the "spontaneous generation" of new concepts. The steady exploration of cultural resources, and in particular of the horizon of possibilities offered by the instruments of science as they are available at a given historical moment, should represent an analogous process allowing to explain the emergence of the new in scientific thinking.

In fact, the instruments of science – in the quite general sense introduced above, that is, including any form of external representation of thinking – appear to offer a suitable basis for such an explanation as they open up – by their very nature as parts of the material world – a range of applications that is wider than that for which they were originally intended. In other words, the knowledge that can be acquired by the application of material means exceeds the cognitive presuppositions necessary for their creation (Hegel's *List der Vernunft*). While the production and tradition of these means can be studied using the traditional methods of the historical sciences, the analysis of the cognitive processes corresponding to the exploration of the range of possibilities determined by these means should refer, as I have pointed out above, to laboratory studies in a similar way as the formulation of evolutionary biology had to rely on the study of changing life forms in breeding.

2.

The discussion in part one has emphasized the need for integrating various disciplines in order to prepare the ground for an historical epistemology. However, the question remains as to how such an integration can actually be accomplished, since it presupposes mastering an enormous array of knowledge that is presently divided into disciplinary structures that may well not correspond to the intellectual structures emerging from a synthesis of this knowledge. It may, for instance, be relevant to such a synthesis to systematically compare – with the theoretical questions of an historical epistemology in mind – historical developments in physics, biology, and chemistry from early modern times until today – but who could actually survey a quantity of knowledge that presently has to be subdivided into chunks each corresponding to an entire academic career?

I would like to address this question by way of a brief reflection on the historical development of the disciplinary structure of science, in order to examine whether perhaps some of the conditions of this development have changed or are changing in such a way as to create more realistic conditions for the synthesis mentioned above. The disciplinary organization of science as we know it today is the result of an historical development, the onset of which can be roughly dated to the end of the eighteenth century. This development depended both on external conditions such as the changing relation between science and industrial production and internal conditions such as the rapid accumulation of knowledge and the stabilization of area-specific knowledge structures. Furthermore, the emergence of disciplines would have been impossible without the social mechanisms mediating between the individual and the social aspects of knowledge systems, such as the educational system and scientific journals. While the genesis of disciplines enhanced scientific productivity in an unprecedented way, the negative aspects of disciplinary specialization can be recognized already in the nineteenth century. They are visible in the increasing difficulty to communicate knowledge across disciplinary borders and in the failure of attempts to create a scientific world view integrating the knowledge produced by the various disciplines.

It now turns out, however, that precisely on the level of the mechanisms that mediate between the individual and the social aspects of knowledge systems, recent technical developments of information processing have created the preconditions for new forms of the integration of knowledge. The electronic representation of knowledge – for instance in the form of electronic texts – opens up quantitatively and qualitatively new possibilities of access to knowledge which will in the long run also change the social structure of its organization and hence the disciplinary structure of science. They thus provide a case in point for the role of scientific instruments emphasized in part one. In fact, if human thinking is conceived as a process to which the relationship between mental structures and their culturally determined external representations is crucial, then it becomes clear that the introduction of electronic information processing may affect the very nature of scientific thinking processes.

Some of these implications can be strikingly illustrated by referring to the *Perseus Project* of Crane and collaborators.[5] By making not only texts of Greek literature electronically available but also images of vases, of archeological sites, maps etc. it furthers the interdisciplinary study of Greek culture in a variety of ways. First of all, the same electronic corpus of material provides access to documents which are still primarily evaluated by different disciplines, such as philology, archeology, or anthropology. Secondly, the electronic access enables a scholar to gain an overview over a much larger body of material in a much shorter time than it has hitherto been possible. In other words, the size of a "digestible unit" of information increases dramatically, and hence for instance the time it takes to test a working hypothesis on a large body of material. But perhaps the most incisive innovation if compared to traditional forms of the representation of knowledge lies in the "active" character of the new forms of access. By transferring certain mental operations such as searching, grammatical analysis, or translation to the computer, problems which traditionally presupposed a technical competence that could only be acquired by a disciplinary training (such as mastery of the grammar of the ancient Greek dialects) now become intellectually accessible also to the non-specialist.

The example of the *Perseus Project* also demonstrates, however, that the new possibilities created by technological development can only be used if they are combined with new forms of the intellectual organization of knowledge. In particular, only the integration of the source material into an hypertext environment enables the user to navigate through an otherwise overwhelming amount of information. Whereas a text is an essentially linear representation of non-linear intellectual structures, a hypertext is a representation of such intellectual structures which itself comprises non-linear features, although it may ultimately also be based on the textual representation of thinking. Almost every traditional book contains features of a hypertext, even if still in a rudimentary form. A scholarly edition, for instance, embeds primary sources into a hypertext structure, which is realized by tables of content, headnotes, footnotes, indexes, bibliographies etc. A more elaborate hypertext structure, as it can be realized in an electronic medium, not only provides a network of internal and external references to the sources, but it also allows to overcome some of the basic limitations of access to a linear text structure, limitations that become the more severe the larger the body of information.

The revolutionary consequences of realizing hypertext structures in a new medium can best be illustrated for the example of an index. Because an electronic hypertext link is not only a direction sign but also a vehicle, an electronic index can be much more efficiently used as an actual access to the text than its traditional counterpart because now "browsing through an index" becomes a feasible way of exploring a large text body. In fact, by way of an electronic index, any part of a text can now be "linked" to any other, semantically related part of the text. Conceptually even more important, however, is the possibility of constructing much more complex index structures than is presently customary. A traditional index has a flat tree structure necessitating a non-semantical ordering – alphabetical ordering – at the top level. Queries by a scholar are, however, as a rule related to semantic structures and must therefore first be mapped, in a rather artificial way, onto this mechanical ordering. An electronic index, on the other hand, may incorporate several levels of a semantic hierarchy and thus directly represent aspects of the conceptual organization of the knowledge to which it provides access.

But let us return to our original question which – in the light of our reflection on the technological conditions of the organization of knowledge – now becomes the question as to how these new technological developments may help to prepare the knowledge base required for the emergence of an historical epistemology.[6] In fact it seems rather obvious that the new information processing technologies radically change the conditions of access to the sources of the history of science. Traditional history of science still lacks the broad base of standard editions, translations, textbooks etc. that are the heritage of some of the more mature historical disciplines. The creation of electronic archives will help to overcome the difficulties related to the availability of primary material and with them the phase in which history of science used to be a field of idiosyncratic expertise, divided into Galileo studies, Darwin studies, Einstein studies, etc. In fact, by making a broad array of primary sources easily available to every interested scholar, electronic archives will bridge many of the obstacles on the traditional route from the location of documents in archives to their intellectual evaluation by the scientific community at large. They will hence radically change the nature of the sometimes narrow-minded disciplinary specialization within the history of science.

A broader text base is in fact a crucial condition for the development of an historical epistemology, as this approach depends on the evaluation of a larger set of source materials than traditional chronological, biographical, or descriptive accounts. The contextualist approach to the history of science, for instance, confronts the history of science with new qualities of documentary sources, such as visual images, and makes it necessary to take a broader selection of sources into account than it has hitherto been customary. Pursuing questions of the cognitive sciences within the history of science, on the other hand, lays an increased emphasis on the use of primary material such as drafts, sketchy calculations, etc., which can be seen as immediate representations of thinking processes. High quality electronic reproductions of the originals will allow to study also those features of such documents that might have been filtered away in a traditional editorial process.

This shift towards archival raw materials is, however, associated with some potential drawbacks of an electronic publication in comparison to, say, a traditional edition. Having to confront electronic reproductions of originals rather than traditionally edited texts requires, it seems, an increased amount of expert knowledge of manuscripts, handwriting, language, etc., and possibly also the need for technical knowledge that is usually only available to computer experts. While these features may indeed counteract the potential of electronic archives to further the development of interdisciplinary competence, they can in fact be more than compensated if the new technologies are not just considered to be a new medium for reproducing sources but as a new and powerful tool for structuring and communicating scholarly information, as is illustrated by the example of the *Perseus Project*.

In exploring these new intellectual possibilities historical epistemology may even play a pioneering role for the impact of the new technologies on the organization of scientific knowledge in general. If electronic media will in the not-too-distant future become the dominating carriers of scientific communications, the question of how to structure scientific knowledge within and across disciplines will determine in a decisive way the use of the benefits of the new technologies. In a sense, the history of scientific disciplines would then return to its point of departure: while the system of scientific disciplines once replaced traditional models of knowledge classification,[7] classificatory enterprises would then take on again a central role in organizing a pool of information whose internal connectivity will have rendered largely obsolete the classical concept of a scientific discipline.

In this context an historical epistemology may contribute its share to the intellectual organization of scientific knowledge, e.g., by identifying the cognitive models shared by different branches of knowledge. As a side-effect, historical epistemology might then also help to definitely reverse the trend in the philosophy of science towards content-independent methodology. This trend, which began with early Neo-Kantianism and continued with analytical philosophy, may be understood as a response also to the disciplinary specialization which shifted the perspective on the unity of science from metaphysics to the realm of methodological reflection. It becomes, however, more and more obvious that it is not in this metatheoretical realm that the problems of a synthesis of scientific knowledge can be solved. As it turns out, these problems are no longer of a purely academic dimension but a matter of the intellectual and practical mastery of nature, a mastery without which the very survival of the human species may be in danger.

Max-Planck-Institute for the History of Science, Berlin

NOTES

[1] See Cohen 1985, pp. 204–205.
[2] The following sketch of the project of an historical epistemology is largely based on discussions over the last years with Peter Damerow, Gideon Freudenthal, Wolfgang Lefèvre, Peter McLaughlin and other members of a research colloquium based at the Max-Planck-Institute for Human Development and Educational Research in Berlin; see, e.g., Damerow, Freudenthal, McLaughlin, Renn 1991.
[3] See Wartofsky 1985.
[4] The analogy between historical epistemology and evolutionary biology sketched in the following is based on the historical assessment given of Darwin's discovery in Lefèvre 1984.
[5] See Crane 1991 for penetrating reflections on the impact of the new technologies on the humanities.
[6] For a discussion of the technical implementation of electronic archives in the history of science as they are advocated in the following, see Renn, Damerow, and Galluzzi 1992, prepared with the assistance of Antonella Krige, Jochen Schneider, Martin Schreiber, Martin Warnke, and others.
[7] See Stichweh 1984.

REFERENCES

Cohen, Robert S. "Constraints on Science," in *A Portrait of Twenty-five Years. Boston Colloquium for the Philosophy of Science 1960–1985*, ed. Robert S. Cohen and Marx W. Wartofsky. 198–205. Dordrecht, Boston, Lancaster: D. Reidel Publishing Company, 1985.

Crane, Gregory. "Composing Culture. The Authoring of an Electronic Text," *Current Anthropology* **32**(3) (1991).

Damerow, Peter, Freudenthal, Gideon, McLaughlin, Peter, and Renn, Jürgen. *Exploring the Limits of Preclassical Mechanics*, New York: Springer Verlag, 1991.

Lefèvre, Wolfgang. *Die Entstehung der biologischen Evolutionstheorie*, Frankfurt/M, Berlin, Wien: Ullstein, 1984.

Renn, Jürgen, Damerow, Peter, and Galluzzi, Paolo. *The Galileo Einstein Electronic Archives Project*, NSF Grant Proposal, 1992.

Stichweh, Rudolf. *Zur Entstehung des modernen Systems wissenschaftlicher Disziplinen. Physik in Deutschland 1740–1890*, Frankfurt am Main: Suhrkamp, 1984.

Wartofsky, Marx W. "Perception, Representation, and the Forms of Action: Towards an Historical Epistemology," in *A Portrait of Twenty-five Years, Boston Colloquium for the Philosophy of Science 1960–1985*, ed. Robert S. Cohen and Marx W. Wartofsky. 215–237. Dordrecht, Boston, Lancaster: D. Reidel Publishing Company, 1985.

MATHIAS WALLNER

AMERICAN CREATIVITY RESEARCH IN A BIPOLAR WORLD: A LOOK AT ONE CHAPTER IN WORLD HISTORY AND HISTORY OF SCIENCE*

1. INTRODUCTION

The end of the cold war not only presents us with new political and economic problems; it also poses questions in the fields of philosophy, history and the sociology of science that have previously received very little attention.

Specifically, the impact of the historical conflict between East and West upon scientific development in both societies has been given little attention by these disciplines, although the problem area "Science and Society" has for quite some time been an object of research in the sociology of science. While each side understood, as a rule, the underlying micro and macro-sociological conditions of its own social system, neither was particularly well-informed about the conditions prevailing in the rival society.

One cause of this deficiency is probably that such influences were so evident in the postwar development of science and engineering that they hardly seemed to merit investigation. After all, the influence of the cold war had often enough accelerated research in strategic areas, a fact true not only of military research.

It is therefore all the more interesting to examine the influence of the East-West conflict on progress in the humanities and the social sciences. Did it guide the course of research and accelerate progress here as well? Are the humanities and social sciences indebted to the cold war for new problems, new areas of knowledge, new approaches and directions of thought, new emphases in research, new disciplines or even a significant expansion in our body of knowledge?

Or put more provocatively: have the fallen Communist system and ideology of the East actually been a primary motivating force in Western humanities and social sciences? This question is interesting especially from the standpoint of science policy and theory of science, because answers to it could help to clarify the possibilities and the limits of

K. Gavroglu et al. (eds.), Physics, Philosophy and the Scientific Community, 253–272.

scientific development within the economic and political structures in the countries of the western world.

However, such answers require empirical data concerning the development of the humanities and social sciences. The present article is intended as an empirical contribution in the form of a case study of the postwar development of an area of research that arose within American psychology, namely research into the phenomenon of creativity.

The development of creativity research is particularly germane to this context because it is a classic example of the influence the cold war has had upon progress in science – "classic" because the object of this research was pivotal to the survival of each of the societies within the context of their rivalry. After all, the collapse of the eastern bloc was caused primarily by its failure to sufficiently promote and exploit creativity in science, technology, economics and politics.

2. THE BIRTH OF CREATIVITY RESEARCH

Certain theoretical barriers to the establishment of creativity research as an independent field of research were evident within American universities at the outset. One of these was the dominance of behaviorism, which banned virtually all analysis of individual epistemic processes.

An even greater theoretical hurdle, however, was to be found in the traditional monolithic understanding of intelligence. Prevailing theory dictated that creativity was not a valid object of independent research, since it was presumed that the concept of intelligence could explain virtually all aspects of mental processes and that intelligence tests were the correct instrument for recording them.

> The intelligence test became the bastion of the tough-minded empiricists in the new science, while the 'creative mind' remained the subject for anecdotal discourse of Hegelian romantics, Bergsonian evolutionists, and tender-minded and scientifically suspect antibehaviorists.[1]

The first attempts to extract creativity from the concept of intelligence can be found in three approaches arising the 1930's.

The first "way beyond the IQ" was taken by L. L. Thurstone, who moved from engineering into psychology and founded a psychometric research laboratory at the University of Chicago. Thurstone localized 20 creativity-related intellectual traits that were not registered by intelligence tests.[2]

Another leading proponent of psychometric factor analysis was J. P. Guilford, who pursued the second avenue to creativity research and became its true Nestor. Guilford developed his factor analyses of personality traits independently of Thurstone in 1940 at the University of Southern California, San Diego, with the intention of isolating individual creative abilities. This research was temporarily interrupted by W.W.II – Guilford was inducted into the US Army Air Corps and led research into criteria for choosing bomber crews.[3]

The third approach to creativity research was founded by H. A. Murray, who worked at Harvard University from 1926 to 1962. His most well known efforts are to be found in a clinical study of 50 Harvard students in which the "Thematic Apperception Test" was employed as an observational instrument for the measurement of the level of fantasy-development.[4] Murray's personality studies emphasized medical, sociological, historical, cultural anthropological and psychological aspects.

This interdisciplinary character was promoted by Harvard President J. B. Conant, who was himself clearly aware of the importance of fantasy and the fantasy-endowed scientist. He had for this reason endeavored in his "General Education Program" both to promote the dissemination of this insight and to support the training of top scientists. The staff teaching natural sciences, humanities and social sciences in this program met several times a year in the "Shop Clubs" Conant organized. In addition to Murray, J. Bruner and T. Parsons also participated in these meetings in the 1930's.[5]

At the "Pareto seminars" organized by L. J. Henderson between 1932 and 1934, Murray met Parsons as well as his student R. K. Merton and other sociologists.

These contacts lent Murray's concept of the creative personality a marked sociological character, which is also revealed in the writings of students of his, who later became proponents of creativity research. Most notably among these are M. I. Stein, D. W. MacKinnon, R. S. Crutchfield and N. Sanford.

They worked during W.W.II alongside Murray on the Assessment Staff of the Office of Strategic Services (OSS) and applied the methods of personality evaluation developed at Harvard to the selection of persons who were to be employed overseas in irregular warfare as spies, counterespionage agents, and as leaders of partisan groups.[6]

The few relevant psychological investigations concerning aspects of creative thought, intelligence and the creative personality that were carried out after the war as of 1946 were at first completely uncoordinated efforts and enjoyed neither very much attention nor financial support. However the situation changed fundamentally in 1949. This period saw the integration of the various disparate individual research projects into a single path of research, as well as the creation of an institutional research context promoted by both government and private industry.

The growing support for creativity research lay not, however, in the academic, but rather in the political and military arenas. This rise in prestige and support was above all due to the growth of the Soviet Union into a superpower, which was able to relatively quickly overcome the devastations of war, extend its sphere of influence to all of East Europe and break the American atomic weapon monopoly, thereby precipitating in 1946 the transition from postwar cooperation to the cold war.

Subsequently, the then-seminal scientific-technological revolution became the focal point, especially in the area of weapons technology, of the East-West conflict.

On the basis of the experiences of the Manhattan Project, leading American political figures believed quite correctly that this would be the decisive field of battle in the confrontation with Communism. The result was the immediate jump in demand for warriors in this battle, i.e., for the producers of the required top scientific and technological performance.

The underlying rationale was as follows: The scientific-technological competition is above all a competition of the intellectual potentials of both sides, whereby it is usually only a few highly gifted scientists who can deliver the level of performance required to gain preeminence over, in the case of the USA, the Soviet Union. These champions of science thus hold the fate of the nation in their hands. In Conant's opinion, ten second-class scientists are just no replacement for one such first-class scientist. It was therefore of greatest importance to discover, promote and exploit the capabilities of as many such talents as possible. Moreover, the United States suffered a great deficiency of scientists relative to the international competition after W.W.II.[7]

Guilford clearly described this situation, which was subsequently to lead to the elevation of creativity research:

> The Second World War had called forth great efforts toward innovation in research and development, culminating in the atomic bomb. The coming of peace that was no peace left us in the cold war, which called for ever-accelerating efforts in a contest of intellects. Inventive brains were at a premium, and there were never enough. We were on the eve of the space age, and rockets were already taking trial flights, stirring our imaginations of things to come. The stage was well set, then, ready for the psychologist to play his proper role in trying to fathom the creative person and his creative processes.[8]

In order to fill the "scientific manpower pool", the United States Atomic Energy Commission funded a scholarship program for gifted scientists just after the war. However, the selection of candidates proved difficult because no exact criteria existed for the evaluation of potential intellectual performance, and because standard IQ-assessment offered no means to "separate the chaff from the grain."

The pragmatic stumbling-block of finding appropriate selection procedures was the major reason that government offices and private foundations alike began making funds available for research into the identification, analysis and furthering of creative personalities among scientists, engineers and officers. Thus arose the material support necessary to unite the disparate efforts of creativity researchers into a systematically organized field of research and to guarantee their further institutional support.

The establishment of creativity research was initiated in 1949 at two Californian universities. With financial support from the Rockefeller Foundation, D. W. MacKinnon and N. Sanford grounded the Institute for Personality Assessment and Research (IPAR) at the University of California, Berkeley.

Here they intended to develop further and to apply the OSS-assessment method as a research tool in order to answer two questions that had long been neglected:

(1) What are the characteristics of persons who are highly effective in their personal lives and professional careers? and

(2) How are such effectively functioning persons produced in our society?[9]

At the same time, a research project studying the training of military leadership personnel under the direction of Guilford was financed by the Office for Naval Research of the US. Navy at the University of Southern California. "Its primary goal was to understand human intelligence in general, including the thinking processes of individuals when they are in the act of creative production."[10]

It was certainly no coincidence that Guilford was voted President of the American Psychological Association in 1949. Shortly afterwards, the Soviet Union broke the American nuclear weapons monopoly, an event that served to spotlight the scientific progress of the USSR, as well as the significance of scientific creativity within the framework of the confrontation of systems.

Immediately thereafter Guilford employed the prestige of his office at the APA-congress of 9/5/1950 to direct attention to the field of creativity, to stimulate new research and to integrate the disparate directions of research in the field. In a speech that can justifiably be viewed as the starting signal of creativity research at American universities, he expressed his regret at the neglect to which this area had long been subjected, defined the creative personality as *the* problem facing psychologists, offering as support for this claim the great need for creative scientists, engineers and other leadership personnel in industry and government.[11]

If we view the ascent of creativity research as described here not merely from an academic standpoint, but rather from within its socio-historical context, it immediately becomes clear that the decisive force behind its formation originated in the competition between the two Æsuperpowers. For it was only through the comparison of the intellectual potential of the USSR and the USA within the context of the given historical conditions that Guilford's appeal for research into creativity could achieve the kind of results that only two decades before were unattainable by earlier proponents like E. D. Hutchinson.[12]

3. TENDENCIES IN THE FIFTIES

In 1950 the US Congress established the National Science Foundation for the purpose of discovering and furthering American scientific potential. One of its first duties was to conceive a scholarship program for postgraduate study, which also absorbed the earlier USAEC-scholarship program, taken over by the NSF in 1952.

NSF-leadership immediately announced the priority of research towards more effective methods of candidate selection according to the principle of "separating the men from the boys." The responsibility for this research was invested in the National Research Council of the

National Academy of Science, which named C. W. Taylor as the first director of the project. Taylor expressed the goal of these studies clearly: "We were seeking future scientists who might make a major difference in the nation's scientific progress."[13]

Because it was at first believed that the strength represented by the greater numbers of Soviet scientists could be paralyzed by achieving an American scientific manpower pool of higher quality, Taylor investigated criteria of quality applicable to scientists, in order to run down as many candidates as possible in this hunt for "Wunderkinder." Scientific talent has since been searched for and promoted like a mineral natural resource.

Taylor continued this "prospecting" for talent in 1954 at the University of Utah, Salt Lake City, where he began to prepare the first National Research Conference on the Identification of Creative Scientific Talent under the auspices of and with the financial support of the NSF. This first conference was then held in 1955.

The theme of the conference alone made evident the narrow focus of creativity research in the early 1950's. At the same time, research elsewhere did not necessarily concentrate exclusively on the traits of the creative personality. The inner logic of the analysis of the multifaceted phenomena creativity was guaranteed to lead into other problem areas. Because the creative performance of the subjects selected was the result of processes, the study of the course and features of such processes was inescapable.

Moreover, the processes under study themselves took place within a particular professional and social milieu. Ultimately the personality traits identified could only be understood as the result of personal background, upbringing and education. Thus the field of research expanded to include the process of creative thought and the social environment of creative people.

One pioneer in the analysis of the microsocial milieu of creative personalities was A. Roe. After completion of a series of personality studies, she gained attention with the publication of a paper dealing with the social background, family relationships and religious beliefs of 64 scientists.[14]

A similarly pioneering role was played by M. I. Stein in the areas of macrosocial conditions of creativity and thought processes of the individual. In a lecture at the Northwestern University Centennial Conference on Anthropology and Psychiatry in 1951, he emphasized that creativity is always defined by an inner and an outer orientation. The latter is the cultural environment of the individual. Stein included here the factors of educational chances, upbringing, the tolerance of the society towards deviations from the status quo or its pressure to conform, and the society's attitudes and philosophies. These are the cultural factors which, according to Stein, promote or hinder creative thinking.[15]

This kind of sociopsychologically and sociologically oriented study of creativity proliferated over the next years. C. R. Rogers criticized American culture at a creativity conference at Ohio University, Granville in 1952. He claimed that it tended to promote stereotypes and conformity in education, recreation, family life, science and industry.[16]

The sociologist M. Tumin also sees the release of creative potential endangered in a society that pressures the individual to achieve a certain social status by means of conformity.[17]

In this vein, R. May complains in 1953 that the typical American is "externalized"; Americans live as if they were guided by a kind of radar that informs them what others expect of them and are characterized by passivity, conformity and apathy.[18]

Similar positions were also clearly expressed at the IPAR. In 1955, the foundation's later Director, R. S. Crutchfield, criticized the pressure to conform evident in many American institutions, which in his view undermines the creative powers of the individual.[19]

In 1958, at the interdisciplinary Symposium at Michigan State University, East Lansing, H. A. Murray came to the drastic conclusion that the United States was suffering from an emotional deficiency, a paralysis of fantasy and an addiction to superficial suffering.[20] H. H. Anderson summed up one result of the conference, stating that all who participated agreed that the social environment can either promote or it can hinder creativity.[21]

These few examples alone serve to show that the German literature is inaccurate when it alleges that American creativity research exhibits minimal interest in the social aspects of creativity.[22]

On the contrary, such claims are easily refuted solely by considering that most American representatives of this field, e.g. H. A. Murray, A. Maslow, C. R. Rogers and others, are fully entrenched in the tradition of humanistic psychology. They focus not upon the isolated individual, but rather upon the whole personality trying to develop within the social environment. Proceeding from this theoretical standpoint, the urgent need to record and evaluate social obstacles to creative thinking was inescapable.

The continually growing interest in the social determinants of creativity also became obvious at the creativity conferences at the University of Utah. Looking at the topics and contents of contributions to the first three conferences in 1955, 1957, and 1957, we can match them with definite problem areas.

The order of importance of the problem areas in 1955 and 1957, arranged according to the number of contributions in each, is as follows: (1) The traits of creative personalities, (2) The concept and the criteria of creativity, (3) The traits and phases of the creative process, (4) The prediction and development of creativity, and (5) The characteristics of the social environment that promote or hinder creativity.

Of the fifteen lectures given at the second conference in August of 1957, only two were directly concerned with the social context of creativity, and two others dealt with it only peripherally. The discussions on this topic were held in a subgroup of the conference whose results were so meager that they were not even able to complete a report.[23]

At the third conference in June of 1959, the situation had changed radically. Here, 5 out of 19 contributions directly addressed the social aspects of creativity, thereby, from a statistical standpoint, catapulting this problem area from last to second place, a result which of course makes no statement concerning the quality of the research discussed.[24]

But these statistical results alone do make one conclusion inescapable: the focus and emphasis within the field of creativity research clearly took a new course in the period from 1957 to 1959! The initially only gradual increase in the number of sociopsychological and sociologically oriented analyses was overshadowed by a dramatic surge of interest in questions of the social and cultural conditions surrounding creativity. It is not an exaggeration to characterize this massive swing of attention towards the social aspects of creativity as a *sociological turning point* in the course of creativity research.

But we must ask ourselves where the causes for this qualitative reorientation of the whole direction of research are to be found.

4. SPUTNIK AND CREATIVITY RESEARCH

Indeed, the path to the sociological turning point in creativity had been prepared long before by the growing realization that the great social need for creatively-thinking individuals was faced with equally great obstacles to their development. In my opinion, however, the turning point was first truly forced upon the scientific community by an event that served to acutely spotlight the dangers of the newly-discovered obstacles to creativity threatening the survival of the nation in its confrontation with the USSR: the launching of the first Soviet Sputnik on 4 October 1957!

This event served to bring home the message that the Soviet Union had managed to catch up to the USA in such areas as weapons and space technology, and in particular in rocket technology and electronics, areas which are well known to be dependent upon the productive capacity of the entire economy.

The resulting "Sputnik shock" had two main effects on creativity research.

First, it led to intensified governmental support and consequently to a substantial quantitative increase in research projects. The exploitation and promotion of the creative potential of the populace was elevated to the position of highest national priority. As T. A. Razik aptly described this process:

> Sputnik catalyzed the realization for Americans that further life and development would depend on having many creative men at work in a constant effort to transcend what had already been done with accomplishments still more novel and powerful . . . In the presence of the Russian threat, 'creativity' could no longer be left to the chance occurrences of the genius; neither could it be left in the realm of the wholly mysterious and the untouchable. Men *had* to be able to do something about it; creativity *had* to be a property in many men; it *had* to be something identifiable; it *had* to be subject to the effects of efforts to gain more of it. . . . Research on creativity became legitimized as a properly serious concern of the military, government, and industry.[25]

Second, Sputnik shock affected the substance of creativity research by raising questions regarding the connection between prevailing political and cultural conditions within society on the one hand, and the possibilities engendered by these factors for exploiting the society's creative potential on the other. Sputnik seemed, namely, to contradict the widely accepted attitude already expressed by H. S. Truman: "I think that creativity grows best where there is freedom of the individual and free government."[26]

If Truman was right, the question became unavoidable as to how an anti-creative system like communism could within one short decade succeed in releasing enough creativity among scientists and engineers to enable them to pull ahead of the free societies of the West in the sensitive area of scientific and technological progress.

N. E. Golovin, technical consultant for space research and technology to the presidential advisor for Science and Technology, had already in 1959 warned of the dangers of overestimating one's own, and thereby underestimating one's opponent's, ability to exploit creative potential. As he explained, "It is probably dangerous folly to discount the creative effectiveness of Iron Curtain physical scientists and engineers on the presumed grounds that their abilities have been stunted by authoritarian political indoctrination."[27]

Rather, one should examine what has been done wrong in the development and exploitation of one's own intellectual potential.

Guilford was among those who posed the uncomfortable question:

> It is obvious that we are in a tight race with Russia with respect to these technological advances. It is almost obvious that we are in competition with Russia also with respect to education, and in fact, with respect to other areas of social and economic progress. In many respects our adversary has shown more initiative and imagination than we have. It is not likely that our potential intellectual resources are inferior to those of Russia. We may well ask ourselves, therefore, what we are doing that is wrong and how we may promote better use of the intellectual resources that we have.[28]

In other words: The search for the obstacles to creativity in education, upbringing, prevailing social attitudes, etc., and their removal by means of corresponding reform movements was more than ever a question of the survival of the American way of life.

Moreover, Sputnik directed attention to the fact that science is a social institution. In this regard N. W. Storer believes that ". . . the advent of Sputnik I and the ensuing 'crisis' in American science and science education finally completed the definition of science as a 'social problem.' "[29]

This judgment, too, points out that Sputnik shock acted as a multiplexed transmitter, its signal suddenly opening the gates of transition, not only instigating reform programs in education and science, but also the full-blown sociological change in course, both in creativity research as well as in the philosophy of science. The ousting of the "accepted view" by the "new approach" is, in my opinion, also a result of the intensifying competition between the two systems because it made ever more apparent the nature of science as a social system.

Thus, a hitherto unseen connection between the two disciplines becomes evident and we realize that the nearly simultaneously occurring transition in creativity research was thus no accident.

5. HOW DO KUHN AND MERTON RELATE TO CREATIVITY RESEARCH?

Increased interest in the macrosociological conditions of creative thinking was reflected in the development of new socio-psychological analyses. But these necessitated strong interdisciplinary cooperation because of the paucity of theoretical and methodological tools for the investigation of the social determinants of creativity available to creativity research at the end of the 1950's. It is therefore quite understandable that the sociologist of science, N. Kaplan, and the historian of science, T. S. Kuhn, were invited to participate in the 3rd Creativity Conference held at the University of Utah.

Kuhn's invitation was extended primarily on the basis of the leaning towards psychology evidenced in his occupation with Gestalt psychology and the works of Piaget during the preparation of his book, *The Structure of Scientific Revolutions*. However his contributions was destined to be disappointed to the psychologists; from the start in the Introduction, he expressed doubt concerning the value of such discussions.[30]

It must of course be admitted that none of Kuhn's works was intended as a substantive contribution to the explanation of scientific creativity and the conditions thereof. He limited his view of the impact of creativity on the development of science primarily to the revolutionary phase of the paradigm change; nor did he explain how innovations arise. He only described the way in which these are accepted by the scientific community.

As M. W. Wartofsky correctly observed:

> . . . Kuhn also left the nature of creativity in scientific thought untouched, though he fully took into account its subsequent role in science. . . . In fact, like Polanyi, Kuhn left the process of scientific change to the domain of the non-rational, if not indeed the irrational.[31]

This criticism refers to the limited nature of Kuhn's direct influence on the development of creativity research. His contribution is not to be found in the area of reaching an understanding of creativity, but rather in acting as a stimulus for further research into the relations among the cognitive and social aspects of creativity, as well as their macrosociological determination.

Kuhn achieved the latter effect with the report of the conference committee on "Environmental Conditions Affecting Creativity", which he compiled together with Kaplan. Here they reach the conclusion that

> Unlike the problems of criteria, predictors, and education in the sciences, the problems of environmental conditions have hardly been studied. There are several useful discussions of the socio-economic conditions that promoted the Scientific Revolution of the seventeenth century, as well as a few recent studies of working conditions, morale, and productivity in individual industrial and government laboratories. But almost no other guideposts exist. The man who wishes to understand, for example, why some cultures, like our own, have been particularly productive of technological innovation but conspicuously backward in contributions to the basic sciences can scarcely find a basis for even an educated guess . . . Yet the ability to answer questions like these will inevitably affect the nation's ability fully to utilize its creative scientific potential.
>
> Under these circumstances our committee's first and most important recommendation is a plea. Research on the general environmental conditions – cultural, professional, and institutional – conducive to first rate scientific research needs major encouragement. We are aware of no area in the social sciences where research is simultaneously so vitally needed and so much neglected.[32]

Although Kuhn and Kaplan's contribution in this area remained humble, their suggestions were to produce significant long-term effects revealed in related studies in the 60's and frequent references to Kuhn and Kaplan by researchers in the area. In 1969, among the problems dealt with in the mushrooming number of publications on creativity, the problem of the "social environment" reached a proportional ranking at 16% of second place.[33] In the same year, C. W. Taylor directly credited the plethora of publications dedicated to the topic "Climate for Creativity" to Kuhn and Kaplan's committee report of 1959.[34] From this we can conclude that, against the historical background of the sensitizing effect of Sputnik shock, their suggestions acted as a theoretical catalyst, both stimulating the actual boom in sociopsychological and sociological creativity studies in the 1960's, and setting the course of their development.

Looking more deeply behind the scenes at Kuhn and Kaplan and their urgings to intensify the investigation of the social conditions of creativity, we discover another personality who has received little acclaim in this context: the sociologist of science, Robert K. Merton! One can safely assert that Merton played a key role in these events. The committee report of 1959 referred implicitly to his pioneering work in the area of the socio-cultural and socio-economic environmental conditions of scientific creativity, namely his dissertation dealing with "Science, Technology & Society in Seventeenth Century England," which he finished in 1935 and published for the first time in 1938.[35]

Considering the topic of this contribution, it is important to examine the historical background of his work. The volume "Science at the

Crossroads", containing articles submitted to the II. International Congress for the History of Natural Science and Technology, appeared in London in 1931, just as Merton had completed his studies in sociology and was in search of a topic for his dissertation and empirical material concerning the historical relations between science, technology and society.

One of these articles, the essay on "The Social and Economic Roots of Newton's 'Principia' ",[36] submitted by Soviet physicist Boris Hessen, was a source of some excitement. Hessen attempted to carry out an historical-materialistic analysis of scientific development which, despite the exaggerations and stigmata of popular economics contained in it, presented a clear challenge to western philosophy and history of science. As a result, the essay was enthusiastically received by representatives of the English Marxist scientific left, such as J. D. Bernal, L. Hogben, J. B. S. Haldane and J. Needham, while it was sharply criticized by anti-Marxist historians such as Ch. Singer, G. N. Clark and H. Butterfield. Merton found Hessen's article "immensely stimulating"[37] and decided in 1932 to select an analogous subject for his dissertation.

It must be noted, however, that Merton's understanding of scientific development was not colored exclusively by Marxist attitudes; he was much more strongly influenced by his teachers, G. Sarton, P. Sorokin, A. N. Whitehead, T. Parsons and L. J. Henderson.

We can see these contradictory theoretical influences reflected in his dissertation in the form of the "Merton Thesis." On the first page, he investigates the connection of science and technology with the religious values that guided the rising bourgeoisie of the 17th Century.

Merton agrees with M. Weber's view of the role of the Protestant ethic in the rise of capitalism in Europe, and sees English Puritanism as a powerful source of development in the sciences. However, it was not Puritanism that was responsible for the selection of subjects of research.[38]

For this reason, Merton also investigated the relations between science, technology and economic development. In this connection, he showed that approximately 30% to 60% of scientific research was influenced by "socio-economic needs" that resulted from developments in the areas of shipbuilding, mining, metallurgy, military technology and the textile industry.[39] It is this part of Merton's dissertation that most closely relates to Hessen's essay. Merton himself admits this fact: "In the discussion of the technical and scientific problems raised by certain economic developments, I follow closely the technical analysis of Professor

B. Hessen in his provocative essay . . ."[40] In his conclusion he writes: "We have already indicated that the preceding three chapters of the present study, despite certain differences of interpretation, are heavily indebted to Hessen's work."[41]

Merton was not generally opposed to Marxist thought; rather, he made use of it to a certain extent while working out his position on the sociology of science. Marx's analysis of the connection between science and society bears out the legitimacy of such an approach.

Merton clearly describes his basic attitude to Marxism:

> I have long since abandoned the struggle to determine what 'Marxism' is or is not. Instead, I have taken all that I find good in Marxian thought – and that is a considerable amount – and neglected conceptions which do not seem to me to meet tests of validity.[42]

Merton's ambivalent attitude towards Marxism can also be discovered in his work in the 1950's on the socially conditioned nature of multiple scientific discoveries. He counts Marx among the theoretical precursors of a theory, still in the stage of germination, concerning the social determinants of discoveries, and finds support for this claim in one of the "Altersbriefe" of F. Engels.[43]

Broadly viewed, we can see Merton's interpretation of the social determination of the appearance of scientific discoveries as an attempt synthesis of Marx, Engels and Hessen on the one hand, and of M. Weber, M. Scheler, K. Mannheim, W. F. Ogburn and D. S. Thomas on the other.[44]

Merton quite correctly perceives the value of investigations into multiple discoveries, carried out according to such a theory, in its nature as a strategic research site for sociology and psychology of science. Results obtained in this way could contribute to the clarification of the mechanism of social and cultural processes in the development of science. In addition, they would complement research into creativity, which still concentrated for the most part on the social status and the psychological traits of scientists.[45]

These remarks bring us full circle – back to the suggestions of Kuhn and Kaplan in 1959, which stem logically and historically from Merton's work on the social determination of science as an institution and of creative work in science. In this manner, Kuhn and Kaplan consciously or unconsciously introduced stimuli for thought, or at least theoretical problem areas, that originated, via Merton as intermediary, from Marxism.

6. APEX AND DECLINE OF CREATIVITY RESEARCH

Thanks to generous financial support, the 1960's and early 1970's represent the most productive period in the development of American creativity research. This is equally true both of the intensity and of the content of the investigations. Particularly, the analysis of the social environment in which creativity can blossom enjoyed its heyday. This development was essentially guided by the maxim "that dividends in creativity can be attained most rapidly through proper change of environmental factors."[46]

Such considerations led to intensified criticism of cultural and political conditions prevalent in the USA at the time. For example, E. P. Torrance criticized the success orientation of American culture,[47] A. Roe the extant cultural institutions in general,[48] and A. Toynbee the "egalitarian deformation of American democracy", as well as a "benighted conservatism" that undermined common creativity.[49]

Social utopias, in which the contours of an "ideal", creativity-promoting society were drafted as alternatives to existing social orders, were being developed with increasing frequency. The most impressive visions in my opinion are those of a "free society" as conceived by M. I. Stein,[50] and those of the "creativogenic society" described by S. Arieti.[51] These theoretical currents culminated in the formation of "Socialpsychology of Creativity", represented most prominently by D. K. Simonton[52] and T. S. Amabile.[53]

In the second half of the 1970's, the intensity with which creativity research was carried out in the USA began to stagnate, nor was it able to reinvigorate in the 80's. The number of publications diminished in relation to the 60's, while at the same time the focal points shifted to, among others, topics like the process and the elements of creative thought, metaphoric and visual thought. In line with this new trend, the significantly more pragmatic and for the most part self-financed applied creativity research emerged more strongly.

I feel that this trend can largely be attributed to the fact that the excessively high expectations for immediate practical results from academic creativity research could not be realized sufficiently.[54]

At present, such research enjoys little financial support and very little attention at American universities, and in the eyes of some psychologists it is already a dead field of inquiry. This seems to me to be an exaggeration. Moreover, this would be most regrettable, because we still know far too little about creativity to be able to employ it effectively in solving the problems of mankind.

Furthermore, the decline of creativity research is only in part due to the insufficient realization of the expectations placed on it. It is much more that, since creativity research was a product of the cold war, the cessation of hostilities can be seen as the fulfillment of the task it was originally given. However, this is not to say that creativity research must necessarily disappear due to the transition from a bipolar to a multipolar world. After all, it is an area confronted today and in the future with a multitude of new tasks important to securing the existence of mankind.

7. CONCLUSION

What, ultimately, is the message of this article? It is certainly not meant simply as a nostalgic look at a chapter in the history of science, during which the cold war caused the financial fountains of the humanities to spew forth in greater quantity than seems to be the case today.

Above all, it should be evident that the confrontation of systems led not only to great dangers for mankind, but that it forced both rival systems to mobilize and concentrate to the limit the inner forces of their respective societies in order to survive. This held true in science as it did in economic, politics, culture and education.

On the other hand, positive impulses for the development of its rival's social sphere also emanated from each historical combatant. As we have seen, at least for a while, even the humanities were able to profit from this competition, even if this profit was to be lost at the conclusion of the cold war. As I see it, this fact is equally as true of philosophy as it is of creativity research.

Given that this is the case – and there seems to be much evidence supporting this claim – dangers become apparent that threaten the existence of the humanities and the social sciences in any society which, due to the removal of great external threats, suffers a reduction in stimuli for its social and intellectual progress. But a multipolar world presents many new threats. If only to be able to recognize these, we are in need of the humanities and the social sciences.

University of Bamberg

NOTES

* This article arose from research carried out in 1985/86 when the author was Research Associate at the Center for Philosophy and History of Science, Boston University, Boston, Massachusetts.

[1] J. W. Getzels, and M. Csikszentmihalyi, "From Problem Solving to Problem Finding", in I. A. Taylor, and J. W. Getzels (eds.), *Perspectives in Creativity*, Aldine, Chicago, 1975, p. 95.

[2] L. L. Thurstone, and T. G. Thurstone, *Factorial Studies of Intelligence*, University of Chicago Press, Chicago, 1941.

[3] See F. Barron, *Creative Person and Creative Process*, Holt, Reinhart, and Winston, New York, 1969, p. 5.

[4] H. A. Murray, *Explorations in Personality*, Oxford University Press, New York, 1938.

[5] See G. Holton, *Thematic Origins of Scientific Thought*, Harvard University Press, Cambridge (Mass.) and London, 1973, pp. 41–42.

[6] See D. W. MacKinnon, *In Search of Human Effectiveness*, Creative Education Foundation, Buffalo (NY), 1978, p. xi.

[7] See N. E. Golovin, "Creativeness in the Sciences and the Problem of National Growth", in M. A. Coler (ed.), *Essays on Creativity in the Sciences*, New York University Press, New York, 1963, p. 13.

[8] J. P. Guilford, "Creativity: Yesterday, Today and Tomorrow", *Journal of Creative Behavior* **1** (1967), p. 7.

[9] D. W. MacKinnon, *In Search of Human Effectiveness*, p. xii.

[10] J. P. Guilford, "Creativity: Retrospect and Prospect", *Journal of Creative Behavior* **4** (1970), p. 150.

[11] See J. P. Guilford, "Creativity", *American Psychologist* **5** (1950), pp. 444–454.

[12] See E. D. Hutchinson, "Materials for the Study of Creative Thinking", *Psychological Bulletin* **28** (1931), pp. 392–410.

[13] C. W. Taylor and R. L. Ellison, "Moving toward Working Models in Creativity", in I. A. Taylor and J. W. Getzels (eds.), *Perspectives in Creativity*, p. 193.

[14] A. Roe, *The Making of a Scientist*, Dodd & Mead, New York, 1952.

[15] See M. I. Stein, "Creativity and Culture", *Journal of Psychology* **36** (1953), pp. 319–322.

[16] See C. R. Rogers, "Toward a Theory of Creativity", in S. J. Parnes and H. F. Harding (eds.), *A Source Book for Creative Thinking*, Scribner's, New York, 1962, p. 64.

[17] See M. Tumin, "Obstacles to Creativity", in *ibid.*, pp., 106–110.

[18] See R. May, *Man's Search for Himself*, Norton, New York, 1953, pp. 19, 177.

[19] See R. S. Crutchfield, "Conformity and Creative Thinking", in H. E. Gruber, G. Terrel, and M. Wertheimer (eds.), *Contemporary Approaches to Creative Thinking*, Atherton Press, New York, 1962, pp. 121–139.

[20] See H. A. Murray, "Vicissitudes of Creativity", in H. H. Anderson (ed.), *Creativity and its Cultivation*, Harper & Row, New York and Evanston, 1959, p. 96.

[21] See H. H. Anderson, "Creativity in Perspective", in *ibid.*, p. 251.

[22] See for example G. Mehlhorn and H.-G. Mehlhorn, *Zur Kritik der bürgerlichen Kreativitätsforchung*, Deutscher Verlag der Wissenschaften, Berlin, 1977, p. 168; E. Kellner, *Wissenschaftliches Erkennen: Plan oder Intuition*?, Dietz Verlag, Berlin, 1987, p. 19.

[23] See C. W. Taylor (ed.), *The Second (1957) University of Utah Research Conference on the Identification of Creative Scientific Talent*, University of Utah Press, Salt Lake City, 1959.

[24] See C. W. Taylor (ed.), *The Third (1959) University of Utah Research Conference on the Identification of Creative Scientific Talent*, University of Utah Press, Salt Lake City, 1959.

[25] T. A. Razik, "Psychometric Measurement of Creativity", in R. L. Mooney and T. A. Razik (eds.), *Explorations in Creativity*, Harper & Row, New York, 1967, p. 302.

[26] H. S. Truman, "The Creative Purpose", in *Creative America*, Text by J. F. Kennedy and others, Ridge Press, New York, 1962, p. 111.

[27] N. E. Golovin, "The Creative Person in Science", in C. W. Taylor and F. Barron (eds.), *Scientific Creativity: Its Recognition and Development*, Wiley, New York and London, 1963, p. 21.

[28] J. P. Guilford, "Creativity: Its Measurement and Development," in S. J. Parnes and H. F. Harding (eds.), *A Source Book for Creative Thinking*, p. 156.

[29] N. W. Storer, "Introduction", in R. K. Merton, *The Sociology of Science: Theoretical and Empirical Investigations*, University of Chicago Press, Chicago, 1973, p. xxiv.

[30] T. S. Kuhn, *The Essential Tension*, University of Chicago Press, Chicago and London, 1977, p. 225.

[31] M. W. Wartofsky, "Scientific Judgment: Creativity and Discovery in Scientific Thought", in T. Nickles (ed.), *Scientific Discovery: Case Studies*, Reidel, Dordrecht, Boston and London, 1980, p. 4.

[32] T. S. Kuhn and N. Kaplan, "Committee Report on Environmental Conditions Affecting Creativity", in C. W. Taylor (ed.), *The Third (1959) University of Utah Research Conference on the Identification of Creative Scientific Talent*, p. 313.

[33] See R. S. Albert, "The Concept of Genius and its Implication for the Study of Creativity and Giftedness", in R. S. Albert (ed.), *Genius and Eminence*, Pergamon Press, Oxford, New York and others, 1983, p. 12.

[34] See C. W. Taylor, "Acknowledgments", in C. W. Taylor (ed.), *Climate for Creativity: Report of the Seventh National Research Conference on Creativity*, Pergamon Press, New York, Oxford, and others, 1972, p. vii.

[35] R. K. Merton, "Science, Technology and Society in Seventeenth Century England", *Osiris*, Vol. VI, Part 2 (1938), pp. 360–632.

[36] B. Hessen, "The Social and Economic Roots of Newton's 'Principia' ", in N. I. Bukharin (ed.), *Science at the Cross Roads*, Kniga, London, 1952, pp. 147–212.

[37] R. K. Merton to M. Wallner, unpublished letter of April 13, 1988.

[38] See R. K. Merton, *Science, Technology & Society in Seventeenth Century England*, Howard Fertig, New York, 1970, pp. 55–136.

[39] See *ibid.*, pp. 137–198.

[40] *Ibid.*, p. 142. Footnote (24).

[41] *Ibid.*, p. 206. Footnote (8).

[42] Cit. in P. Sztompka, *Robert K. Merton: An Intellectual Profile*, St. Martin's Press, New York, 1986, p. 26.

[43] See R. K. Merton, *The Sociology of Science*, p. 354.

[44] See *ibid.*, p. 371.

[45] See *ibid.*, pp. 372–373.

[46] C. W. Taylor, "Some Knows, Needs, and Leads", in C. W. Taylor (ed.), *Creativity: Progress and Potential*, McGraw-Hill, New York, 1964, p. 183.
[47] See E. P. Torrance, "Education and Creativity", in *ibid.*, pp. 98–102.
[48] See A. Roe, "Psychological Approaches to Creativity", in M. A. Coler (ed.), *Essays on Creativity in the Sciences*, p. 179.
[49] See A. Toynbee, "Is America Neglecting Her Creative Minority?", in C. W. Taylor (ed.), *Widening Horizons in Creativity*, Wiley, New York, 1964, pp. 3–9.
[50] See M. I. Stein, "Creativity in a Free Society", *Educational Horizons* **41** (1963), pp. 129–130.
[51] See S. Arieti, *Creativity: The Magic Synthesis*, Basic Books, New York, 1976, pp. 312–325.
[52] See for example D. K. Simonton, *Genius, Creativity and Leadership*, Harvard University Press, Cambridge (Mass.) and London, 1984.
[53] T. S. Amabile, *The Social Psychology of Creativity*, Springer, New York, Berlin and others, 1983.
[54] Newer publications also evidence an equally high level of academic quality, but seem somewhat underdeveloped from a pragmatic point of view. See for example, S.G. Isaksen (ed.), *Frontiers of Creativity Research: Beyond the Basics*, Creative Education Foundation, Buffalo, 1987; F. Flach (ed.), *The Creative Mind*, Bearly Limited, Buffalo, 1988; R. J. Sternberg (ed.), *The Nature of Creativity: Contemporary Psychological Perspectives*, Cambridge University Press, Cambridge and New York, 1988; D. K. Simonton, *Scientific Genius: A Psychology of Science*, Cambridge University Press, Cambridge and New York, 1988; R. Ochse, *Before the Gates of Excellence: The Determinants of Creative Genius*, Cambridge University Press, Cambridge and New York, 1990.

KAREL BERKA

RATIONAL AND NONRATIONAL ELEMENTS IN THE HISTORY OF SCIENCE*

It is the aim of my contribution to analyze the relevance of nonrational elements in the history of science, more exactly the dialectics of scientific and unscientific roots of modern science. As a prerequisite I shall discuss, first of all, some general problems concerned with the rational reconstruction of the origin of modern science in respect to its unscientific or even counterscientific roots. Against the one-sidedness of two opposing views, namely "that science is fundamentally an irrational enterprise", and – in contradiction to it – that "scientific behavior . . . is the best example we have of rationality",[1] I want to stress the interplay of rational and irrational motives and ideas which is, in my view, vital for the advancement of scientific knowledge. By focussing attention on unscientific aspects in the actual history of science and by emphasizing the historical nature of scientific rationality, I do not share the relativism of Th. S. Kuhn and P. K. Feyerabend connected with the assumed incommensurability of theories and the rigid dichotomy of the cumulative and non-cumulative process in the development of science.[2] I am, however, fully aware that some of my views are controversial and will deserve a deeper justification from the philosophical point of view and by case studies as well. What I can offer here is only a brief outline of some substantial topics that support the basic idea of my approach, namely, that even outdated, metaphysical or counterscientific doctrines, religions, magical and mystical beliefs and opinions have positively influenced the rise and development of scientific knowledge.

Let me start with obvious questions already raised by various historians and philosophers of science, especially by I. Lakatos.[3]

Is it possible to conceive a historical reconstruction of scientific achievement of the past as a rational process? In what sense can we speak about the rationality of science, about the rational development of scientific knowledge? Have we at our disposal some universal norms of rational behaviour valid for scientists in all epochs or have we rather to accept the view that rationality is relative to a given period and that perhaps every individual has his or her own idea of rationality? Where are the borderlines between science and pseudo-science, science and

K. Gavroglu et al. (eds.), Physics, Philosophy and the Scientific Community, 273–285.

metaphysics, science and myth? How can irrational doctrines be rationalized in order to attain unbiased, objective and true knowledge in the pursuit of scientific activities?

These and similar questions have to be answered historically. The concept of scientific rationality itself, which can be considered as a special instance of rationality side by side with other kinds (e.g., rational behaviour in decision situations), is equally well historically determined. In addition, this concept is related to concrete social conditions of a given epoch and the peculiarities of the scientific problem in question, hence never absolutely divorced from nonrational or irrational features. In order to avoid simplifications that either absolutize the scientific rationality of the present epoch or overvaluate the relativity of subjective moments and nonscientific impacts of the past, one has to admit the existence of inconsistencies and contradictions in the development of scientific rationality.

Taking into account the difference between the method of presentation and the method of inquiry, we have to compare very carefully past and contemporary views, past and present standards and norms in scientific activities. When looking backwards to some epoch in the history of science, we cannot avoid the impact of present views and doctrines. Even the most faithful insight into the past is thus burdened by our present views. Our exegesis will therefore be more or less authentic. The reconstruction of the actual history will oscillate between a literal interpretation and a conceptual reinterpretation. This bifurcation already implies a danger of misinterpreting the past. We are accustomed to interpret older doctrines with respect to our own ideas, often as anticipations of modern conceptions. However, did the scientists of the Renaissance understand a certain concept the same as we do today? Perhaps. But there is not enough evidence to show that this agreement is not just a coincidence without the same intellectual intensions. We are, of course, able to understand and explain ideas developed many centuries ago, but we can hardly assume that our interpretations will faithfully represent what people of past epochs really had in mind. Every epoch has its special conceptual climate, which is only approximately similar to an analogous atmosphere of our epoch.

We cannot introspect the real thought process of someone else. What we can do is to compare the conceptions of a scientist with those of his predecessors, contemporaries and followers, and then outline a model of his mental evolution. We have to admit that a scientist was influenced by various conceptual impulses which can be inconsistent, or very unlikely, or even evidently false. There need not be an inner harmony between what a scientist says about himself, how he explains his scientific procedures, what he published, what he intended to achieve and what he actually attained. We must even deal with his attempt to hide the real motives of his activities, to conceal the conceptual sources of his work. He can pretend to be better than he was in reality. He will often suppress his own mistakes, his unrealized wishes, his intuitions. Since every scientist lives in a concrete society under specific political and ideological conditions, he may well adopt the *Zeitgeist* of his epoch. As a citizen, he is often forced to conceal his real opinions simply because he could be accused of heresy, blasphemy or deviation in respect to the political or ideological views prevailing in the society in which he lives. We have further to mention the fact that a good scientist need not be a good man, i.e., one with unshakeable, firm moral views and behaviour. He can be fond of awards and titles and for this reason willing to make compromises with mighty and influential people. He can officially proclaim doctrines he in reality abhors only in order to have the possibility of satisfying his scientific ambitions. There need not be a unity of words and deeds. Just recall the advice of a heavily smoking physician that smoking is very unhealthy.

Man is not only a "rational animal", but rather a creature of emotions, desires, hopes and wishes. Each of us combines in his thinking, feeling and acting rational and nonrational elements. Very often people later on highly esteemed for their work have traits in their character that seem to be nearly unexplicable, especially when taking into account their intellectual level. We are, therefore, surprised that famous mathematicians make mistakes in simple counting, that excellent logicians sometimes draw invalid conclusions, that great scientists have held out-dated, even reactionary ideas.

Remember e.g., the famous mathematicians A. L. Cauchy and E. Galois: one an ardent monarchist and the other one a consistent anti-monarchist. Take, as another example, G. Frege's very progressive conceptions in logic, his conservative views on non-Euclidian geometry and his nationalistic political attitude. Or what shall we say about the anti-Semitic racist criticism of Einstein's theory of relativity by German physicians in the Third Reich, defended even by J. Stark, the Nobel Price winner for the year 1919? This discrepancy in character, this confusion in thought or, as it is sometimes expressed, this schizophrenia in behaviour, is evidently an even more concomitant phenomenon in the period of transition from the medieval picture of the cosmos to the modern image of the universe.

A conscious veiling of the real course of past events is often the result of the epigones of some eminent person, the work of his ardent and faithful followers. Possibly in good faith, but more likely for apologetic reasons, they hide or simply "reinterpret" certain facts that according to their view could show their master in an inappropriate light. The history of Aristotelianism is a very instructive example of such an uncritical, and in its consequences harmful, approach to the heritage of this scholar who has for a very long time influenced the scientific and philosophical development of mankind in a positive as well as negative sense. Something similar can be said in the case of Marxism.

The understanding of the historical development of science is burdened by facts that question the rationality of scientific conceptions or at least the actual behavior of their bearers. Without taking into account the concrete historical situation in which a scientific conception had been evolved we cannot clearly decide about its character, whether it belongs to science or to pseudoscience, to science or craft, to science or practical art.[4] It is thus problematic, e.g., to draw a demarcation between astronomy and astrology, medicine and witchcraft in this sense. It is clear that there is a real difference, if one tries to explain the assumed discrepancy between the astrologer Kepler and his great astronomical achievements on the one hand, and on the other, a similar relation holding between astrology and astronomy at the end of the 20th century. The degree of scientific rationality connected with astrology in the days of J. Kepler, whether he really believed or only pretended to believe in this activity, was rather high, whereas today it is very near to zero.

On the assumption that there cannot be a full agreement in the assessment of scientific rationality that would be historically valid for every period in the history of science and acceptable to all the various kinds of human civilizations, one can adopt Laudan's conception that "contrary to common belief, it can be rational to raise philosophical or religious objections against a particular theory or research tradition, if the latter runs counter to a well established part of our general *Weltbild* – even if the *Weltbild* is not 'scientific' (in the usual sense of the word)".[5]

This standpoint implies the problematic question of the influence of nonscientific world-views on science, represented chiefly by religion, theology and metaphysics. Is there any historical evidence that these world-views had a positive impact on the development of science? The answer will depend less on empirical facts and theoretical arguments than on one's own unscientific, i.e., metaphysical, prejudices, because when divorcing scientific and nonscientific rationality, it does not suffice simply to say that nonscientific views contradict the attained level of scientific knowledge.

How then to explain the scientific results obtained in past periods by scientists with an outspoken religious world-view? When dealing with the Middle-Ages, the answer to this puzzle was clear. Since the prevailing world-view of this epoch was theology, no real advancement of science was possible. After convincing evidence of scientific conceptions unknown before, which testified to the view that there was, in reality, no gap of many hundred years between the classical and Hellenistic period of ancient Greece and the rise of modern scientific knowledge in the Renaissance, that there was no absolute stagnation in scientific activities, this strict denial of all unscientific impacts on science – in the given case, those of religion – had to be modified. The most convenient way out was found in the medieval doctrine of double truth and in the explanation that new scientific ideas occurring under the hegemony of theology were merely absorptions of ancient conceptions or intuitive anticipations stemming from heretical deviations of their representatives.

This rationalization of the growth of scientific knowledge under feudal social conditions and the ideological rule of theology could, however, not satisfy another historical situation, which is, in respect to the opposition of scientific and nonscientific motives in the development of modern thought, even more paradoxical. How to reconcile positive scientific results in the period of the rise of modern science with an outspoken religious world-view? There is no doubt that such a giant of modern natural science as I. Newton has written more treatises on religion and theology than on physics or "natural philosophy". For Newton the existence of God was not a hypothesis which he did not need for his work, as Laplace later explicitly said. "Newton was confident . . . that certain empirical facts open to anybody's observation, unqualifiedly implied the existence of a God of a certain definite nature and function. God was not detached from the world that science seeks to know".[6]

How to handle this case? The usual way of getting rid of nonscientific elements is an "appropriate" reinterpretation. The influence of religion, which I took as an extreme case of nonscientific motives instead of magic which is more characteristic for other cultures than the European one, can be explained – and this explanation is really rational – as an expression of the dependence of the scientist on the general intellectual atmosphere of the epoch. A scientist cannot rid scientific activities as a whole from all nonscientific impulses. Like every other man, he also is influenced by the *Zeitgeist* or *Weltanschauung* of the epoch. This dependency can be, of course, weakened in a modern interpretation by holding that the nonscientific impulses did not negatively influence his scientific activities. It can be also declared as a contingent expression of the spiritual trends of his epoch.

When thinking about the impact of religion upon Newton, we must be aware of its socio-historical context. This context changes from epoch to epoch. The growth of scientific knowledge in ancient Greece – by contrast to India, Egypt or China – was determined by the fact that philosophy and not religion became a leading component of the ideological superstructure of this epoch. In the Middle Ages the feudal ideology was a combination of religious and scientific views, of Christian doctrines and a modified Peripatetism. The thinkers and scientists of the Renaissance revolted against this ideology without losing belief in the existence of God. This belief was acknowledged in principle, till the time of I. Kant, who made the attempt to prove that the question of God's existence leads to an antinomy. The impact of religion on man survived till our epoch. Faith, religious feelings did not weaken their influence on people's personal life for a quite simple reason: because of human fear of death and man's irrational desire for salvation and immorality.

For the actual rise and development of scientific thought in the modern era a change in the prevailing philosophical, i.e., ontological and metaphysical, ideas was more effective and important. It is well known that the revolution in the general outlook was led by a renewed spirit of Pythagoreanism, Platonism and Neoplatonism against a Thomistically grounded peripatetism. How to evaluate this change in respect to its progressiveness? When comparing medieval peripatetism – and *a fortiori* ancient Aristotelism – with the above-mentioned philosophical movements, the first impression is quite clear: Peripatetism with its empirical background and qualitativism is more appropriate to common sense. Neoplatonism with its quantitativism in close connection with number mysticism and the hermetic-caballist tradition is remote from the everyday experience of people. Which of these metaphysical approaches is, therefore, more appropriate to natural investigations? Without anticipating what consequences that have resulted from the wedding of natural philosophy and universal mathematization, a soberly thinking man would rather prefer a detheologized peripatetism for its unspeculative empiricism, which could easily be confirmed by the human senses.

The Aristotelian and the modern approach to nature were based on certain metaphysical assumptions. Aristotle and his followers believed in what they could learn by direct observation. Their naive empiricism was aimed against conceptions not supported by empirical evidence. Aristotle could not measure quantities. For this reason, he adopted, even in physics, a standpoint similar to that in biology. A quantitative and infinite standpoint was for him admissible only in mathematics. He did not assign a mathematical character to nature. In contradistinction to Plato he not only refused his mathematical idealism, but also his conception of space – *chora*.[7] He rejected the views of Demokritos, especially his conception of the void, as well.

Aristotle refuses to admit that space is something separable from the bodies that occur there. Against the views of Plato and Demokritos, which anticipate the concept of absolute space of Newtonian physics, he uses the concept of place – *topos*. The place of bodies – which is for him the starting point for the description of the motion of bodies – is a certain relationship between bodies in a common system. The existence of places follows from the possibility of changing places. The place is, of course, differentiated from the bodies that occur in it. It cannot be a body since in this case one and the same place would be occupied by two bodies. Such an assumption is for logical reasons absurd.

For similar reasons Aristotle could not admit the existence of the void of atomism. Such a space would include every body and would be prior to all of them. The void of the atomists was, of course, a metaphysical concept and its identification with the physical concept of the vacuum is not appropriate. The objections against the void, in scholastic peripatetism, incorrectly interpreted in an anthropomorphic sense as "horror vacui", could not be, therefore, refuted by the experiments of Guericke or Toricelli.

The refutation of the concept of the void was confirmed by unacceptable consequences that would follow from the assumption of this view, e.g., the discontinuity of the cosmos, the identity of all velocities etc.

Aristotle's empiricism which suited only the sublunar realm could not, for methodological reasons, serve the revolutionary change of the modern world-view with its underlying mathematization or rather geometrization.

For Aristotle, abstraction and idealization were appropriate and useful procedures only in mathematical sciences. When investigating natural phenomena another approach is suitable. Such phenomena have to be handled in their natural milieu, they have to be investigated according to their actual behaviour and not how they would behave under idealized conditions. Aristotle assumes a conceptually hypothetical situation, but then uses his observations to show that such a situation is not natural. He assumes, e.g., what would happen if he admits the existence of the void, but based on everyday experience together with theoretical conclusions, he shows at the same time that such an idealized case is, in respect to the real conditions we find in nature, actually nonexistent.

The scientists of the modern era, especially Galileo, adopted another standpoint. They grounded their success on the denial of the predominance of sensory perception and everyday experience. They became well aware that a strict empiricism is a serious obstacle to scientific investigations. Taking into account all the plain sense date accessible to every layman, they could not assume that the earth is moving, whereas the sun is fixed. Natural phenomena have to be studied in isolation and then mathematically expressed. Galileo emphasizes idealized conditions. He knows very well that the air influences the free fall of bodies, but ignores this fact in order to simplify observations and experimentation. He accepts the simplification of the actual course of natural events in its idealized circumstances, which in reality do not exist, because such an idealization can bridge the gap between natural phenomena and mathematical entities.

The "new science" of Galileo and before him the Copernican revolution in astronomy, with the position adopted by J. Kepler, already anticipated by the ideas of G. Bruno, Nicholas of Cusa and the impetus conception of the Parisian nominalists, represents in its nonscientific aspects a step backwards from Aristotle to Plato, from a conceptualistic metaphysics to a realistic one.

These relatively less progressive metaphysical roots of modern science were based not only on the mathematical ideal of Pythagoras, Plato and their ancient and medieval followers. The mathematization of modern science with its emphasis on quantitative aspects, on exact numerical data based on the well-known assumption that "the book of nature is written in geometrical characters", was not only a metaphysical presupposition. The evidence for the instrumental nature of mathematics and its extensive applicability in scientific investigations was also the result of the acquaintance with the mathematical works of ancient authors, predominantly of Archimedes and Apollonios. An important role was played even by methodological views of simplicity and harmony already elaborated in the Middle-Ages.

Neither can we disregard the influence of the Scripture, however. If modern science tries to explain all natural phenomena by "numbers, figures and motions", then this conception obviously reflects the biblical statement that God has built the world according to number, weight and measure.[8] The view that God is a mathematician, which was later on modified into the idea that he is rather a physicist, could have been easily united with Platonism and Pythagoreanism.

One can even assume that the Renaissance scientists, especially G. Bruno, were under the influence of "Jewish mysticism, Neo-platonic cosmological and theological speculations, numerological prescriptions, and theories of Natural magic".[9] According to the testimony of Burtt, J. Kepler was a sunworshipper who drew the following analogy between the Trinity and the new astronomy: "The sun is God the Father, the sphere of the fixed stars is God the Son, the intervening ethereal medium . . . is the Holy Ghost".[10]

Is it not amazing that the great revolution in science, started by the conceptual upheaval of Copernicus, was actually in a very relevant manner influenced by so many nonscientific ideas? Quite naturally, therefore, there has often been raised the question: on what empirical and/or theoretical evidence could Copernicus and Kepler have been persuaded that their heliocentric picture of the universe is really correct. From the very beginning of this conception in the sixteenth century, not only theologians, because of the biblical conception of creation, but also the majority of scientists and philosophers did not accept it for various empirical and theoretical reasons. Before experimental verification many years later, Copernicus and his followers could present as convincing arguments only that it "saves the phenomena" in a simpler way and brings unity into the whole universe by overcoming the difference between the sublunar and the supralunar, and by assuming that the cosmos is mathematical in its structure.

The driving motives of the new astronomy are, thus, evidently expressions of a nonscientific rationality, which has its roots in the Middle Ages. The postulates of simplicity, harmony and unity reflect such older sayings as e.g., "Natura semper agit per vias brevissimas", "Natura nihil facit frustra", "Natura simplicitatem amat" etc.

Perhaps J. Kepler, in spite of his clear-cut Platonism accepted even Occam's razor in this respect, as long as it was directed against his realistic opponents. This diversity of influences, often burdened by inconsistent motives, is nothing uncommon for the thoughts and behaviour of the pioneers of modern science. It is a controversial point in the modern historical reconstructions of this epoch, whether to portray Kepler "as a weird schizophrenic who happened to have been a modern physicist part of the time but was a medieval astrologer at others" or "to present Kepler's view as a unified and integrated melding of metaphysics, physics, geometry and theology, not an arbitrary and insulated juxtaposition".[11]

The second alternative, which is from our viewpoint more convenient, is elucidated by G. Holton as follows: "Conceptions which we might now regard as mutually exclusive are found to operate side-by-side in his intellectual make-up . . . when his physics fails, his metaphysics comes to the rescue, when a mechanical model crashes down as a tool of explanation, a mathematical model takes over, and at its boundary in turn there stands a theological axiom."[12] Holton's conception is finally summarized by his statement that Kepler "in one brilliant image, saw the three basic themes or cosmological models superposed: the universe as physical machine, the universe as mathematical harmony, and the universe as central theological order."[13] This unified picture of Kepler, which has, of course, its lacunae, points to the mutual relations holding between scientific and nonscientific moments in the development of modern scientific thought.

In general, it is admitted, at least in respect to the history of scientific activities in the Middle Ages, that in this period theology not only determined the content, but also the goals of knowledge and the scientific interests of medieval scholars. Many important and, for the future development of science, relevant conceptions and theories were elaborated in this epoch not because of their interest for nature, but for theological reasons. So for example, John Philoponos, living in the early decades of the Middle Ages, and evidently still in close contact with ancient views, developed his physical doctrine in order to support the Christian doctrine of creation against the standpoint of Simplikios.[14]

How nonscientific preconceptions were integrated into a scientific system and achieved their own scientific career can be illustrated by the approach to the concept of infinity, which played a very important role in the rise of modern scientific thought.

Ancient Greek thought was largely characterized by the idea of finitism, expressed by the concept of *peras* as opposed to the *apeiron* of Anaximandros. The infinite, as Aristotle explicitly says "exists when one thing can be taken after another endlessly, each thing taken being finite".[15] In regard to the present distinction of actual and potential infinity, Aristotle and, with the exception of Anaxagoras and Zeno of Elea, who anticipated the idea of actual infinity, all other ancient philosophers adopted the concept of potential infinity. For methodological reasons, they gave a clear preference to finite procedures, especially when dealing with proof and definition.

The finiteness of the world, a conception developed in the Aristotelian and Ptolemaic picture of the Cosmos, is a positive value. This property characterizes the regularities of the Universe, its order against chaos. In opposition to it, the infinite is evaluated as a negative term, being connected with the borderless and unlimited.

The distinction of two levels of the Cosmos – the sublunar and the supralunar with its different kinds of substances and motions, as it was elaborated by Aristotle, was later transformed by Plotinos and neoPlatonism: the human world and nature are finite, whereas the heavens, the divine, is infinite. With Neoplatonism still another conceptual shift changed the relationship of the finite and infinite: the finite was conceived as something imperfect. This was obviously a logical consequence of the application of the infinite to the divine hypostasis in Plotinos.

With Christianity and the development of theology, the scholastic tradition accepts this neoPlatonistic distinction together with the corresponding valuation. For God is infinity, a positive property: infinity expresses the absoluteness of God, who is considered to be an "infinitum esse". Hence, the concept of actual infinity, according to Augustine, is accepted as the basic notion of theology. For man and nature, by contrast, the infinite is something imperfect, rather a negative property, being a chaotic potentionality of material things. Nature cannot be infinite, since it was created by God.

The sharp borderline between God and the world, the infinite and the finite, was already questioned in the XIIIth and XIVth centuries. The mystic Eckhart was the first Christian thinker who attributed the infinite creativity of God to Nature as well. A qualitative change in the assessment of the distinction between the finite and infinite in the theological, metaphysical and cosmological levels of this contradictory couple of terms is due to Nicolas of Cusa in the XVth century. In his famous work *De docta ignorantia* Cusanus characterizes the universe as a finitely infinite universe, as an infinite sphere.[16] By this picture, he wants to express that the world is not actually infinite as God, but only potentially infinite, respectively, that nature is infinite, but not absolutely, as is God.

This view had its social background and revolutionary consequences in science and politics. The conception of a non-finite universe reflects the endeavour of the late Middle Ages and the early Renaissance to bridge the gap between Nature and God, to divinize the world and human conditions, and at the same time, to secularize God, so that in this respect there is no distinction between God and Nature. Later on this view enabled B. Spinoza to speak about "Deus sive Natura". Thus, the dualism of the sublunar and supralunar spheres has been overcome. By smoothing out the difference assumed between celestial and terrestrial phenomena the foundations of the geocentric systems of Ptolemy together with those of the physics of Aristotle were in their essential features endangered. In its political consequences the *coincidentia oppositorum* of Cusanus implied that man is not absolutely powerless, not a creature living in fear and awe in respect to the Absolute and Infinite. The Medieval concept of *ordo*, of a hierarchical universe and society which was a feudalized transformation of the ancient ideal of *peras* lost its firm foundations. This was another aspect of this revolutionary idea.

"The dissolution of the Cosmos," being "the most profound revolution achieved or suffered by the human mind since the invention of the Cosmos by the Greeks,"[17] was rather the result of views proclaimed by G. Bruno, than by Copernicus or Kepler. G. Bruno said very clearly that the *astronomia nova* must replace the idea of a closed and finite world by that of an open infinite universe. This view was adopted neither by Copernicus nor by Kepler.

The idea of the actual infinite, which was without any doubts in its origin a metaphysical, respectively a theological concept, played a very important role not only in the scientific revolution of Copernicus and his followers. This concept has, in its mathematical version, exercised an essential impact on the origin and development of the infinitesimal calculus since Newton and Leibniz and it was not irrelevant to Cantor's conception of set theory.

How shall we characterize the actual course of the development of ideas, when a property of the divine has been transformed into a property of mathematical entities? Under what conditions was such a conceptual transformation from the domain of theology and metaphysics into astronomy and later on into mathematics possible at all? These and other related questions can be explained only by a realistic reconstruction of the history of scientific cognition. That nonscientific and even counterscientific conceptions have positively influenced the development of scientific knowledge is a fact that cannot be denied. By stressing their fruitful impacts on the growth of science neither irrationalism nor epistemological anarchism is thereby accentuated or supported.

NOTES

* In memory of a round-table discussion with R. S. Cohen and other colleagues held during the 8th International Congress of Logic, Methodology and Philosophy of Science, Moscow, August 17–22 1987.

[1] Kuhn, Th. S., "Notes on Lakatos", in: Buck, R. C. and Cohen, R. S. (eds.), PSA 1970. *In Memory of Rudolf Carnap, Barton Studies in the Philosophy of Science*, vol. VIII, Dordrecht/Boston 1971, p. 143f.

[2] Cf. Shea, W. R.: "Do Historians and Philosophers of Science Share the Heritage?" in: Shea, W. R. (ed.), *Nature Mathematized*, Dordrecht/Boston/London 1983, p. 17f.

[3] Cf. Lakatos, I.: "History of Science and Its Rational Reconstructions", in: Buck, R. C. and Cohen, R. S. (eds.), *PSA 1970*, pp. 91–136.

[4] Cf. Kuhn, Th. S.: "Logic of Discovery or Psychology of Research," in: Schilpp, P. A. (ed.), *The Philosophy of Karl Popper*, La Salle 1974, vol. II. p. 804.

[5] Laudan, L.: *Progress and Its Problems. Towards a Theory of Scientific Growth*, London-Henley 1977, p. 127.

[6] Burtt, E. A.: *The Metaphysical Foundations of Modern Physical Science*, Garden City 1954 (1st ed. 1924), p. 283.

[7] Cf. North, J. D.: "Finite and Otherwise. Aristotle and Some Seventeenth Century Views," in: Shea, W. R. (ed.), *Nature Mathematized*, p. 114.

[8] Cf. Koyré, A.: *Metaphysics and Measurement*, Cambridge, Mass. pp. 2, 91.

[9] Lindholm, L. M.: "Is Realistic History of Science Possible? A Hidden Inadequacy in the New History of Science," in: Agassi, J. and Cohen, R. S. (eds.), *Scientific Philosophy Today*, Dordrecht/Boston/London 1982, p. 169.

[10] Burtt, E. A.: *op. cit.*, p. 60.

[11] Lindholm, L. M.: *op. cit.*, p. 177.

[12] Holton, G.: *Thematic Origins of Scientific Thought: Kepler to Einstein*, Cambridge, Mass./London 1975, p. 70.

[13] *Ibid.*, p. 86.

[14] Cf. Floss, P.: *Proměny vědění* (*The Metamorphoses of Knowledge*), Praha 1987, p. 152.

[15] *Arist. Phys.* III c.6.

[16] Cf. Floss, P., p. 190.

[17] Cf. North, J. D., *op. cit.*, p. 144.

RUDOLF HALLER

DIRT AND CRYSTAL: NEURATH ON THE LANGUAGE OF SCIENCE

For a long time I have been a defender of the philosopher Otto Neurath. Why? First, he was underrated by the history of philosophy; second, he was misunderstood by many, even those who were in sympathy with him or with one or the other part or aspect of his work. Thirdly, I think that his contribution to philosophy is not only of historical value but has the mark of being an important step in the development of modern empiricism (and this is not to take account of his many-sidedness in such different fields as sociology, economy, politics, history, architecture, visual education, language construction, etc.).

Perhaps the main reason why taking up the study of Neurath anew will be desirable and necessary however, is that many of his ideas are flourishing today in other contexts. Therefore to look back and to identify their origin will help us to understand the philosophy of our time better than would be possible without this perspective.

Until recent times it was commonly held that logical empiricism was a failure, that most of the tenets of those philosophers turned out to be not tenable but false, that especially their views of science and their theory of science could not withstand the heavy attacks of the so-called antipositivists. Therefore most philosophers agreed and agree to the happy pronouncement that logical empiricism is dead. The true story of all of this, however, is different, and some have even claimed that logical positivism has been murdered (and some have even proudly taken responsibility for this happy event). However, it is not the case that there was only an *attempt* at revolution in philosophy which tried to overthrow the traditional one. There actually was such a revolution and philosophy in truth has been changed ever since. And surely it also is not the case that the revolution failed because it was murdered.

As we know, in spite of the fact that there happened to be an assassination – namely of Moritz Schlick in 1936 – the philosophical movement stemming from the heyday of the Vienna Circle was never murdered, as I have pointed out already on a number of different occasions.[1] Contrary to many shortsighted judgments concerning the philosophy of this century, there is no special reason to pick out of the

K. Gavroglu et al. (eds.), Physics, Philosophy and the Scientific Community, 287–299.

successful philosophical movements precisely those that were shown to be based on tenets which could be falsified. True: Statements like beliefs may turn out to be false, arguments may be rejected, and theories like other assumptions may lose their explanatory power when what they predict does not occur. But the great project of logical empiricism was a composite of *different* projects, based on different philosophical assumptions. Indeed, one of the main goals of work *on* the Vienna Circle during the past years has been that of bringing to light the philosophical *differences* to be found within the Circle. Only against the background of a more accurate picture of the individual contributions as well as of the general views held by the members of the group will we be able to judge their achievements. The rediscovery of the almost forgotten Vienna Circle is well under way.

Perhaps the greatest surprise in the course of this work was the rediscovery of the work and the works of Otto Neurath, starting with the Vienna Circle Collection in 1973, the edition of the collected works of Neurath in 1981 and the Schlick–Neurath-Symposium, the centenary-celebration of both of these philosophers a year later.

In the course of these studies the shape of Neurath's philosophical silhouette has been changed remarkably. Where in earlier reports and writings on the history of logical empiricism his name was mentioned mainly in connection with either the debate about protocol-sentences or with regard to his opposition to Schlick's foundationalism, it can now be recognized that to take into account only this small part of his work is neither just nor sufficient. Since only part of his work has won recognition, we had to start anew. Fortunately this task has been taken up in different quarters of the world; although it would be interesting to write about it, one has to consider if this is the place to report about the new wave of Neurathiana and the rediscovery of the Vienna Circle.[2]

Here and in the following I have to restrict myself to bringing into focus only those parts of his vast works which will contribute to an understanding of his ideas concerning language and science.

First, Neurath's philosophy of science, continuing his early work in the period of the *First Vienna Circle*, tries to unite the Machian project of a dynamic theory embracing all the sciences with some of Duhem's main ideas.

Certainly, one of Mach's problems was how we should deal with the different sciences under the pretext that they make up one system or a unity. To free the idea of the unity of the sciences from the charge that it is a sham or mere object of wishful thinking, Mach tried to find a bridge which would permit the scientist to change from the theory of physics to psychology and vice versa, without a complete change of ontology. According to Mach this is possible if we take elements as the basic stuff which may be viewed differently under the *aspect of different directions* of research. We know this view very well from Russell who, in the *Analysis of Mind*, stated that physics and psychology are not distinguished by their stuff or ontology, which are the elements, out of which Mind and Matter are only logical constructions. Our interest primarily is what it is that directs theory building towards different ordering.

As concerns Duhem's ideas, I should like to pick out only one: the principle of theory choice, which gives credit to the fact that a theory and therefore also a theoretical hypothesis are not and cannot be uniquely determined by experience. When combined with a tolerant application of Mach's principle of economy, this opened the door for an interpretation of the foundations of scientific knowledge that was rather more liberal than purely and rigorously empiricist. As soon as decisions enter the process of theory-building it is true that no law expressed in a theory can collide with experience. This after all, was for Duhem the strongest reason for the rejection of the idea of an *experimentum crucis* as between conflicting hypotheses.

Perhaps Neurath was already aware of the importance of practice in theory as well as in life. As Hilary Putnam rightly observed (or may have learned from Wittgenstein): "The primary importance of ideas is that they guide practice, that they structure whole forms of life:"[3] What Neurath like Mach was always asking for was that these ideas should not be overrated since they themselves very often are nothing else than what history, usage, or custom has brought about.

Since I do not like to repeat too often what I have already repeatedly said about the nature, origin, and success of Neurath's general views, I cannot go here into any detailed account of the development of his conception of the scientific world-view, which from the beginning was 'naturalistic' in character. By this I mean a view of a sort that, first, *denies* that there are *genuine philosophical questions* which can be answered only by a special discipline, namely, the most general one. If, however, there are no genuine philosophical questions and no genuine philosophical answers, as Neurath thinks, then science alone can give the answers we need to improve our practice and to provide the predictions necessary for this purpose. Such a naturalistic conception is, secondly, bound to avoid questions which cannot be put into the framework of science. The upshot of this conception will then consist in the claim that all *meaningful questions* which deserve to enter scientific discourse have to be tackled by one of the disciplines that are part of the entire enterprise called *science*. But Neurath does not accept a real departmentalization of science into disciplines, meaning: theories separated one from the other. According to his aim of *uniting* what is divided, he takes the unity of science not as a hierarchical pyramid, not as a hypothetic-deductive axiomatic system, as Carnap did, but as an *encyclopedic* undertaking where under every heading we find numerous references to other fields of research, different accounts and entries.

It is not easy to see the main point of the gigantic project of an encyclopedia of unified science, particularly because the actual outcome does not at all fit the program that prompted it. To get a better understanding of the linguistic apparatus working in this project, I think it will be useful to look at the structure as well as at its purpose. A very early version of Neurath's general point of view is to be found in a paper by him on Wilhelm Wundt's voluminous *logic*.[4] This paper, published in 1910, starts with a quotation from Comte, thus indicating where Neurath's interests at this stage already lay.

At the very beginning of this article Neurath turns to one of his favourite topics, the questions of universal science, *scientia generalis* or *scientia universalis*, a project which was dealt with already in the works of Descartes and Leibniz. In the same vein Neurath is not only *interested* in the *idea of a universal science*, but determined that to bring about the development of mankind it is necessary to establish the idea of such a universal science as mankind's highest goal. If science embodies the best methods of a rational pursuit of knowledge, then no other aim can compete with it. If science is one, then no division should divide it. Neurath thinks that it is in the nature of hypothetical systems that they are *theoretical*. "One cannot see if a statement is practical or theoretical"; the distinction between practical and theoretical sciences should be *abandoned*, only the *objects* of science should serve as a *different specifica*. That is, only the properties of the objects should be of interest, not how they are to be applied (GA, p. 29).

As early as 1910 we see him, therefore, developing some of his later methodological principles. One concerns the idea that, contrary to a widely held opinion – not only *quantitative concepts* may enter into the exact sciences. We can equally well use *qualitative concepts*, referring to properties of objects in an extensional way without any loss of exactness. What Neurath is suggesting is that *relations* of the kind important for economics are to be dealt with in *a calculus of relations*.

Another topic which we find in this early piece concerns the much discussed topic of *is* and *ought*. Here Neurath tries to *refute the distinction in regard to science* itself. The first step is to say that *imperatives* or *rules* are not true or false, are not *descriptive statements at all*, but *suggestions*, suggestions to be used *to influence the behaviour of other persons*. The descriptive content of a statement has nothing to do with its suggestive power. The warning of Max Weber that we should not let value judgments enter into science is, Neurath says, just a result of a misunderstanding of science. This is an interesting and important point. Because Neurath distinguishes the question of the sense of rules and imperatives from the question of the role of value-judgments versus empirical judgments. From the fact that the analysis of rules and imperatives makes manifest that it is the *use* of the words and *not* the 'content' or sense of a sentence which functions as suggestions, Neurath argues that we cannot accept the exclusion of value judgments from science. It is wrong to confront value judgments with empirical judgments since "every value judgment is an empirical judgment".[5] Value judgments as well as the effects of the use of imperatives or rules can therefore be the objects of scientific investigations.

According to Neurath we could analyze all *value judgments* on the model of likings and dislikings, like descriptive judgments on the model of accepting *p* or rejecting *p*. And Neurath thinks that to investigate the acts of people on the basis of their likings and dislikings would be the best way of bringing about *a scientific theory of economic preferences*.

When he summarizes his article – which after all was some sort of review of the work of Wilhelm Wundt – Neurath stresses another point that later on became quite important in philosophy of science, namely, *the meaning and importance of the history of science for the philosophy of science*. As long as theory is not universal it will profit from history of science. It is remarkable that Neurath is quite aware of the fact that "we have not [yet] developed a special technique for the analysis of trains of ideas". And he then emphasizes the point that most historians "operate with dichotomies . . . as if they were the most obvious" (GA, p. 88).[6] This quotation shows that Neurath is concerned with a general description of scientific theories based on history. Thus from the very beginning of his work he sees the aim of all investigation as lying in the realization of a *universal science*.

Since the history of science can show how scientific ideas and scientific theories are changing, the question how to view the change of theories was important. Very early Neurath found out two things: first, when we look at a scientific system like physics, we realize that its *growth* does not follow the simple rule that the results of scientific research *are added* one to the other in such a way as to enrich the storehouse of truths: rather, "the whole system itself is very often questioned. Therefore it is crucial to be aware of the problem, *how* long it is *useful* to keep the total system of a theory to explain a certain fact by using auxiliary hypotheses in order to accomplish the task, *or* when it is rather advisable to *reorder* and to change the whole system" (GA, p. 44f.).

In saying this Neurath shows a clear awareness of the main problem concerning a criterion for scientific progress and theory-change. When do we have to give up a theory and how long may we defend an old one? As we know, these questions turned into the questions of Kuhn, Feyerabend and contemporary philosophers of science.

The second point concerns Neurath's hopes and beliefs that the progress of science will lead to an *integration*. Thus he prophesies that the borders of scientific disciplines will not any more bring about an isolation of the researchers,

> but that a general, comprehensive research-work will determine the general principles of scientific work and so will provoke us to understand that the sciences *are* a unity of their own. The individual will then be able to gain an easy survey of the whole system of science, while today he is confronted with a chaos; what is common to all the sciences, will then be formulated, and *organization* of scientific work will be possible (GA, p. 45).

With these somewhat emotional words Neurath summarizes what will be the course of his intellectual career. It is mainly the *goal of the unity of science* that rightly is associated with the name of Neurath. But until a short time ago, it was not known that Neurath had formulated this program already in 1910.

In his article "On the Theory of the Social Sciences" Neurath mentions as a further aim – one which is exactly what was later on stated as the aim of the *Encyclopedia of Unified Science* – that there should come into existence a *unified view* of the world. It is this view which should prevent the atomization of our theories into individual sections and should provide us with a more satisfactory, non-atomized picture, so that the world will be seen as a whole again.

Those who follow the lifelong struggles of Neurath will agree that with these very general points the main topics of his philosophical work emerged very early and are to be found already in his earliest writings.

In at least two of these early writings Neurath was also concerned with the methodology of the history of science. He states at the beginning that one of the obstacles to an informative and up-to-date history of science lies, *first*, in the lack of systematic reconstruction of the theories, the system of hypotheses, and *second*, in the diversity of opinions on the principles of classification. "A continuous cooperation of those doing research-work in the history of science" – he claims – "is still missing."

He suggests, one of the ideas needed to get nearer to a genuine cooperation is a diversification of the underlying models guiding the research. What is meant by this can perhaps be brought out with the following example. If we want to reconstruct the history of some special field – the history of a period, say of physics, we always seem to be tempted to use one or two dichotomies to describe what was going on and what changes took place in a certain period of time in the given special field of science.

But in doing this, says Neurath, the historian of science is not at all doing justice to the real facts of the history of science. He introduces a classification, distinguishing what we might call the *A-type* and the *B-type*, the two being quite often independent of each other. Contrary to this tempting method of thinking in dichotomies – which Neurath declares to be unrealistic and to be the *result of an adversarial spirit – Neurath suggests a theory of theories*, that is, a metatheoretical science of scientific hypotheses and theories, where

> the theories would have to be dissected into their elementary components whose combination could then be fixed by a kind of formula ("On the Classification of Systems of Hypotheses").

Given such elements, Neurath thinks, *we could work out their possible combinations* and apply this richer framework to the description of individual cases in history. In using the total set of possible combinations, the description would be more adequate than would be that which flows from the fixed desire to stick to the device of simple dichotomies.

If we now pass from these early views of Neurath to his mature conception of an encyclopedic undertaking that is designed to comprise a metatheoretic philosophy of science, we still will recognize the old ideals of the unification of science by stressing the need for cooperation among the members of the scientific community. This community is seen as a republic, none of those members should have special privileges and each of whose members should have the same right to propound his/her own views, hypotheses and theories, all of them having the same duty to criticize the others, if for instance they use nebulous formulations, meaningless words or unfounded statements. Sometimes in this connection Neurath mentions half-seriously the advantage of an *index verborum prohibitorum*, containing a list of all words which should be avoided in scientific discourse or, what amounts to the same, in unified science.

Only in the sciences can we look for such a purification of our language. Therefore there *can* be a republic of scientists – dubbed by Kuhn the 'scientific community' – but *there cannot be a republic of philosophers*. For in the camp of philosophers, meaning metaphysicians, there *cannot* emerge cooperation, a republic built on cooperation is here not conceivable. If, however, it would be possible to unite philosophy and science in such a way that the latter provides the answers to those questions that can be tested in an appropriate way, then all meaningful problems fall on the side of science. This, by the way, was a different view of philosophy than the one Wittgenstein propounded in the *Tractatus* (4.11), because Neurath did not accept the view that "Philosophy" should mean something which is either "above" or "below" the natural sciences. "Philosophy", if taken seriously, should appear on the index of forbidden words.

But what kind of language is Neurath prepared to accept? On what kind of meaning criterion is he relying when he decides which words should be included on the index? Concerning the second question, Neurath for a long period seems to remain undecided as between a kind of corroboration criterion and the old means of banishing any "metaphysical" expression from meaningful talk. More and more however, he preferred the term "isolated sentence" for those series of words that are not connected and cannot be connected with the set of scientific propositions. The latter are first of all the *accepted* scientific propositions of our time, since we do not have any better. This should not mean that such an isolated sentence (or proposition) will remain excluded from the scientific discourse for ever. The "isolated statement" of today may fit into the scientific discourse of tomorrow. But how can we find this out? Certainly not in advance. The process of acceptance and rejection is a matter for the republic of scientists and depends on their *decisions*. If they change the totality of statements by changing some of them, then they always have the last word – at least for a certain period. It is clear that, according to this view, it is the present scientific community that decides which statements belong to scientific discourse at any given time.

Concerning the first question, namely, what language should be taken as the basic language of science, Neurath's answer may be a surprise: "It is the actual store of expressions which people have in common all over the world."[7] From this basis we may build an encyclopedia. If, however, we think that there will be an accepted procedure for selecting for instance those statements that are "in harmony with protocol statements", we may be wrong. Because Neurath warns us not to rely on truth-claims, because even "truth" depends on acceptance and there does not exist an arbitrator to decide between the truth of conflicting protocols. On the same line of argument also the choice of a universal jargon – the language of science – cannot be justified in advance: "No judge is in the chair who says which of us has THE TRUTH."[8]

The starting-point of this discussion can be found in Neurath's criticism of the Wittgensteinian concept of a language which is such that it can picture reality like a painter can paint a tree. Against such a view Neurath's opposition was strong and uncompromising in spite of the fact that in general we see him eager to reach compromises. He strongly denied the possibility that language can picture states of affairs or that signs may have the same 'multiplicity' as facts or reality.

Against these metaphysical-sounding formulations Neurath maintained that *even if* we would speak of "sentences and state of affairs", we should have to use *sentences* which means "sentences about sentences" and "sentences about non-sentences" to make clear what is meant by the one and the other. After all, we do have the possibility to speak about our language and to speak about it with our language.

If our encyclopedia contains a series of words which is in conflict with a statement '*p*', then we are not helped by the statement that '*p*' is true if and only if *p*. Neurath therefore suggests that we should use instead "accepted" and "non-accepted" or even better "encyclopedic sentence" *versus* "non-encyclopedic sentence". To use the semantic vocabulary for an ordinary language leads to unnecessary and dangerous complications which we avoid when taking only *sentences* into account. When we take a proposition, a sentence, let's say within a new hypothesis, this sentence, Neurath says:

> is confronted with the totality of existing statements already accepted. If it agrees with them, it is joined to them; if it does not agree, it is called 'untrue' and is rejected; or, the existing complex of statements of science is modified so that the new statement can be incorporated . . . There can be no other concept of 'truth' for science.[9]

Here as elsewhere decisions enter. But as we have seen from his later work, the mildness with the use of the "truth-vocabulary" vanished.

"Truth" therefore belongs to those expressions, which – like "reality", "correspondence" – have to be abandoned. The famous ship-metaphor – made famous by Quine who used it instead of quoting Neurath's naturalized epistemology – is applied in the argument that "we cannot find an absolute immutable basis for science: . . . various discussions can only determine whether scientific statements are accepted by a more or less determinate number of scientists and other men."[10]

I think it may seem strange that the most caustic critic of Wittgenstein within the Vienna Circle shares some of Wittgenstein's main points and shares in some sense a general view of language, which departs strongly from that of Carnap, his closest friend within the Circle. While Carnap – after sharing Neurath's move towards physicalism – worked for four years on the large project of the *Logical Syntax of Language*, a project opposed by Neurath, Neurath started with the much larger project of an encyclopedia of all the sciences. The main reason for investing his entire energy into this enterprise was the growing suspicion that the classical model of scientific systems could not be taken as a model for the vast variety of the actual sciences.

To achieve the goal of representing the scientific world view, one cannot rely on neatly defined concepts in all fields and stages of the sciences. Neither can we assume that the theories we accept at a certain time will not contradict others which also contribute to our world view. Emphasizing the fact that quite often propositions belonging to the corpus of different disciplines, like physics, geology, zoology and sociology, are used to refer to the *same kind of object*, Neurath thinks that only unification will, step by step, lead to a unified language. The program of physicalism was a sign-post pointing the way to how propositions of the different disciplines should find comparable referents.

From an encyclopedia proper, which may comprise the total mass of accepted scientific propositions (Neurath speaks consistently of "sentences"), there are to be distinguished those encyclopedias which Neurath called "model-encyclopedias". What is the task of the encyclopedia proper – as we have heard? It is to represent the total mass of scientific propositions accepted by the scientific community at a certain time; it is the task of a model-encyclopedia to try to present a type of encyclopedia which is clean, so to speak. Because we have to pay attention to the fact that the encyclopedia is bound to contain a great number of the impressions indirectly present in all the empirical sciences. It is not sufficient for us to have overcome the absolutism of systematic philosophy: "We have to beware also of the pseudo-rationalism which treats the empirical sciences as if they were a matter of logical relations between clearly distinguished propositions, independent one of the other, and constructed on the basis of perfect definitions."[11] The deeper reason for this are the cluster-concepts. In the mass of accepted propositions there are also those that are not consistent with many other propositions which nevertheless belong to one or the other of the accepted theories. One reason why this may happen is because the terms used in some of the theories are not clearly defined, that up to now a whole bundle of concepts remains in a mess, that even contradictions are not excluded. The region where we find formalizations that are not precise, not cleared from contradictions, Neurath calls "Ballungen" – a term which may be translated "clusters".

What leads Neurath to insist that even logical analysis will not lead to a language of science that has been purified from all inconsistencies and containing only exact defined terms? And further, to the view that we have to rely on a common mode of using language as the basis of the interpretation of any scientific theory?

To the first point. In a letter to Neurath of October 7, 1928 – at a period when they had not yet changed to the friendly "Du" – Carnap acknowledges that what Neurath had suggested to him belongs to "the most important things". Among others there are two points that Carnap emphasizes. First, that one should work on a logic, "a method of concept formation . . . which takes into account the fact that we are constantly faced with crystals and dirt (*Kristalle und Dreck*), a method which states the conditions that scientific concepts and statements should meet, as long as there does not exist an 'ideal language'". The second point in this early acknowledgement of Neurath's influence on Carnap concerns the importance of historical and sociological problems for a full understanding of science and society, another point Carnap promised to take into account – but almost never did.

I think this passage makes clear why Neurath, after his *dissociation from the semantic turn of Carnap*, feels free to resume the old topic of the "dirty" language which nevertheless is the basis for understanding and interpreting scientific discourse. As a historian and sociologist Neurath was aware of the fact that many sciences are not in a state that could be compared with mathematics or mechanics. Taking the actual state of science, that is, of that part which is accepted by the scientific community, leaves open what forms a rational reconstruction in Carnap's sense might take. It is precisely such attempts that Neurath called model-encyclopedia. But it would be completely wrong to think of them as the ideal form of a theory. There is doubtless more than one such form. Actually many of them are conceivable in regard to *one* theory.

As long as there is no method to decide which of the model-encyclopedias is preferable, all models are on the same level, as are the different disciplines – they are all waiting for acceptance. It is this fact which Neurath tried to reconcile with our common life: it is the practice of life, he says, that forces us to accept one or other of those encyclopedias.

There is no way to step behind language, the language of our common world, and there remains always the practice of life which brings about decisions. Even if science for Neurath seems to have the final word on everything empirical, there remains the fact that science cannot be done and cannot be judged beyond the practice of life.

University of Graz

NOTES

[1] Cf. R. Haller, "New light on the Vienna Circle", in: *The Monist* **65** (1982), pp. 25–37.

[2] Thomas E. Uebel (ed.), *Rediscovering the Forgotten Vienna Circle.* Dordrecht-Boston: Reidel, 1991 (Boston Studies in the Philosophy of Science 133).

[3] H. Putnam, "The 'Corroboration' of Theories", in: R. Boyd *et al.* (eds.) *The Philosophy of Science*, Cambridge: MIT Press, 1991, p. 134.

[4] O. Neurath, "Zur Theorie der Sozialwissenschaft", in: *Jahrbuch f. Gesetzgebung, Verwaltung und Volkswirtschaft im Deutschen Reich* **34** (1910), reprinted in: *O. Neurath, Gesammelte philosophische und methodologische Schriften* (eds.) R. Haller & H. Rutte (abridged: GA), Wien 1981, vol. 1, pp. 23–46.

[5] Cf. Neurath, “Zur Theorie der Sozialwissenschaft”, GA, p. 42.
[6] O. Neurath, “On the Classification of Systems of Hypotheses”.
[7] “The Orchestration of the Sciences by the Encyclopedism of Logical Empiricism”, in: O. Neurath. *Philosophical Papers* 1913–1946 (eds.) R. S. Cohen & M. Neurath (abridged: PP) (Vienna Circle Collection vol. 116) Dordrecht: Reidel, 1983, p. 242.
[8] O. Neurath, “Universal Jargon and Terminology”, in *PP*, p. 229.
[9] O. Neurath, “Physicalism” (1931), in: *PP*, p. 53.
[10] O. Neurath, *PP*, p. 181.
[11] O. Neurath, “Le developpement du Circle du Vienne et l’avenir de l’empirisme logique”, 1936 (German translation in GA).

JAAKKO HINTIKKA

WHAT IS ELEMENTARY LOGIC? INDEPENDENCE-FRIENDLY LOGIC AS THE TRUE CORE AREA OF LOGIC

1. IS ORDINARY FIRST-ORDER LOGIC THE TRUE ELEMENTARY LOGIC?

My ultimate aim in this paper* is to show that conventional wisdom is seriously wrong when it comes to logic. I want to challenge the traditional ideas as to what the most basic part of logic is like and how it ought to be studied. What is more, I will actually prove that these widely accepted ideas are mistaken.

Now what is the most basic part of contemporary logic, the true elementary logic? Most philosophers, and most logicians, would undoubtedly answer: first-order logic, also known as quantification theory, lower predicate calculus or – *nomen non est omen* – elementary logic. This part of logic was first developed explicitly by Frege. Indeed, the often-repeated claim that Frege is the true founder of modern logic is largely predicated on the assumption that first-order logic, that versatile creation of Frege's, is the true ground-floor part of logic.[1]

The belief in the status of quantification theory (first-order logic) as the true elementary (basic) logic is not unanimous. For instance, some philosophers and mathematicians have argued that second-order logic is the appropriate universal medium of mathematical theorizing.[2] I shall postpone my comment on the claims of second-order logic to a later stage of my line of thought. For my present purposes, it suffices to assume that first-order logic is a part of the genuine basic logic.

This assumption looks so trivial that it is seldom explicitly stated by philosophers. It is codified in virtually all introductory textbooks of formal logic which are nothing but expositions of first-order logic preceded by an ancillary presentation of propositional logic. Occasionally, however, the claims of first-order logic are stated in so many words. "If I don't understand first-order logic, I don't understand anything", a well-known philosopher-logician once said to me in discussion, undoubtedly echoing the views of a large number of other philosophers, not only of some of his Harvard colleagues.

K. Gavroglu et al. (eds.), Physics, Philosophy and the Scientific Community, 301–326.

The claims of first-order logic to a special status are often reinforced by, and sometimes probably based on, the idea that it is part and parcel of the logic of our natural language. The most influential (though cautious) proponent of this view is probably Noam Chomsky. His version of the time-honored idea of logical form (Chomsky's LF) is essentially the idea of a logical form of a first-order formula.[3] Some of Chomsky's followers have in fact argued for an even more intimate relation between first-order logic, as illustrated by a recent book title *Logic as Grammar*.[4] The "logic" intended here is essentially first-order logic.

Elsewhere I have shown that the claims of first-order logic to be the true *Sprachlogik* are to a considerable extent unfounded.[5] For instance, anaphoric pronouns do not behave in natural languages like variables of quantification theory, conventional wisdom notwithstanding, but more like quaint definite descriptions.[6] I shall not review my earlier arguments here, however. It suffices for my present purposes to note that the claims of first-order logic to a specially central role in logic cannot be based on its especially close relation to natural language and its logic.

Another line of argument for the sufficiency of first-order logic as the true basic logic relies on its alleged capacity of accommodating all mathematical reasoning. This type of argument proceeds step by step. On the one hand, it is argued that all mathematics can be carried out in terms of set theory.[7] On the other hand, it is argued (or assumed) that set theory itself can, and should, be thought of as a first-order axiomatic theory.[8]

Once again, this is too big an issue to be argued adequately here. It suffices to point out that it is by no means obvious that all mathematics should be approached via set theory. Not only are there serious alternatives to the status of first-order axiomatic set theory as the true logic of mathematics, such as categories and topoi,[9] as well as a suitable Bourbaki-type approach.[10] I shall in fact try to show on another occasion that, independently of the claims of such rivals, set theory is not the natural foundation of mathematical theories. Hence, the self-sufficiency of first-order logic as the basic logic cannot be argued by reference to set theory, either.

However, the claims of first-order logic as the true elementary logic might seem to be safe enough without such supporting arguments from linguistics or foundations of mathematics. It is an incontrovertible fact that even those philosophers who plead for the claims of suitable extensions of first-order logic to the title of general logic do usually consider first-order logic as the core area of general logic. It is this belief that I want to shake.

The first main point I shall argue for is that the usual formulation of first-order logic incorporates completely arbitrary restrictions. As soon as you understand the usual form of first-order logic, you *ipso facto* understand logical ideas that take you beyond it. They should therefore be incorporated in our basic general logic on a par with the ideas of traditional first-order logic. The only reason why they have not been codified in our ground-floor general logic is a number of arbitrary notational conventions which have no foundation in the true order of things or perhaps rather in the true order of logic. Hence, even though the conventional first-order logic is part of the true elementary logic, it is not all of it.

2. QUANTIFIER DEPENDENCE AS THE ESSENCE OF FIRST-ORDER LOGIC

In order to argue for this claim, we have to ask: What is needed to understand first-order logic? What is its conceptual gist? The obvious answer is: quantifiers. First-order logic does not have the alias "quantification theory" for nothing.

But this answer is seriously incomplete. There is more to first-order logic than the license to speak of *all* members of a domain and of *at least one* of its denizens. By means of these two ideas taken separately, we can obtain little more than the old syllogistic (monadic) logic. Something else is needed to provide first-order logic the escape velocity needed to propel it beyond the trivial realm of syllogistic reasoning.

Where does this extra force come from? The answer is not trivial, although there is not likely to be serious disagreement about it. The "secret" of quantification theory, the idea that enables us for instance to formulate highly interesting mathematical theories by its means, is the notion of *dependent quantifier*. Because of the availability of this idea, we can say in an applied quantificational language, not only that everything is so-and-so or that something is thus-and-so, but that for each individual there is at least one individual related to it in such-and-such a way. Thus it is the notion of a dependent quantifier that enables us to deal with functional relationships by means of first-order logic. It is not merely that we can admit relations into first-order languages; the logic of relations would be seriously handicapped without the idea of quantifier dependence. This is the reason why the order of dissimilar quantifiers matters, so that $(\exists x)(\forall y)Rxy$ and $(\forall y)(\exists x)Rxy$ have to be distinguished sharply from each other. It is also the reason why on the semantical plane we need the notion of satisfaction over and above the notion of truth.

The notational implementation of quantificational dependence is carried out by means of the notion of scope. In first-order logic, each quantifier (Q_1) comes fully equipped with a pair of parentheses following it which define the scope of (Q_1). Thus in $(Q_1)(-)$, $(-)$ is the scope of (Q_1). Any other quantifier which occurs within the scope of (Q_1) depends on (Q_1). Any quantifier which occurs outside the scope of (Q_1) is independent of (Q_1).

The idea of independence is of course familiar to my readers. For instance, what is said in

(1) $(\exists y)Rya$

is that there exists an individual which is related to the individual a in a certain way. What is said in

(2) $(\forall x)(\exists y)Ryx$

is that for *each* individual, call it x, there exists an individual which is so related to it. The choice of the value of "y" which makes (2) true depends on the choice of the value "x".

The insight that the true gist of first-logic lies in quantifier dependencies is not new. Among others, it is discussed in a most illuminating way by Warren Goldfarb, who argues persuasively that this insight was one of the motive forces of Hilbert's work in logic, culminating in his epsilon-calculus.[11]

The functions that serve to codify truth-making choices of the values of dependent quantifiers are known as Skolem functions.

3. QUANTIFIER INDEPENDENCE AND ITS CODIFICATION

So far, everything that has been said has been relatively unsurprising. How can it be, then, that the innocent-sounding remarks I have made point beyond the traditional first-order logic? The answer is exceedingly simple. To understand customary first-order logic means essentially to understand quantifier *dependence*. To understand quantifier dependence is the same thing as to understand quantifier *independence*: they are the two sides of the same conceptual coin. Hence to understand first-order logic presupposes understanding the idea of quantifier independence.

Now how are quantifier dependence and independence expressed in logic? Bracketing is one partial way of signalling quantifier dependence, but neither the only possible way nor an exhaustive one. In the light of what has been said, we are not taking a single step beyond what is needed to master the ordinary first-order logic if we amplify its notational apparatus by indicating explicitly quantifier independence. Because of the presence of the old device of parentheses, independence needs to be expressed only for ordered pairs of quantifiers where the latter of the two quantifiers would otherwise depend on the former, i.e., where the latter occurs within the scope of the former.

For the purpose I propose to introduce a simple slash notation. It can be explained as follows:

Assume that F is a well-formed formula of the usual first-order logic containing a quantifier (Q_0) which occurs within the scope of quantifiers $((Q_1), (Q_2), \ldots (Q_k)$, plus possibly some others. Then we obtain a well-formed formula F* by replacing, in F, the quantifier (Q_0) by

(3) $(Q_0/Q_1, Q_2, \ldots, Q_k)$.

Naturally, more than one quantifier in F can be independent of others in this way.

The intended interpretation of F* is determined by the idea that in it (Q_0) or, rather, (3), is to be taken to be independent of (Q_1), (Q_2), . . . , (Q_k). Otherwise, F* is to be understood in the same way as F.

How this intuitive idea can be implemented by an explicit semantical treatment will be discussed later. (See Section 8 below.) The result of carrying out the possibility of independent quantifiers (and of extending this idea back to propositional logic), starting from the usual first-order logic, will be called independence-friendly (IF) logic. More explicitly, such a logic will be called an IF first-order logic. Languages, including interpreted languages, whose logic is an IF logic, will be called IF languages.

Somewhat fuller explanations of the syntax and the semantics of IF first-order languages will be provided later. For my purposes in this paper, including the definition of IF first-order languages, it suffices to consider only such formulas whose negation normally do not contain independencies other than that of existential quantifiers.

The relation of IF first-order languages to ordinary first-order languages is thus like that of a game with imperfect information to the corresponding game with perfect information. As an example, you can think of the relation of the game of bridge to a simplified form of bridge in which all the players know the distribution of cards. What is important to realize is that if you understand one game, you automatically understand the other one, too. (Just think of how actual bridge games are analyzed by reference to diagrams showing all the hands.)

What has been said can be illustrated by means of particular examples. For instance, consider an ordinary first-order sentence like

(4) $(\forall x)(\exists y)\mathrm{R}xy$.

Its meaning is clear to everyone who understands first-order logic. But what would happen if the existential quantifier $(\exists y)$ were made independence of $(\forall x)$? In other words, how are we to understand the following?

(5) $(\forall x)(\exists y/\forall x)\mathrm{R}xy$

Consider, for a moment, what (5) says. Unlike the situation in (4), in (5) the truth-making choice of the value of "y" does not depend on the value of "x". Hence it might as well be made prior to the latter choice. In both cases, the formula which is to be made true is the same, viz, $\mathrm{R}xy$. Thus it can be seen that the force of (5) is the same as that of

(6) $(\exists y)(\forall x)\mathrm{R}xy$.

Thus, anyone who understands ordinary first-order logic will readily see, not only what (5) means, but that it is logically equivalent with the ordinary independence-free first-order formula (6). In other cases, the meaning of IF sentences is equally easy to understand, but they no longer reduce to ordinary first-order logic. The simplest example of such a formula is the following:

(7) $(\forall x)(\forall z)(\exists y/\forall z)(\exists u/\forall x)\mathrm{S}[x, y, z, u]$.

In (7), the truth-making choice of the value of "y" depends on the value of "x" but not on that of "z", and *vice versa* for "u".

It can be seen in a simple fashion that (7) cannot be expressed in a linear form without the independence indicator. For in such a representation, the quantifier $(\forall x)$ must precede $(\exists y)$, for the latter depends on the former. Likewise, $(\forall z)$ must precede $(\exists u)$. Again, $(\exists y)$ must precede

$(\forall z)$, for it is independent of the latter. By the same token $(\exists u)$ must precede $(\forall x)$. These requirements may be symbolized as follows

(8) $$(\forall x) \supset (\exists y), (\forall z) \supset (\exists u)$$
$$(\exists y) \supset (\forall z), (\exists u) \supset (\forall x)$$

However, they are easily seen to be incompatible with the requirement of linearity.[12]

A knowledgeable reader has long since noted that IF logic I am outlining comprises what is known as the logic of partially ordered quantifier prefixes, popularly but inaccurately known as the logic of branching quantifiers.[13] For instance, (7) is tantamount to what is known in the trade as the Henkin quantifier. It can be written in a self-explanatory notation as

(9) $$\begin{matrix} (\forall x)(\exists y) \searrow \\ \qquad\qquad\qquad S[x, y, z, u] \\ (\forall z)(\exists u) \nearrow \end{matrix}$$

Such a notation enhances further the case with which IF formulas can be understood, by displaying graphically the relative dependencies and independencies of the several quantifiers.

The idea of branching enables me to illustrate in yet another way the extent to which the principles needed to understand IF logic are already needed to understand the normal independence-free first-order logic. In a perfectly good sense, branching quantifier structures have always been part and parcel of the first-order notation. The branching of quantifiers *to the right* has always been admissible and understood without any special explanations. Quantifiers in different branches (branching to the right) are simply subordinate quantifiers whose scopes do not overlap. If one understands rightwards-branching quantifiers without any special instructions, by the same token one can appreciate the meaning of leftwards-branching quantifiers without being told how they are to be understood. The fact that quantifier structures branching to the right reduce to the usual linear independence-free quantifier prefixes (e.g., as in a prenex normal form) does not invalidate my point, for we do not understand them by means of a translation to a linear form.[14]

However, the IF logic envisaged here is in several respects richer than the logic of partially order quantifier prefixes. For one thing, the ideas of dependence and independence can be extended to the propositional connectives &, v.[15] They, too, can be dependent on each other and on quantifiers. For them, independence can be indicated by the same slash notation as in the case of quantifiers with one exception (or, rather,addition). Since different occurrences of the ampersand or of the wedge are not always distinguished from each other typographically or by their location in a sentence, it may be necessary to use subscripts to keep them apart from each other.

For instance, we may have expressions like the following:

(10) $(\forall x)(A[x]\ (v/\forall x)\ B[x])$

(11) $(A_1[x]\ (v/\&)\ A[x])\ \&\ (B_1[x]\ (v/\&)\ B_2[x])$

(12) $(\forall x)(\forall y)((A_1[xy](v/\forall x)A_2[x,y])(v/\forall y)(B_1[x,y](v/\forall x)\ B_2[x,y]))$.

It easily turns out that the first two of these reduce to ordinary first-order notation but the third one does not.[16]

In a similar way, quantifiers may or may not be independent of connectives within the scope of which they occur. Such independence can likewise be indicated by the slash notation.

One can extend the idea of independence even to negation. Notationally, we can simply extend the slash notation to negation. What the interpretation of this notation is has to be discussed separately. I shall not do so in this paper.

Thus it can be seen that understanding quantifier independence is an integral part of understanding ordinary quantification theory. If the conventional first-order logic is the ground-floor logic, or part of it, then so is the IF first-order logic.

4. THE SCOPE OF SCOPE

But this result poses a problem. How come, in view of the fact that the idea of independence is implicit in ordinary first-order logic, that is not indicated in the usual formulations of this logic? The right answer, it seems to me, is the notorious Montague reply: historical accident.[17] Or, perhaps I should say instead: arbitrary choice of notation which unnecessarily and artificially restricts what can be expressed in first-order logic. Above, I hinted at this fact by saying that in the usual first-order logic we have a way of signalling dependence but not a general way of signalling independence. This point deserves to be spelled out more fully.

The joker in the pack of the basic concepts of conventional first-order logic is the notion of scope. It is one of the received ideas which are taken virtually for granted but which soon turns out to involve a mare's nest of problems. I have spelled out some of those problems elsewhere.[18] A sense of the problem situation can perhaps be conveyed to the reader by a quick comparison between customary formal languages and natural languages. In the former, scope is usually indicated by a pair of parentheses (brackets) attached to a quantifier or to a propositional connective. In a natural languages, there are no parentheses available that could be used in the same way. (Customary parentheses are not scope indicators, but a discourse device, indicating as it were that certain contributions are not a part of the speaker's or writer's main line of thought.) How can natural language get along without the usual scope indicators?

A full answer cannot be given here. I have provided one on an earlier occasion. Instead, a few more specific comments can be made. One function that parentheses have in the usual formal languages is to indicate the syntactical limits of binding. This is in fact how the customary quantificational notation is usually explained. Let (Qx) be a quantifier. The explicit or implicit parentheses attaching to it, as in

(13) $(Qx)(S[x])$

demarcate (so we are told) that segment of the formula in which the variable "x" is *bound* to that (occurrence of the) quantifier (Qx). Sometimes this "scope" is compared with the segment of a sentence or discourse in a natural language in which anaphoric pronouns are "bound" to a head.[19]

Be this comparison between the variables of quantification and anaphoric pronouns as it may, we can see at once that the usual quantificational scope notation is arbitrarily restrictive. Accepting for the moment the idea of binding, we can see that there are a number of unwarranted assumptions built into the customary use of parentheses. Given a quantifier (Qx), the pair of parentheses indicating its scope are supposed to demarcate a segment of the formula in which variables are bound to it. However, the usual notation presupposes that a quantifier scope (i) begins immediately after the occurrences of the quantifier in question;

and (ii) is continuous (without *lacunae*). Moreover, (iii) quantifier scopes are assumed to be nested, that is to say, assumed never to overlap only partially. If the scopes of two quantifiers overlap, then the scope of one of them must lie completely within the scope of this other.

None of these three assumptions is lent any *a priori* support by the very explanations that are given of the meaning of parentheses and thus are supposed to enable us to understand them. For instance (cf. (i)) there is no law, logical, legal or divine, that says that the part of a sentence in which a variable is bound to a given quantifier (Qx) must being immediately after (Qx). This requirement is made awkward already by the fact that the conventional parentheses notation has systemically speaking other functions than to indicate the limits of binding. In fact, its first and foremost function is to show the relative logical priority of the different quantifiers. Such collateral functions may make it awkward to assume that the "scope" of a quantifier is always and everywhere adjacent to it.

Likewise (ii) represents a substantial assumption to require that the binding range of a quantifier is continuous. It is even fairly obvious that the corresponding assumption is false in natural languages.[20] It is fairly easy to find examples of a discourse where an anaphoric pronoun momentarily changes heads, only for the speaker to resume later to refer to the original head.

Furthermore, (iii), and most importantly, there is no *a priori* reason to assume that the ranges of binding for different quantifiers must be nested. In terms of dependence, this would mean assuming that quantifier dependencies are always transitive. Why should they be? From game theory, it is known that informational dependencies of the kind we are dealing with here need not be transitive.[21] But if they are not, quantifier "scopes" will not be partially ordered, contrary to what they are assumed to be in the customary notation for quantification theory.

If a quick example is needed for the difficulties that beset the notion of scope as applied to natural languages, a variant of the so-called Bach-Peters sentences will serve the purpose.[22] Consider the following perfectly understandable English sentence:

(14) The boy who was fooling her kissed the girl who loved him.

How are the quantifier scopes supposed to be ordered in (14)? There is no reasonable way of making them linear in this usual independence-free first-order logic, no matter how you analyze the definite descriptions.

However, when it comes to logic it might seem that I am here quibbling about a minor point of notation. In order to see whether the point is in reality so small, let us see what happens if we give up the third assumption (iii). If we do so, we need some notational convention to indicate which pair of parentheses or brackets goes together with which quantifier. This can be done by using different kinds of parentheses or more generally, by co-indexing parentheses and quantifiers.

Then we can construct well-formed formulas like the following:

$$(15) \qquad (\forall x)((\exists y)[Axy) \;\&\; (\forall z)((\exists u)[Bzu) \;\&\; R[y, u]]].$$

Of course, (15) could be written more pedantically (and less perspicuously) as

$$(16) \qquad (\forall x)_1({}_1(\exists y)_2({}_2Axy)_1 \;\&\; (\forall z)_3({}_3(\exists u)_4({}_4Bzu)_3 \;\&\; R[y, u])_2)_4$$

But the $64,000 question still remains: Does the liberalization of the use of parentheses represent more than a minor change in the notational conventions, perhaps something of the same order of (un)importance as the use of the Polish notation?

The answer is that the notational change has indeed significant consequences. It has to all practical purpose the same consequences as the step to an IF first-order language.

This can perhaps be seen from an example. Let us assume that we have a language that includes the pairing function $z = p(x.y)$ (expressing the fact that z is the ordered pair $\langle x, y\rangle$) and the left and the right unpacking functions $l(z)$ and $r(z)$. Thus, e.g.

$$(17) \qquad (\forall x)(\forall y)(\forall z)((x = l(z)) \;\&\; (y = r(z))) \leftrightarrow (z = p(x, y))$$

Then we can write out sentences like the following:

$$(18) \qquad (\forall x)((\exists y)[(x = l(y))) \;\&\; (\forall z)((\exists u)[(z = l(z)) \;\&\; S[l(y), r(y), l(u), r(u)]]].$$

This is seen to be of the same form as (15). But a moment's reflection also shows that it has the same force as the Henkin quantifier sentence (7) or (8).

Thus the unavoidable liberalization of the use of parentheses once again leads us to IF first-order logic. This logic is therefore installed more and more firmly as the true basic logic.

5. THE FAILURE OF AXIOMATIC METHODS IN IF LOGIC

What has been established so far? It may have disturbed some of my readers that we have to dethrone conventional first-order logic from its place of honor as our basic general logic and to install in its stead the IF first-order logic in the way I have indicated. But this is not the end of my story. The most important shock (or, for the happy few, the reassurance) is still to come. We encounter it when we actually begin to study the IF first-order logic sketched above, to gain a general view of it, to develop a metatheory for it, etc. How can we to do so? Here comes the real shocker: *We cannot study IF first-order logic satisfactorily by means of the most common tools of customary logical theory*. For what are those tools? On the formal (syntactical) level, the most important conceptual tool is the deductive-axiomatic method. What that means is that we try to enumerate recursively all valid logical truths as theorems derivable from formally characterized axioms by means of purely formal rules, the so-called "rules of inference". As Gödel showed in 1930, all logical truths of the ordinary first-order logic can be so derived, i.e., there are complete axiomatizations of first-order logic.

It quickly turns out however, that the IF first-order logic is not axiomatizable. Perhaps the easiest way of seeing this intuitively is to see what happens if you try to treat higher-order logics simply as many-sorted first-order logics.[23] You can capture everything*salva* satisfiability in first-order terms except the idea that there must be a higher-order entity (of a suitable type) corresponding to each extensionally possible class of (n-tuples of) lower-order entities. For instance, there must be (on the standard interpretation of second-order logic) a second-order entity corresponding to each class of individuals.

This requirement can be expressed by means of one single second-order universal quantifier. Hence the decision problem (for satisfiability) for a higher-order logic can be reduced to the decision problem for second-order sentences of the form

(19) $(\forall X)S[X]$

where the initial universal quantifier is the only higher-order quantifier.

Now it can be shown (this is implicit in Enderton's results)[24] that each sentence (19) is equivalent with the contradictory negation of an IF first-order sentence. Hence the decision problem of validity for higher-order logic reduces to that of IF logic. From this it follows that IF logic cannot be axiomatizable.

6. QUALMS ALLAYED

It is not that the failure of the axiomatic-deductive techniques in (a special case of an) IF logic has gone unnoticed. In fact, Quine has relied on this very failure in his attempt to discredit in effect IF logic as being an integral part of logic in contradistinction to mathematics.[25]

Quine's argument is seriously inconclusive. He tries to conclude from the possibility of representing the branching quantifier formula

$$(20)\qquad \begin{matrix} (\forall x)(\exists y) \\ (\forall z)(\exists u) \end{matrix} \Big> S[x, y, z, u]$$

in a second-order form

$$(21)\qquad (\exists f)(\exists h)(\forall x)(\forall z)S[[x, f(x), z, h(z)]$$

that (20) is in reality a higher-order (and hence, according to Quine's lights, "mathematical") statement. But precisely in the same way we can represent any first-order statement in a second-order form. Does that show that all of first-order logic is really "mathematics"? If Quine's argument is valid, it would show just that, providing us with a neat *reductio* refutation of Quine's position.

What is wrong with Quine's argument is that it confuses the question as to what is needed to understand a branching-quantifier sentence with the question as to what is needed to master the set of logical truths of the theory of branching quantifiers. It is virtually a confusion between the notions of truth and logical truth. The axiomatizability (or non-axiomatizability) of a branch of logic pertains to the latter, that is, to the treatment of logical truths, not to the former, that is, to the treatment of truth *simpliciter*. Yet it is the latter question alone that is relevant to the question of understanding a language. To understand a sentence is to know what things are like when it is true, not what they are like when it is logically true. Thus it is the question of understanding of actual truth of sentences that is crucial here. One could even turn the tables here on Quine and recall his reservations about any sharp distinction between analytic and synthetic truths. What is so especially important about the axiomatization of logical truths if there is no important distinction between them and the synthetic (empirical) ones? Complete axiomatization fails quickly outside pure logic anyway. Why should there be any problem about that in logic?

I suspect that the likes of Quine are suffering from methodological insecurity. If we cannot study logic by means of the axiomatic-deductive method, they seem to think, there is nothing equally clear-cut and familiar that we can resort to – or so it seems. The antidote to this insecurity is obviously to show that there are plenty of ways of handling and even mastering IF logic, even though its logical truths are not recursively enumerable. Indeed, that very nonaxiomatizability makes it possible to use IF logic for what I consider the most important task of logical theory: the discovery of successively stronger and stronger logical principles.[26]

This argument might seem to be too sophisticated, too *ad hominem*, to be convincing. It is nevertheless based on a solid insight.

The right method of freeing philosophical logicians from this methodological insecurity would be to show how IF logic can be successfully studied notwithstanding the absence of a completely axiomatization. On another occasion, I shall try to do so and even to indicate how IF logic can be employed to facilitate the discovery of new logical principles.

7. THE FAILURE OF TARSKI-TYPE TRUTH-DEFINITIONS IN IF LOGIC

We have discovered that axiomatic-deductive methods cannot serve to build an exhaustive theory of IF first-order logic. But perhaps semantical (model-theoretical) methods can do the trick? Well, what are those methods? Ninety-nine percent of my readers will answer: They are methods based on Tarski-type truth-definitions.[27] And now comes the next shock: *Tarski-type truth-definitions are not applicable to IF first-order logic*, either.

Why not? The reason is that Tarski-type truth-definitions rely on another widely (but not universally) accepted principle of language theory, a principle whose role has been thought to be to keep, if not the body and the soul of language theory together, than at least to keep syntax and semantics together. It is most commonly called the *principle of compositionality*.[28] Among its aliases is "Frege Principle". In one of its formulations it says that the meaning of a complex expression is a function of the meanings of its constituent parts. In practice, what it amounts to is an inside-out principle that implies a kind of semantical context-independence. If the meaning of a complex expression depends only on the meanings of its parts, it can never depend on its context in a still more comprehensive expression. Thus much of the actual force of the principle of compositionality lies in ruling out semantical context dependencies.[29]

Whether the principle of compositionality can always be honored in one's semantical theory, come hell or high water, will not be discussed here. Suffice it to observe that in an obvious sense a Tarski-type truth-definition presupposes compositionality. In such a truth-definition, the concepts of truth and satisfaction for a complex expression are defined recursively in terms of the truth and satisfaction of certain simpler expressions. In brief, a Tarski-type truth-definition works its way from inside out, and hence cannot accommodate any real semantical context-dependence.

But the very idea of quantifier independence violates the principle of compositionality. For the force of an independent quantifier (Q_1/Q_2) depends on another quantifier (Q_2) which occurs, not within the scope of the former, but outside this scope, in other words in the context of (Q_1/Q_2). This clearly violates compositionality, and it is the ultimate reason why Tarski-type truth-definitions do not work in an IF first-order logic.[30] By the same token, any approach to formal semantics, such as Montague semantics, which relies on compositionality, is incapable of handling IF quantifiers, at least without some special measures, such as taking refuge in higher order logic or some comparable counsel of despair.

Thus the role of IF first-order logic as the true elementary logic has a general methodological moral. It suggests that we should not put much trust in the principle of compositionality. This suggestion pertains in the first place to the semantics of formal languages. However, to the extent we can argue that an IF logic is a better framework of semantical representation for natural languages than ordinary first-order logic, to the same extent the injunction against compositionality applies also to the semantical treatment of natural languages.

Although the point will not be argued here, there is in fact ample evidence to suggest that independence of the kind we are dealing with here is a pervasive phenomenon in natural languages. Hence the status of compositionality in IF logics is relevant also to the semantics of natural languages.

In general, the role of informational independence in natural languages needs a closer study than can be devoted to it here. In particular, we need a fully argued-for answer to the question: Why did the presence of informational independence in natural languages elude so long the attention of linguists, logicians, and philosophers? The answer that seems to be current is: Because it is a relatively rare and marginal phenomenon. There is reason to think that the true answer is almost diametrically opposite.[31] It seems to me that informational independence is one of the most common and most spontaneously understood aspects of the semantics of natural languages. It escaped attention partly because of its very familiarity and of the very ease of its interpretation. An important contributing factor is that independence is not uniformly marked syntactically in natural languages. However, this absence of syntactical expressions of independence admits of an interesting explanation, and hence does not count as an argument against the ambiguity and naturalness of the phenomenon of independence in natural languages.[32]

In any case, the failure of compositionality in formal languages already has interesting consequences. One thing it shows is that the so-called substitutional interpretation of quantifiers does not work even in elementary logic, now that we have seen that this elementary logic has to include IF first-order logic.[33] The reason is that the substitutional interpretation is usually taken to conform to the principle of compositionality. If so, the question whether x, y, z and u satisfy the formula $S[x, y, z, u]$ as it occurs in (7) or in its independence-free variant

(22) $(\forall x)(\forall z)(\exists y)(\exists u)S[x, y, z, u]$

depends only on its substitution-instances $S[a, b, c, d]$. But the relevant substitution-instances are different in (7) and in (22) and they are not shown by the formula $S[x, y, z, u]$ alone. You also have to know the dependencies and independencies between the quantifiers to which x, y, z and u are bound.

This failure of our most cherished methods in systematic logical theory undoubtedly is not easy to countenance. Sixty years after Gödel's results we have perhaps become prepared to come across logics that cannot be axiomatized, but even then it is a surprise to see that our most basic logic is among them. And several logicians and philosophers have been so impressed by Tarski-type truth-definitions that they consider them the alpha and the omega of logical semantics. For them, it is bad enough a shock to find that Tarski-type truth-definitions cannot be used (at least not in any direct manner) in such exotic looking logics as the theory of branching quantifiers or infinitely deep logics. So much the worse a shock if Tarski-type truth-definitions cannot be used in the most fundamental region of logic which does not go beyond concepts which we have learned to understand and to master so completely that we are virtually taking them for granted.

8. GAME-THEORETICAL SEMANTICS TO THE RESCUE

In spite of the failure of Tarski-type truth-definitions in a first-order IF logic, this logic can easily be treated semantically. This fact should give us a pause, for it shows strikingly that Tarski-type truth-definitions do not constitute a universally applicable approach to logical semantics nor indeed the most basic approach.

One approach that is applicable to IF logics is what I have called game-theoretical semantics (GTS).[34] On an earlier occasion, I have suggested that, far from being a contrived and artificial approach to semantics, it is merely a systematization of the ways in which mathematicians (and other folk, for that matter) spontaneously speak and think of (dependent) quantifiers.[35] (Subsequently, of course, this systematization has been extended to the semantics of expressions other than quantifiers.)

The need to resort to GTS in IF logic further strengthens the justification of this way of looking at GTS. That GTS had not been formulated explicity any earlier can be attributed to its very naturalness. It was literally too obvious a way of thinking to have caught the attention of mathematicians and logicians. It is nevertheless significant that whenever Tarski-type methods failed, logicians spontaneously resorted to game-theoretical conceptualizations well before the explicit development of GTS. Examples are offered by the theory of partially ordered quantifiers, game quantifiers, and Gödel-type functional interpretations.

Epistemologically, GTS can be thought of as a way of relating the truth-conditions of a sentence S to the processes by means of which the truth or falsity of S can be ascertained.[36] They are in GTS conceived of as games between an initial verifier or "myself" and an initial falsifier or "nature". The game associated with S will be called G(S).

It starts with S, and its rules can be gathered from the basic ideas just adumbrated. They can easily be formulated for the ordinary first-order logic.

A background explanation is nevertheless needed. The game G(S) is played on a given model M of the language (L_s) of S. That M is a model of (L_s) implies that all the nonlogical symbols of (L_s) have been interpreted on M. This means that all the atomic sentences of (L_s) have been assigned a definite truth-value. The same goes for all the atomic sentences of any extension of (L_s) obtained by adjoining to it a finite number of names of the members of the domain do(M) of M. Thus the function of a game-theoretical truth definition, like a Tarski-type truth-definition, is in effect to extend the concepts of truth and falsity from atomic sentences to all others.

The game rules can be formulated as follows:

(R.v) $G(S_1 \vee S_2)$ begins with a choice of S_i ($i = 1$ or 2) by myself. The game is continued as in $G(S_i)$

(R.&) $G(S_1 \,\&\, S_2)$ begins with a choice of S_i ($i = 1$ or 2) by nature. The game is continued as in $G(S_i)$

(R.E.) $G((\exists x)S_0[x])$ begins with a choice of a member of do(M) by myself. Let "*b*" be a name of the individual chosen. Then the game is continued as in $G(S_0[b])$.

(R.U) Likewise, except that nature chooses *b*.

(R.~) $G(\sim S_0)$ is like $G(S_0)$, except that the roles of the two players, as defined by these rules, are reversed.

(R.A) G(A) is won by myself (and lost by nature) if A is true. If A is false, nature wins and myself loses.

(R.true) S is true iff there exists a winning strategy for myself in G(S).

(R.false) S is false iff there exists a winning strategy for nature in G(S).[37]

It is easy to see that in ordinary first-order logic this truth-definition agrees with the usual (Tarski-type) truth-definition, if it is assumed that all (mathematically possible) strategies are allowed for myself. If myself's strategies are restricted to recursive ones, we obtain a different concept of truth and falsity. This is one of the ideas on which

Gödel's famous functional interpretation of first-order logic and first-order arithmetic is based.[38]

In the rules so formulated it was tacitly assumed that the games G(S) are games with perfect information. What happens in IF logic is simply that this assumption is given up. The slash notation will then be a way of specifying the information sets of the different moves (in the usual game-theoretical sense of the term). For instance, the information set of

(24) $(Q_0/Q_1, Q_2, \ldots, Q_k)$

(where $Q_0, Q_1, Q_2, \ldots, Q_k$) are quantifiers or connectives) consists in moves connected with all the quantifiers and connectives within the scope of which (24) occurs, *except for* $Q_1, Q_2, \ldots, Q_k$.

Through this simple stipulation GTS can be extended to IF languages.

Thus we have explained completely how an IF first-order language is to be interpreted. Each sentence of such a language has been assigned a definite truth-value.

9. THE GENERALITY OF THE PHENOMENON OF INFORMATION INDEPENDENCE

At the same time, game-theoretical semantics shows that the phenomenon of independence is not a curious, isolated feature of the behavior of quantifiers. It is pervasive phenomenon which can come up in virtually any part of the semantics of formal as well as natural languages.

In order to see this, you only have to realize the general explanatory strategy involved in a game-theoretical approach to our *Sprachlogik*.[39] This strategy consists in associating a rule for semantical games with a wide variety of different logical and nonlogical symbols (words, constructions, formal symbols, etc.). The idea is to capture the semantics of the symbol in question by means of this game rule.

This strategy can be shown to work in a wide variety of different cases. A game rule is naturally associated with the usual truth-functional connectives, as illustrated by (R.v), (R.&) and (R.~) above. Likewise, a game rule can be associated with many of the logical or nonlogical symbols whose introduction gives rise to well-known extensions of first-order logic, including the symbols for necessity, knowledge (knowing that), belief, obligation, etc. Indeed, the possibility of so doing is implicit in the usual possible-worlds semantics for these notions. Indeed, operators like necessity are in that semantics in effect universal quantifiers ranging over alternative worlds.[40]

It is not my purpose in this work to discuss in detail the game-theoretical treatment of the different extensions of first-order logic. In part, such a treatment is typically quite obvious. In part, it is (or will be) actually worked out elsewhere. Even in cases where the role of game-theoretical ideas is not obvious, there are explicit examples of how they can be used in a highly interesting manner. The so-called functional interpretations of first-order logic and arithmetic, launched by Gödel in 1957, are cases in point,[41] as was first pointed out by Dana Scott in 1968.[42]

In a different direction, when GTS is applied to natural languages, a game rule not only can but must be associated with a number of further words, including pronouns[43] and prepositions.[44] Furthermore, it can be argued that a move (and hence an application of a game rule) must be associated even with our most primitive expressions, proper names and words expressing primitive predicates.[45] The background of this *prima facie* surprising phenomenon is possible-worlds semantics, where the value (interpretation) of any symbol, including a proper name, must be specified for each relevant possible world.

The same reason is operative in formal languages, where a move thus can be associated with the interpretation of a proper name in each model or "world" we may have to consider.

We do not have to consider the details of the rules governing such moves. The only fact that is relevant here is the existence of a game rule governing the move in question. For as soon as the semantics of a symbol is handled by means of a game rule, the notion of informational independence is *ipso facto* applicable. In the theory of natural languages, this leads to the expectation that informational independence is a widespread phenomenon, which rears its head in a wide variety of different categories. That this expectation is in fact fulfilled, has been argued extensively by Jaakko Hintikka and Gabriel Sandu.[46]

In formal languages, this means that the slash notation can be applied to expressions of all sorts of different logical types, with an interpretation which is obtained automatically as a corollary to any satisfactory game-theoretical treatment of the semantics of the formal language in question.[47]

We have already relied on this line of thought in applying the slash notation to propositional connectives.[48] Of the multitude of extensions of IF first-order logic, I mention here only IF (first-order) epistemic logic. It turns out to be especially important because of its applications to epistemology[49] and to the logic and semantics of questions and answers.[50] In such extensions, there is a game rule associated with each of the additional ingredients of the language, e.g., with the epistemic operator "K_x" which is to be read "*x* knows that". Hence we must consider additional types of independencies. For instance, an existential quantifier can be independent of K, as in

(23) $K_b(\forall x)(\exists y/K_b)S[x, y]$

It turns out that such existential quantifiers independent of a sentence-initial K play a major role in the theory of questions and answers in natural languages.[51]

Thus IF first-order logic can stake a claim to being the true basic logic also in the sense that it gives rise to various extended IF logics precisely in the same way in which ordinary independence-free first-order logic can be extended in different directions.

But are there restrictions as to how far IF first-order logic can be extended? It turns out that the main restriction is to be found in a surprising direction: negation. As a preparation for a study of this restriction, I shall offer here a couple of comments on the game-theoretical truth-definition. The rule (T.~) defines the concept of negation – or, perhaps I should say, *a* concept of negation – for IF first-order languages, by defining uniquely (jointly with the other rules, of course) what it means for a negated sentence to be true. In the usual independence-free first-order logic this concept of negation agrees with the usual one. However, it is easy to see that in the IF first-order language the game-theoretical concept of negation defined by (R.~) is not the contradictory negation. For this reason, it has to be handled with care. For instance, even though the law of contradiction holds for it, the law of excluded middle does not do so in general.

A closer examination nevertheless shows that the failure of *tertium non datur* is practically the only major change caused by the step from ordinary first-order logic to IF one. Practically all the "nice" metatheoretical results that do not depend on the presence of contradictory negation hold in IF first-order logic, such as compactness, upwards Löwenheim–Skolem theorem, axiomatizability of logically false sentences, separation theorem, Beth's theorem on definability. Some of these results hold even in a stronger form. For instance, in the separation theorem the separating formula can always be chosen to be an ordinary first-order one.

The exploration of all these opportunities is too large a task to be undertaken in one paper, however.

NOTES

* Even though this paper does not come as close to Bob Cohen's central interests as I hoped, its history perhaps makes it a suitable tribute to him. It was written at his request and presented under the auspices of Bob's Boston Colloquium for the Philosophy of Science as my unofficial inaugural lecture at Boston University, where I had moved to, to a large extent because of Bob's persuasiveness. I owe him much more than this modest paper can even symbolize.

1 The gradual development of the conception of first-order logic is an intricate subject of which the final truth has not yet been told. Meanwhile, you can have a glimpse of the problems from studies like Gregory H. Moore, "The Emergence of First-Order Logic", in William Aspray and Philip Kitcher, editors, *History and Philosophy of Modern Mathematics* (Minnesota Studies in the Philosophy of Science, vol. 11, University of Minnesota Press, Minneapolis, 1988, pp.95–135). Moore's paper is to be read with caution, however, for he is unaware of some of the most important conceptual points concerning the idea of first-order logic. For one thing, he does not even mention Henkin's distinction between standard and non-standard interpretations of higher-order logic. Yet it is possible to reconstruct a higher-order language, with a suitable non-standard interpretation, as a many-sorted "first-order" language, as far as logic is concerned. Cf. also below, especially sec. 16.

For some reasons for the ambivalence of Frege's status, cf. Jaakko Hintikka "The Place of C. S. Peirce in the History of Logical Theory", forthcoming.

2 Cf., e.g., Stewart Shapiro, "Second-order Languages and Mathematical Practice", *Journal of Symbolic Logic* **50** (1985), pp. 714–42; Georg Kreisel, "Informal Rigor and Completeness Proofs", in Imre Lakatos, editor, *Problems in the Philosophy of Mathematics*, North-Holland, Amsterdam, 1967, pp. 138–86.

3 See here Jaakko Hintikka, "Logical Form and Linguistic Theory", in Alex George, editor, *Reflections on Chomsky*, Basil Blackwell, Oxford, 1989, pp. 41–57.

4 Norbert Hornstein, *Logic as Grammar*, The MIT Press, Cambridge, 1984; cf. Robert May, *Logical Form*, The MIT Press, Cambridge, 1985.

5 See note 3 above and also Jaakko Hintikka and Gabriel Sandu, *On the Methodology of Linguistics: A Case Study*, Basil Blackwell, Oxford, 1991.

6 See Jaakko Hintikka and Jack Kulas, *Anaphora and Definite Descriptions*, D. Reidel, Dordrecht, 1985.

7 Cf. statements like the following: "Among the many branches of modern mathematics set theory occupies a unique place: with rare exceptions the entities which are studied and analyzed in mathematics may be regarded as certain particular sets or classes of objects." (See Patrick C. Suppes, *Axiomatic Set Theory*, Dover, New York, 1972, p. 1.)

8 The way in which first-order logic came to be considered the right foundation for set theory is described in Gregory H. Moore, *op. cit.*, note 1 above.

[9] See, e.g., S. MacLane, *Categories for the Working Mathematician*, Springer-Verlag, Berlin, 1971; Jon Barwise, editor, *Handbook of Mathematical Logic*, North-Holland, Amsterdam, 1977, chapters A8 and D6.

[10] See, e.g., Jean Dieudonné, *A Panorama of Pure Mathematics, As seen by N. Bourbaki*, Academic Press, New York, 1982 (with further references to the literature).

[11] Warren Goldfarb, "Logic in the Twenties: the Nature of the Quantifier", *Journal of Symbolic Logic* **44** (1979), pp. 351–68.

[12] This line of thought actually gives you a general method of deciding as to when a first-order formula with informationally independent quantifiers reduces to the linear form.

[13] Cf. here the brief bibliography of Jaakko Hintikka and Jack Kulas, *The Game of Language*, D. Reidel, Dordrecht, 1983, pp.300–3.

[14] It can be argued that, on the contrary, what the cash value of first-order sentences is, is brought out more clearly by pushing quantifiers deeper and deeper into the formulas, as in the distributive normal forms. This normally increases left-to-right branching, not decreases it.

For distributive normal forms, see Jaakko Hintikka, *Distribute Normal Forms* (Acta Philosophica Fennica, vol. 6), Societas Philosophica Fennica, Helsinki, 1953. A good exposition is also contained in Veikko Rantala, *Aspects of Definability* (Acta Philosophica Fennica, vol. 29, nos. 2–3), Societas Philosophica Fennica, Helsinki, 1977.

[15] See Gabriel Sandu and Jonko Väänänen, "Partially Ordered Connectives", *Zeitschrift für mathematische Logik und Grundlagen der Mathematik* **38** (1992), pp. 361–372; Gabriel Sandu, "On the Logic of Informational Independence", *Journal of Philosophical Logic* **22** (1993), pp. 29–60.

[16] See the papers referred to in note 15.

[17] According to the story (I was not an eyewitness), Richard Montague was once criticized on the grounds that his universal grammar did not explain the properties of actual languages (humanly possible languages) as distinguished from logical languages and from computer languages. "Why is it that we speak English and not ALGOL?" – Montague's sincere reply was, "Historical accident."

One can challenge Montague's reply, but in the case of the usual scope conventions of formal logic this notorious answer can be argued to be literally true; cf. below.

[18] Jaakko Hintikka, "Is Scope a Viable Concept in Semantics?", in *ESCOL '86: Proceedings of the Third Eastern States Conference on Linguistics*, ESCOL, Columbus, OH, 1987, pp. 259–70.

[19] This is in effect the way quantifier scopes are explained in textbooks of logic. In Hintikka and Kulas (note 6 above), it is shown that it embodies a seriously mistaken view of the way pronouns and quantifiers operate in natural languages.

[20] This point is argued in Hintikka and Kulas, note 6 above, especially pp. 137–8.

[21] The information set of a move may be any subset of the set of earlier moves. There is hence no reasons in general why the information sets of successive moves should be ordered even partially.

As it happens, in the special case of ordinary quantifiers, all possible types of informational independence can be shown to reduce to partial ordering. However, this special case is not representative of the general conceptual situation.

[22] Bach–Peters sentences were first introduced as *prima facie* counterexamples to certain types of generative grammars. Independently of this original use of theirs, they can be

employed also as counterexamples of certain types of rules for translating from natural languages to a logical notation.

[23] Cf. here Jaakko Hintikka, "Reductions in the Theory of Types", *Acta Philosophica Fennica* **8** (1955) pp. 57–115.

[24] H. B. Enderton, "Finite Partially-Ordered Quantifiers", *Zeitschrift für mathematische Logik und Grundlagen der Mathematik* **16** (1970), pp. 393–7.

[25] W. V. Quine, *Philosophy of Logic*, Prentice-Hall, Englewood Cliffs, N.J., 1970, p. 91.

Quine bases his negative attitude to branching quantifiers on the alleged fact that the loss of completeness makes a big change in the resulting logic. But this is a strictly circular argument. On the contrary, the preservation of such important results as the separation theorem in an IF logic can be used to argue that the inevitable step from ordinary first-order logic to an IF first-order logic is not a momentous step from the vantage point of a deeper and more sophisticated logical theory. Cf. below, secs. 10 and 12.

[26] This task of logical theory has not been given its proper due. It is perhaps considered most often in a context of an "experimentalist" and antirealist view of mathematics. But, as Gödel's example shows, it can be combined with a realist and even Platonist attitude to mathematical truth. Indeed, semantical (and in that sense "realistic") considerations are especially well suited to sharpen our ideas about possible stronger principles.

An example of how the methods used in this work (cf. sec. 8 below) help to motivate stronger logical principles is provided by the second-order translation to be outlined in sec. 11. The translation of

$$(\forall x)(\exists y)S[x, y]$$

is

$$(\exists f)(\forall x)S[x, fx].$$

If you simply require the two to be equivalent, you receive a form of the principle of choice.

[27] These methods go back to Alfred Tarski, "The Concept of Truth in Formalized Languages", in A. Tarski, *Logic, Semantics, Metamathematics*, Clarendon Press, Oxford, 1956, ch. 8. (The Polish original was drafted in 1930 and published in 1933; the German version appeared in 1935.)

[28] Cf. here Barbara Partee, "Compositionality", in F. Landtman and F. Veltman, editors, *Varieties of Formal Semantics*, Foris, Dordrecht, 1984, pp. 281–313; Barbara H. Partee, Alice ter Meulen and Robert E. Wall, *Mathematical Methods in Linguistics*, Kluwer, Dordrecht, 1990, section 13.1 (pp. 317–38).

[29] See here Jaakko Hintikka, "Theories of Truth and Learnable Languages", ch. 10 in Hintikka and Kulas (1983), note 13 above, pp. 259–92.

[30] Cf. here Jon Barwise, "On Branching Quantifiers in English", *Journal of Philosophical Logic* **8** (1979), pp. 47–80.

Of course, linearity can be restored by moving to higher-order translations. But it can be argued that the price of this move is far too heavy if one's aim is to develop a psycholinguistically viable theory of what is going on in ordinary discourse.

[31] See here Jaakko Hintikka and Gabriel Sandu, "Informational Independence as a

Semantical Phenomenon", in J. E. Fenstad *et al.*, editors, *Logic, Methodology and Philosophy of Science VIII*, Elsevier, Amsterdam, 1989, pp. 571–89; Gabriel Sandu, "On the Logic of Informational Independence", *Journal of Philosophical Logic* **22** (1993) pp. 29–60.

[32] See here Jaakko Hintikka, "Paradigms for Language Theory", *Acta Philosophica Fennica* vol. 49 (1990), pp. 181–209.

[33] This "interpretation" has been entertained *inter alia*, by Stanislaw Lesniewski, W. V. Quine (with reservations), Ruth Marcus and Saul Kripke. Here we finally have a conclusive argument to show why it is not a viable account of the semantics of quantifiers.

Cf. W. V. Quine, *Ontological Relativity*, Columbia U.P., New York, 1969, pp. 63–7, 104–8; Ruth Marcus, "Modalities and Intensional Languages", *Synthese* **13** (1961), pp. 303–22; Saul Kripke, "Speaker's Reference and Semantic Reference", in P. A. French *et al.*, editors, *Midwest Studies in Philosophy* vol. 2, University of Minnesota Press, Morris, MI, 1977, pp. 255–76; D. Grover, J. Camp and N. Belnap, "A Presentential Theory of Truth", *Philosophical Studies* **27** (1975), pp. 73–125; T. Baldwin, "Can There Be a Substantive Theory of Truth?" *Récherche sur la philosophie et le langage* **10**, Université des Sciences Sociales de Grenoble, Grenoble, 1989.

Cf. here also D. Gottlieb, *Ontological Economy: Substitutional Quantification and Mathematics*, Oxford University Press, Oxford, 1980.

In my judgement, the time has come, not only to kill the so-called substitutional interpretation, but to bury it.

[34] See here Hintikka and Kulas (1983), note 13 above; Esa Saarinen, editor, *Game-Theoretical Semantics*, D. Reidel, Dordrecht, 1979.

[35] Cf. Jaakko Hintikka, "On the Development of the Model-Theoretic Viewpoint in Logical Theory", *Synthese* **77** (1988), pp. 1–36.

[36] As such, they are systematizations and further developments of certain types of "language-games" in the sense of Ludwig Wittgenstein. (For them, see Merrill B. Hintikka and Jaakko Hintikka, *Investigating Wittgenstein*, Basil Blackwell, Oxford, 1986.) The failure of the *soi-disant* Wittgensteinians to put his important and promising concept to systematic use is but one example of their failure (or refusal) to recognize the constructive potentialities of Wittgenstein's ideas.

[37] Clauses (R. true) and (R. false) can be viewed as the true core in the often misused idea that a proposition is true if it "can be verified" or if "there exists a method of verifying it". Here the existence of a winning strategy explicates the idea of "can be verified" or "there exists a method of verification", which was left confused or misconstrued in the earlier uses of the same idea.

There might seem to be an alternative to (R. false) namely to define S to be false iff it is not true. This would make it impossible to deal with the falsity of propositions directly by means of game rules, however. Below, it will be seen what other consequences the choice between the two conceptions of falsity has; see especially sec. 15.

[38] See Kurt Gödel, "On a Hitherto Unexploited Extension of the Finitary Standpoint", *Journal of Philosophical Logic* **9** (1980), pp. 133–42. (Translation of Gödel's original 1956 article in *Dialectica* **12**, pp. 280–7, with a bibliography.) See also Kurt Gödel, *Collected Works* vol. 2 (edited by Solomon Feferman *et al.*), Oxford University Press, Oxford, 1990, pp. 217–53.

[39] This is the leading idea of the applications of game-theoretical semantics to natural languages. For examples of this strategy, see Hintikka and Kulas (1985), note 6 above.
[40] This is the gist of the semantics of model logics originally developed by Kanger, Montague and Hintikka. However, necessity-type operators do not interact with ordinary quantifiers quite in the same way they interact among themselves. The reason is that each possible alternative world (i.e., each value of a necessity-type "universal quantifier") has a domain of individuals (i.e., range of values of ordinary quantifiers) of its own, possibly different from others. This fact has important consequences. For instance, not all types of inter- and independencies can any longer be reduced to partial orderings.
[41] See note 38 above.
[42] Dana Scott, "A Game-Theoretical Interpretation of Logical Formulae", McCarthy Seminar, Stanford, 1967, published in the *Yearbook 1991 of Kurt Gödel Society*, Vienna, 1993, pp. 47–48.
[43] See Hintikka and Kulas, note 6 above, especially chapter 2.
[44] *Op. cit.*, especially pp. 172–9.
[45] *Op. cit.*, especially pp. 57, 159.
[46] See note 31 above.
[47] In other words, the slash simply temporarily exempts a quantifier, connective or some other expression from the scope of another quantifier (or similar) within whose scope it would otherwise be.
[48] Cf. note 15 above.
[49] For instance, different constructions in terms of *knows* cannot be analyzed without independence-friendly logic. Cf. here Jaakko Hintikka, "Different Constructions in Terms of 'Knows'", in Jonathan Dancy and Ernest Sosa", editors, *Companion to Epistemology*, Basil Blackwell, Oxford, 1992, pp. 99–104.

The need for independence-friendly logic has further repercussions in epistemology and philosophy of science.
[50] What is true of subordinate questions with *knows* is true of questions in general (cf. note 49): independence-friendly logic is needed to cope with their logic and semantics.
[51] For instance, (23) is the logical form of such wh-questions in natural languages as have an outside universal quantifier, e.g.

Whom does each person admire most?

There are also natural-language questions whose analysis requires independent propositional connectives, e.g.

Whom does each person admire more, his or her father or his or her mother?

THOMAS E. UEBEL

PHYSICALISM IN WITTGENSTEIN AND THE VIENNA CIRCLE

Various disputes about the invention of physicalism have recently come to the attention of Wittgenstein and Vienna Circle scholars.[1] Despite appearances, these matters are not merely of biographical, extra-theoretical interest but also hold lessons for the historiography of analytic philosophy in the 1930's and '40's. I will try to show that the disputants can be largely disarmed by close attention to their texts. Since there obtained a far wider plurality of doctrines in the Circle than is commonly recognized now – or was recognized even then – the disputes are moot. My resolution of the disputes is offered in tribute to Professor Cohen for three reasons. I will expand and substantiate one of his editorial suggestions in the English translation of the works of one of the disputants, and put to use an earlier remark of his on a striking parallel between Marx and Wittgenstein in support of a thesis congenial to his own work: one of the physicalist doctrines in contention may be viewed as the convergence of two philosophical traditions too often thought mutually exclusive.[2]

I

First, an overview of the main claims and the interpretative situation. Carnap's 1932 paper "*Die physikalische Sprache als Universalsprache der Wissenschaft*" (*The Unity of Science*) provoked several so-called priority claims. Thus Wittgenstein claimed that Carnap had stolen his ideas concerning physicalism, ostensive definition, the nature of hypotheses, and the distinction between the "formal mode" and the "material mode of speech" (here I am only concerned with the first).[3] Wittgenstein's displeasure was obviously heightened by Carnap's footnote which credited Neurath with pioneering work towards physicalism.[4] This long footnote in turn was the upshot of an intervention by Neurath sometimes construed as yet another but still earlier priority claim against Carnap.[5] In addition, note must also be taken of the neglected claims by a student member of the Circle, Heinrich Neider, and by its official leader, Moritz Schlick.[6]

K. Gavroglu et al. (eds.), Physics, Philosophy and the Scientific Community, 327–356.

Of the authors who mention this matter only six go beyond remarking on the irony of the situation and explicitly seek to adjudicate the claims.[7] Alberto Coffa called the acknowledgement Carnap granted to Neurath "grossly exaggerated" and Tom Oberdan characterized it as "quite generous."[8] These judgments stand in sharp contrast to Rudolf Haller's, who once remarked in passing that it "seems to be the case that the idea of physicalism – i.e. the possibility of reducing all empirical-scientific statements to those of physics, or of a physicalist language – goes back to [Neurath], even though both Wittgenstein and Carnap claim 'priority' for this conception." Returning to the matter recently, however, Haller somewhat modified his claim about Neurath and differentiated Wittgenstein's and Carnap's versions of physicalism.[9] Next there is the still different conclusion reached by Jaakko and Merril Hintikka, two researchers with access to Wittgenstein's still unpublished manuscripts and notebooks: though physicalism was discussed in the *Tractatus* only implicitly and indeed rejected "in favour of the phenomenological language", it is "understandable" that Wittgenstein should have been upset, for by 1929 he had adopted the view that the physicalistic language was the basic language and regarded this idea as fundamental to his – compared to the *Tractatus* – new outlook. This suggests that Wittgenstein's priority claim was justified. Indeed, more recently, Jaakko Hintikka claimed that "Wittgenstein's own statements show the identity of his views concerning physicalistic languages with those of Carnap."[10] Finally, there is Brian McGuiness' diagnosis – which seems to take the physicalistic doctrines at issue to be identical – that each disputant was concerned with different aspects of it: Wittgenstein "gave the impulse", Neurath "pronounced the importance" and Carnap "began to work out the details"; and Friedrich Stadler's – which follows Haller's differentiation of some of the physicalisms at issue – that one cannot speak of an "explicit appropriation" of their concept of physicalism from Wittgenstein.[11]

I will deal with the assessments of Wittgenstein's and Neurath's real and purported claims in sects. VI and VII, after laying out the relevant material. In my discussion I shall *not* challenge, but rely on, the Hintikkas' account of Wittgenstein's physicalism.[12] I shall reach a *different* conclusion, however, by paying close attention to parallel developments in the Vienna Circle instead. I shall argue that the main contenders provided different arguments and held incompatible views under the name "physicalism." The focus of my argument will be that Wittgenstein's was not the only discussion of the private language problem around and in the Vienna Circle in the '30's.[13]

II

Carnap's *Unity* must be read against the background of his earlier *Aufbau*. The *Aufbau* provided a *rational reconstruction* of scientific knowledge. A rational reconstruction makes no claim to reflect either the way knowledge is arrived at in practice, nor how it is justified in practice; rather, it presents how scientific claims could be understood and justified "in principle." (It does not claim psychological or sociological, but only logical "reality.") Carnap held that all scientific statements could be reconstructed either in the language of physics or in a phenomenalist language. However, in order to reflect what he deemed the order of epistemic priority – first "auto-psychological" objects are cognized, then physical objects, then other minds, and finally "cultural" objects – Carnap preferred a language with an auto-psychological base and adopted the position of *methodological solipsism*: empirical knowledge is built up from an individual's phenomenal experiences.[14]

The *Aufbau* attempted a rational reconstruction of a very radical sort. Carnap held that every statement of science could be (1) translated – via explicit definition – into a statement of the experiential given of an individual subject, and (2) that every statement so reconstructed only spoke of the structure, not the content of the experience. (It did so because Carnap tried to reconstruct even sense-data out of undifferentiated whole experiences and to give a structural description of them.) By combining reductionism with this "structuralism" Carnap sought to save empiricism from the challenge of sceptical doubt about "intuition." Already in 1918, Schlick had noted that empiricism was in trouble if it had to rely on the experiential, intuitive content of experience.[15] For intuition was (a) inevitably vague – when is an impression just of the same sort as a previous one? – and (b) private – the very ground of scientific knowledge was therefore outside of the realm of intersubjective confirmability. By exhibiting the *structure* of experience Carnap sought to restore to the empirical ground of science (a) the very firmness or definiteness that Schlick had suggested scientific theories possess in virtue of their concepts being defined implicitly by others (in terms of their nodal place in the network of scientific terms); and (b) restore its public character, for the structure, unlike private experiential contents, could at least in principle be compared.[16] The reduction of all scientific statements to the auto-psychological base amounted to the defense by proof of the thesis of the unity of science (both as expressing that there is only one object domain and one kind of methodological access to it).

Recent Carnap scholarship, of course, requires us to add that Carnap was not really interested in the phenomenalist reduction of all physical object talk, nor in establishing certain foundations for scientific knowledge claims.[17] His interest lay rather in establishing the objectivity claim of science by means of the structural explication of its concepts. This point is happily conceded. Yet it must be noted that, given his assumption of the epistemic priority of auto-psychological objects, Carnap could not have explicated objectivity as structure in terms of the physicalist language alone and have this explication answer epistemological interests, i.e., make the attainment of objective knowledge intelligible. If objectivity lay in the structure, the form of representation, and if human cognition proceeded from the immediate given, then the private content of representation was required to serve as the conduit for an epistemic subject's participation in that objectivity. Given Carnap's assumption, the physicalist language would have remained unintelligible without its reduction to the language of immediate experience.

Carnap's *Unity* repudiated some, but by no means all, of the *Aufbau's* underlying conception. Some terminological preparation is required before proceeding. For Carnap, the "thesis of physicalism" amounted to the thesis expressed in the original title of his paper: "The Physical Language as the Universal Language of Science." The thesis of the universality of the physical language meant that all scientific statements – including the protocol statements – were translatable into the physical language.[18] To prove this universality would also be to prove the thesis of the unity of science[19] – albeit in the opposite direction from that of the *Aufbau*. In addition to this difference, Carnap's 1932 proof proceeded in the explicitly "metalogical" vein.

The "thesis of metalogic" (or "thesis of syntax") held that metalinguistic discourse about the logical form of linguistic expressions was possible without paradox, or, as Carnap put it, that "all propositions of philosophy which are not nonsense are syntactical propositions, and therefore deal with linguistic forms."[20] Accordingly, Carnap drew the distinction between talk in the correct "formal mode" and in the potentially misleading "material mode" of speech: the former was the language proper to philosophical inquiry and spoke only of intra- or inter-linguistic matters ("syntax"), whereas the latter illegitimately took word-world relations as its topic.[21] Both the thesis of the unity of science and the thesis of physicalism thus must be distinguished from the thesis of metalogic. The thesis of physicalism constitutes a specific application to the languages of the empirical sciences of the metalogical principle of the efficacy of metalinguistic discourse. It represents the proper formal mode expression of the materialistic version of the unity of science thesis (in the material mode expression: of the materialistic unity of the scientific object domain).[22]

Carnap defined "science" as "the system of intersubjectively valid statements."[23] This meant only those statements belonged to science that different people could verify by reference to their own "direct" experience, expressed in their own protocol languages.[24] On this basis, Carnap proved the universality of the physical language for science by showing that the protocol language is but a sub-language of the language of physics. Physicalism demanded that

[formal mode:] statements in protocol language, e.g. statements of the protocol, can be translated into physical language.[25]	[material mode:] given, direct experiences are physically basic facts, i.e. spatio-temporal events.

Carnap's argument proceeded in the formal mode – which permitted the discussion of translation relations between different languages – though for ease of exposition he sometimes slipped into the material mode.[26] First, Carnap assumed the standpoint of the objector and, given his verificationist meaning criterion, came to the unsurprising conclusion that if physicalism were *not* true

> Every protocol language could therefore be applied only solipsistically; there would be no intersubjective protocol language. . . . Our fictitious supposition that the protocol language and the physical language speak of completely different facts cannot be reconciled with the fact that the physical descriptions can be verified empirically.[27]

Later Carnap assumed the physicalist standpoint and found that in this case

> every statement in the protocol language of S can be translated into a physical statement and indeed into one which describes the physical state of S's body. In other words, there is a correlation between S's protocol language and a very special sub-language of the physical language. This correlation is such that if any statement from S's protocol language is true the corresponding physical statement holds intersubjectively and conversely.[28]

Clearly, this is at best an in principle proof. (Carnap next argued that various objections concerning the different "meanings" of the language of experience and the physical language turn "upon a confusion between what is expressed by a proposition and the images we associate with the proposition,"[29] but basically he rested his case with his in principle proof.)

Note that Carnap's argument is clearly a verificationist one, directed against the thesis of the *untranslatability* of private protocol languages. "Physicalism" meant just this for Carnap: all languages could be translated into the physical language. (In broad outlines, this view was already endorsed in the *Aufbau*; thus it might be thought that Carnap's position here in 1932 is simply continuous with that of 1928, but I shall show in Sect. V that the details matter.)

III

The stage is now set to consider the priority claims Carnap's paper set off. Wittgenstein had met with members of the Vienna Circle (Schlick, Waismann, Feigl, Carnap) and discussed his views in 1927–8, but from 1929 to 1932 he only met with Schlick and Waismann when he was in Vienna; it was Waismann who took the role of communicating the latest results of Wittgenstein's thinking to the Circle at large.[30] His notes from these conversations have been published since; the entry for December 20, 1929 contains the following two paragraphs under the heading "Solipsism":

> I used to believe that there was the everyday language that we all usually spoke and a primary language that expressed what we really knew, namely phenomena. I also spoke of a first system and a second system. Now I wish to explain why I do not adhere to that conception any more.
>
> I think that essentially we have only one language, and that is our everyday language. We need not invent a new langauge or construct a new symbolism, but our everyday language already is *the* language, provided we rid it of the obscurities that lie hidden in it.[31]

This is what the Hintikkas call "Wittgenstein's physicalism": that the language which speaks of physical objects and things is the only one we need to use. It stands opposed to the view that in order to attend to philosophical, i.e., epistemological purposes we have to dig deeper and reach or construct a phenomenal or even solipsistic language. In the Hintikkas' interpretation, the *Tractatus* dealt with a "language of immediate experience" and so implied just such a "two languages model" as the one that Wittgenstein repeatedly rejected in the manuscripts of October 1929 (partly published later as *Philosophical Remarks*).[32] Wittgenstein's new preference was first revealed in public in the conversation with Schlick and Waismann.

What were Wittgenstein's reasons for this change of view? Since Waismann's notes are extremely obscure on this point I return to the Hintikkas' exegesis.[33] They note that Wittgenstein was not all that firm in his new conviction that a form of representation for the immediate given was neither possible or necessary; moreover, he continued to be interested in and tried to understand immediate experience.[34] Still, as reconstructed from his notebooks and manuscripts, in October 1929 Wittgenstein came to his physicalism by reflecting on the view that the meaning of a proposition consists in its verification, and on what is required for such verification to take place:

> the basic sentences of our language must be compared directly with (virtually superimposed on) the facts they represent. But since language itself belongs to the physical world, such comparisons must take place in the physical world. Hence only what there is in the physical world can be represented directly in language . . . [Wittgenstein] is moving from the public (physicalistic) character of language to conclusions concerning the nature of the objects that that can be represented in language.[35]

Clearly, Wittgenstein was *still only on the way* to his private language argument of *Philosophical Investigations*, on which his conviction of the primacy of the public language rested in his mature thought. Indeed, the Hintikkas note that in 1929 Wittgenstein "did not yet have available to him the conceptual tools for proving his point", namely that the idea of a phenomenological language has to be rejected.[36] The decisive consideration, according to the Hintikkas, is that "the public character of rule-following [derives] from the subordination of rules to language-games ('practices')"; this primacy of language-games in turn first emerged in Wittgenstein's ms. 116 – that is, not before the fall of 1937.[37] Wittgenstein's 1929 arguments for the primacy of the everyday language, by contrast, is a strongly verificationist one. Like Carnap's 1932 argument, it retains the Cartesian conception of the individual epistemic subject as confronting an immediate given – reined in only by the idea that for verification to be possible it is a public language that the subject must be speaking.

Wittgenstein's priority claim concerning physicalism then amounted to this: his result that the physical language was basic had simply been appropriated by Carnap and dressed up in scientistic garb (whole terms like "primary" language were retained).[38] Carnap did not accept Wittgenstein's claim,[39] but the lingering memory of the dispute may have moved him to have the translator, Max Black, contribute an introductory essay to the English version of *Unity* that discusses the relation between the *Tractatus'* phenomenalism and Carnap's physicalism. Before assessing Wittgenstein's claim, we must consider some parallel developments to his physicalist turn in the Vienna Circle itself.

IV

In the English translation, in fact, Carnap also retained the footnote in which he credited Neurath with the invention of physicalism and added an additional reference to the works of Neurath mentioned.[40] This long footnote was the upshot of an intervention by Neurath, sometimes construed as yet another bland priority claim Carnap had to fend off.[41]

What happened? Neurath's repeated protests[42] were sent in response to Carnap's earlier letter which informed Neurath that Carnap had revised his *Unity* paper – originally written "one and a half years earlier" – such that "the agreement between our views has thus become still closer." Carnap enclosed a copy and noted: "If in response to the revised version of my essay you want to change remarks in your essay that concern this matter, write to [the publisher]."[43] What upset Neurath primarily was his impression that his paper ("Sociology"[44]) would appear in *Erkenntnis* "behind" Carnap's *Unity* – which, as he realized, now formulated the thesis of physicalism much more precisely[45] – when it was in fact written before Carnap's revised version.[46] Moreover, Neurath felt he had contributed to Carnap's conversion to physicalism in the sense adumbrated in the revised version of *Unity*. To judge by his footnote – he credited Neurath with "many valuable suggestions"[47] – Carnap recognized Neurath's priority claim only to a certain degree. Their correspondence bears out this judgment: Carnap resolutely defended the originality of his argument and alleged the absence of any argument for Neurath's prior "theses"[48] – a response which left Neurath distinctly unsatisfied.[49]

I will evaluate the cognitive content of this dispute in Sect. VII below. But we may already note that Neurath claimed priority for the conception of physicalism, not for Carnap's argument. And so we may also ask: how does Neurath's claim relate to Wittgenstein's? Wittgenstein obviously thought he overruled both Carnap's and Neurath's claim to authorship of physicalism.[50] Recalling his conversation with Schlick and Waismann in 1929 we can see why: it was in December of 1929 that he declared the physical language to be basic. Does the shadow of Wittgenstein's accusation fall also on Neurath?

V

At this point we must turn to still another priority claim. In a 1975 conversation with Rudolf Haller and Heiner Rutte, Heinrich Neider – once a student member of the Vienna circle – outlined his claim to physicalism.

> I said to [Carnap]: "You will have to drop the auto-psychological base, for sentences about the auto-psychological are not intersubjectively verifiable, and sentences which are not intersubjectively verifiable do not belong to science. Philosophy can consider them in explaining how these sentences come about, and once they are formulated they must be intersubjectively verifiable." [*sic*] We talked for a long time and then Carnap asked: "This is really a very correct observation." . . . Neurath was delighted. Neurath was not present then, but the next evening we were at Neurath's and told him about it. He said: "Of course, finally!", for with this the bridge to materialism was built which he valued so much as the philosophy of the workers' movement.[51]

Retrospective priority claims need particularly careful assessment, of course, especially when, as in the present one, neither the date of the crucial conversation nor the precise form of the crucial argument is given. Circumstantial evidence – changes in the positions of Carnap and Neurath – leads me to grant it plausibility, to date it roughly and to spell out the argument which Neider left underdescribed.

As I noted, Carnap held that a constitutive system like the *Aufbau* could also have been erected on the basis of the physical language. (When he opted for the phenomenal language as basic, Carnap only sought to accommodate his conception of the epistemic priority of the immediately given.) Indeed, Carnap also remarked that "science . . . needs both an experiential and a materialistic derivation of all concepts."[52] What was his reason for this? It was that science in fact used a materialistic system of concepts. In the *Aufbau* then, Carnap did *not* give as a reason for this the need to support the intersubjective intelligibility of scientific statements.[53] Just that, however, was his declared reason for requiring a materialistic derivation of the concepts of science in a lecture to the *Verein Ernst Mach* in 1930: "[For] scientific statements to be intersubjectively transferrable, the basic concepts must be taken to be physicalistic ones."[54] Similarly, in his first contribution to *Erkenntnis* in 1930 Carnap noted:

> The positivist [phenomenalist] and the materialist system do not contradict one another. Both are correct and indispensable. The positivist system corresponds to the epistemological viewpoint because it proves the validity of knowledge by reduction to the given. The materialist system corresponds to the viewpoint of the empirical sciences, for in this system all concepts are reduced to the physical, to the only domain which exhibits the complete rule of law and makes intersubjective knowledge possible.[55]

But did not Carnap already account for intersubjectivity – and so for the objectivity of knowledge – in the *Aufbau*?[56] The key to Neider's argument lies in seeing how Carnap's claim to have done so is utterly undermined.

As it happens, Carnap's program for sustaining the intersubjectivity of his rational reconstruction of knowledge on the basis of methodological solipsism appears to differ from its implementation. In *Aufbau* §66, Carnap wrote: "even though the material of the individual streams of experience is . . . altogether incomparable . . . all streams of experience agree in respect of certain structural properties."[57] The idea suggested here is that certain structural properties of all possible experience (explicated as the analytical framework of reduction-chains in constitutive theory) were the same for all individual subjects. Intersubjectivity was explicated as the sharing of analytical frameworks. Just this form of intersubjectivity was not reconstructed when Carnap turned to "demonstrate the precise method" by which the *Aufbau* promised to sustain it. In §§146–9 he explicates intersubjectivity as arising from "intersubjective correspondences," namely, the *agreement of a subject's own determinations* of spatio-temporal relations in the world around him *with his reconstructions of another's determinations* of the spatio-temporal relations of objects in the world – of another that is already 'constituted' by the solipsist subject. "The intersubjective objects are abstraction classes of intersubjective correspondences."[58]

There is a world of difference between the two explications of "intersubjectivity." The first conception traded on the neutral reconstructive perspective of a third observer who could compare the analytical frameworks according to which the experience of two subjects is organized – and required a non-solipsistic language. (So it was not the case that the conceptual edifice of science could just as well be built up from a solipsistic starting point.) Only the first conception of intersubjectivity could explain the intersubjective practice of science at least *in prin-*

ciple, for it traced the parallel between two subject's frames of reference all the way to the base in their distinct immediate givens. By contrast, the second conception dispensed altogether with a real existing other; intersubjectivity was explicated as the subject's ability to map his own reconstructions of other's perceptions of the world onto his own – only a solipsistic base was assumed. Obviously, from the methodologically solipsist base no other conception of intersubjectivity could be provided, but such criticism is still pertinent to the problem at issue. Another's phenomenal evidence statements about his given remained wholly inaccessible to the second method of intersubjectivity-by-analogy, which remained restricted to intersubjective correspondences concerning the external world. This means that there could be *no intersubjective verification* at all, given that the ultimate evidence statements were held to be phenomenal ones.

Just this absence of intersubjectively verifiable evidence statements was the point of Neider's argument. Neider's argument to the effect that the basic evidence statements of science must be formulated in the physical language clearly pinpointed one of the weak spots of the *Aufbau*, namely its inability to underwrite actual rather than imagined intersubjectivity. Carnap's position as of 1930, reported above, suggests that he had already switched to a conception of intersubjectivity of the broad sort adumbrated in his 1932 paper: it takes the first reconstructive route, not taken in the *Aufbau*.[59] Carnap's change of mind therefore places Neider's argument prior to his article and lecture of 1930.

Now consider Neurath. Still in 1928, Neurath had endorsed "profess[ions of] materialism" explicitly only as an expression of a proper anti-bourgeois attitude and placed its acceptance by "exact thinkers" in the future.[60] In the published version of his address to the Circle's first conference in Prague in September 1929, however, he clearly and repeatedly affirmed materialism: he spoke of the "scientific world-conception on a materialistic basis," of "sociology on a materialistic basis," etc.[61] Related to this, in a newspaper article of October 1929 Neurath asserted that "all concepts are formed in the same fashion: on the basis of certain rules of control, everything that is asserted is reduced to singular experiences *which everybody can test*."[62] Considering that, for Neurath, the difference between (updated) materialism and physicalism was but a difference in name,[63] it is clear that he too had drawn the conclusion from Neider's argument. This would date Neider's argument to October 1929 or earlier.

But why was the result of this argument published only in 1932, if it was put to Carnap in 1929? Here we must note that the first version of *Unity* was completed in September 1930.[64] Moreover, the abstract of Carnap's 1930 Verein Ernst Mach lecture suggests that the thesis of physicalism was already adumbrated then: it asked "How is 'unified science' possible?" and answered "every (intersubjectively discernible) state of affairs can be expressed in the physical language."[65] In a letter of 1933, finally, Carnap characterized the second, "new phase" of the development of the Vienna Circle by two acquisitions: "syntax mainly since 1931 (my first draft: Jan. 1930)" and "(*mainly since 1929?*) physicalism, unified science."[66] Carnap's adoption of Neider's argument – if that's what it was[67] – simply took a long time to appear in print.

Still, one could only grant Neider's priority claim if one were prepared to neglect the metalogical revisions that the published version of *Unity* incorporated: though the exact degree to which Carnap's argument followed Neider's verbal presentation cannot be assessed, it is most unlikely that the distinction between the formal and the material mode of speech was drawn by Neider.[68] Moreover, the final *Unity* argument talked, as Neider's argument apparently did not, about the protocol language as a sub-language of the language of physics. As we know from his correspondence with Neurath, however, it was this – its development within the strictures of metalogic – that Carnap understood to be the central feature of his argument.[69] Compared to Carnap's mature *Unity* argument and thesis of physicalism, Neider's argument and thesis constituted but a rough precursor.

VI

Wittgenstein also claimed to have drawn the distinction between the formal and the material mode of speech in "the last few propositions of the *Tractatus*."[70] In his *Logical Syntax* of 1934, which represents the full working out of his metalogical approach, Carnap – perhaps in response to these charges – frankly admitted Wittgenstein's precursor role in this respect, but also pointed out two important differences between his own and Wittgenstein's conception: the *logical* form of *sentences* can be *expressed* and *formalized*.[71] But I shall not pursue this theme further since it leads ultimately to the remaining priority claims of Wittgenstein's as well – and to the question of different Wittgensteins "in transition"[72] and away from my topic here: what about Wittgenstein's priority claim concerning physicalism?

It is important, as Haller has pointed out, that for Wittgenstein it was the *everyday* physicalistic language that was primary: there is no indication that he accepted Carnap's view that the "system language" of physics was the primary language, as announced by the latter's definition of physicalism, nor that he embraced the unity of science view that there was only one kind of object. Thus "Wittgenstein's physicalism was only similar to that of Carnap and Neurath."[73] Yet even granted this point, it may still be thought that Carnap merely scientized Wittgenstein's findings of 1929 (engaged in "cheap cannibalization," as Wittgenstein put it uncharitably).[74] Thus it is important to note, first, the independent impetus towards physicalism he received (via Neider and Neurath) and, second, a still further difference between Carnap's and Wittgenstein's conceptions.

From the record of the conversations with Schlick an Waismann it is clear that Wittgenstein already thought the "two languages model" *unwarranted* in the fall of 1929 – there was only one language, the language of everyday which spoke of physical objects, not sense-data. By contrast, Carnap maintained a methodologically solipsistic protocol language alongside the physical language: Carnap *retained* the "primary" language for purposes of epistemology, and so still held to the "two languages model." This position of Carnap's is apparent not only from the passage of his paper of 1930 that I quoted earlier, but is also made clear in the paper so offensive to Wittgenstein.

[formal mode:]	[material mode:]
The simplest statements in in the protocol language are are statements needing no justification and serving as foundation for all the remaining sentences of science.	The simplest statements in the protocol language refer to the given, and describe directly the given experience or phenomena, i.e. the simplest states of which knowledge can be had.[75]

The very point of Carnap's paper was, after all, to show that this "primary" language was fully *translatable into* the physical language, for otherwise the primitive protocols would stand *wholly* outside of the realm of intersubjectivity, and so of science. Now it might be thought that the distinct nature of the protocol language had thus become unimportant for Carnap, especially since he also admitted the possibility of non-phenomenal protocols. But this is to mistake the import of Carnap's concession.

> If A utters a singular psychological sentence such as "Yesterday morning B was happy", the epistemological situation differs accordingly, as A and B are not the same person. Consequently, we distinguish between sentences about other minds and sentences about one's own mind . . . this distinction cannot be made among the sentences of intersubjective science. For the epistemological analysis of subjective, singular sentences it is, however, indispensable.[76]

Carnap is quite definite then about the epistemological role that the primitive protocol language still has to play, despite its translatability into the physical language. They form the termini of the epistemological reduction in virtue of their incorrigibility. In fact, only in their original form are the protocols *not* in need of justification; once they are translated into the system language of science they become revisable (on pain of unrevisability entering into science itself) and need support. In case of a conflict of a person's protocol with "the facts," we can only change the translation of this protocol into the physical language, but not the original protocol itself: "a protocol sentence, being an epistemological point of departure, cannot be rejected."[77] (Only in the still later "On Protocol Statements" did Carnap briefly entertain the idea that these primitive protocol sentences themselves could be declared false and eliminated, but at this point he abandoned his earlier insistence on primitive protocol statements altogether for a preference for Popper's "basic" sentences, indicating a general turn towards conventionalism.[78])

What we find then is that Carnap and Wittgenstein had quite *different* ideas of what physicalism amounted to. Whereas Wittgenstein rejected the phenomenal language (but retained interest in "phenomenology"), Carnap retained the possibly phenomenal, in any case, methodologically solipsist, primitive protocol language for epistemological purposes. The Hintikkas' conclusion that Wittgenstein had reason to be upset about Carnap's physicalism for it incorporated his insight of 1929 is mistaken (as was Wittgenstein's own fear that his own conception of physicalism had been plagiarized by Carnap). Haller's detection of their differences is reinforced: in fact, Carnap was considerably less radical in his physicalism than Wittgenstein.

VII

I turn to Neurath's so-called priority claim. As noted, he did *not* lay claim to Carnap's material-formal mode argument. But was he happy with Carnap's reworked version of physicalism, and did he just go along with Neider's argument? Like Wittgenstein, Neurath held an "only one language" view. But his reasons did not derive from Wittgenstein. From 1931 to the end of 1932, Neurath inveighed against Carnap's "two languages model" on five occasions, with variations of the same argument: the conception that phenomenal protocols – however much they were deemed translatable into the physical language – could serve as termini of testing procedures *had* to ge given up.[79]

Neurath argued against Carnap's methodological solipsism with what amounts to a *private language argument*. In fact, the conflict between Carnap and Neurath is usefully conceptualized as one between *two kinds of private language arguments* of different strength, and the priority dispute about physicalism between Carnap, Neurath and Wittgenstein as turning on what follows from *three distinct private language arguments*.

Is this name and my interpretation an undue anachronism? Well, it isn't unless we think that only Wittgenstein has the right to call his considerations of the private language problem in *Philosophical Investigations* a private language argument.[80] (Given the multiplicity of interpretations of that argument, it is dubious anyway whether the singular article "the" is still justified.) Consider: What was Carnap concerned with in establishing the universality of the physical language? To show that if the phenomenal protocol language did not have a physical interpretation, then it was no more than a methodologically solipsistic, but essentially private language – and to rule it out on these grounds!

The background for Neurath's private language argument is given by Carnap's retention of a primitive protocol language in the sense of providing a private knowledge base for science (as documented above).

> Carnap . . . distinguished two languages: a 'monologising' (phenomenalist) one and an 'intersubjective' (physicalist) one. . . . However, in my opinion . . . on the contrary *only one language comes into question from the start*, and that is the physicalist. One can learn the physicalist language from earliest childhood. If someone makes predictions and wants to check them himself, he must count on changes in the system of his senses, he must use clocks and rulers; in short, the person supposedly in isolation already makes use of the 'intersensual' and 'intersubjective' language. The forecaster of yesterday and the controller of today are, so to speak, two persons.[81]

If physicalist statements like instrument readings need themselves to be translated, in order to be meaningful statements, into phenomenal terms directly related to a scientist's experience, then no touchstone at all would be available by which the correctness of predictions could be established: What could such phenomenal terms refer to but my experience *now*? Neurath suggested that once on a solipsistic base, there was no preventing solipsism of the moment. The incoherence and impossibility to conceptualize experience in this solipsistic fashion showed that the correctness of predictions can be established only by reference to the spatio-temporal determinations of physical states of affairs that the language spoke about. Now, if language use was so controllable then that language was already intersubjective (and intersensual).[82]

What makes this a private language argument? It shows, like Neider's and Carnap's arguments, that a phenomenal protocol language won't do. The important point, however, is that for Neurath, already the checking of predictions by the same individual requires that the language used be physicalistic. Only the physicalistic language afforded the constancy of language use on which the checking of predictions even by the predictor himself depended.[83] Neurath's argument highlights the social nature of language use and cognition which was *perforce* left out of Carnap's account. To emphasize this, consider the following objection to Neurath's argument. According to his argument, only reference to non-phenomenal states of affairs allowed for the constancy of an individual's language use; but how could such checks of constancy be effected? Surely the individual himself could not do so. Whether his language was phenomenalist or physicalist, his understanding of his own language use still remained prone to the vagaries and fleetingness which had convinced Schlick already in 1918 that the objectivity of knowledge was not to be found in its reliance on ostensively defined terms. How could the essentially intersubjective nature of the physicalist language make a difference? The answer is that it was not the physicalist character of the language all by itself which allowed for constancy of language use, but the intersubjective control of an individual's language use that the physicalist language made possible. That its referents were public objects or events, ensured that *others* could correct an individual's language use, and that *thereby* an individual speaker could assure himself of its constancy. A phenomenal protocol language whose physicalist interpretation is at best established "in principle," but for which no translation manual has been given, simply will not be able to do this job.

Here I can only give this one example of Neurath's argumentation; the discussion of its merits against Carnap's rational reconstructionism and for Neurath's own naturalism must be provided elsewhere.[84] Similarly, I cannot provide a comparative study of Neurath's and Wittgenstein's *later* private language arguments, their interesting parallels and differences. Already, however, this much is clear: in the early '30's, *different private language arguments were in play to support different conceptions of physicalism.*

Neurath's private language argument did not merely follow Carnap's verificationist version, for it was not the understanding of others, but the *self*-understanding of an individual epistemic agent that was at issue. Neurath's argument also went beyond Neider and Carnap in that it called into question what Neider thought unusable for the reconstruction of scientific practice, yet did not hold unintelligible in itself: the phenomenal protocol language. Finally, Neurath's conception of the primary physicalist language also seems to have differed from Carnap's early on: unlike Carnap who required the translation of solipsist protocols into the system-language of physics (or considered their translation to belong to a sublanguage of the system-language proper – which is unclear), Neurath's protocols were formulated in the everyday physicalist language that "one can learn . . . from the earliest childhood" – explicitly so still before his "Protocol Statements", which explicitly declared the primacy of the everyday language over the system-language.[85]

Like Wittgenstein's priority claim then, Neurath's must be viewed with reservation, for even Carnap's conception of physicalism wasn't exactly his, but only a weaker version of it. (In fact, Neurath became aware of this soon enough: from mid-1932 onwards he referred to his own version as "radical physicalism," and reemployed his private language argument in further arguments against Carnap.) Yet Carnap did not, after all, seem to have noted these differences himself.[86] So how, it may be objected, could he have resisted Neurath's priority claim? Here we must remember that, for Carnap, the thesis of physicalism was inextricably linked with, though distinct from, his thesis of metalogic: Unlike Neurath, who disputed the need for and existence of a private protocol language and so made the physicalistic language universal by default, Carnap asserted that all scientific languages were translatable into it. This metalogical rendering of the thesis of physicalism further distinguished his from Neurath's and provided the ground on which to resist Neurath's priority claim as far as he did.[87]

Now compare Neurath and Wittgenstein. Whereas Carnap's private language argument turned on what is required for intersubjective verification, Wittgenstein's of 1929 and Neurath's of 1931 turned on what is required for one single individual to verify his statements. Yet Neurath's argument also did not proceed from consideration of what is required to verify a statement by comparing it to reality, as Wittgenstein's reasoning did at the time, but instead proceeded from what is required to compare a later statement (protocol) with an earlier one (prediction). (Thus "statements are always compared to statements."[88]) Moreover, as of 1931/2, Neurath's position was also more radical then Wittgenstein's, who, as the Hintikkas noted, was still attracted to the idea that "the world we live in is the world of sense-data" – it's just that "the world we speak about is the world of physical objects."[89] To judge from his private language argument, Neurath would have rejected the idea of such a phenomenal life-world for its want of constancy. Finally – and this point of his conception is shared with Carnap – Neurath's physicalism differed from Wittgenstein's in respect of its status: physicalism was a proper scientific (metatheoretical) doctrine, not merely a helpful though strictly speaking meaningless, "elucidation", as pronouncements on the status of languages may at this point still have been for Wittgenstein.

Strictly speaking then, Haller's early claim that Neurath holds the priority for physicalism must be rejected for the same reason we rejected the Hintikkas' claim for Wittgenstein's priority: the conceptions of physicalism in play were quite distinct doctrines. Haller's modified assessment is correct, however, in respect of the broad sense shared by Carnap's and Neurath's physicalism: Neurath was "the first to propose 'physicalism' as the correct point of view and physicalization as the correct method for getting rid of those things that cannot be brought down to earth."[90] *Pace* Coffa, Carnap's laudatory footnote was justified. Moreover, Neurath was also the first consistently radical physicalist, asserting the revisability of all scientific statements already in lectures and discussions in March 1931.[91]

VIII

That the priority disputes occurred despite their mootness shows that even the participants and contemporary onlookers of the protocol sentence debate were not always clear about the identity of the positions espoused by their fellow travellers. This conclusion is confirmed when yet another player is brought into the game, Moritz Schlick. Having accepted Carnap's rejection of Wittgenstein's claims in 1932, Schlick joined the fray in 1935 and claimed that *he* had proposed physicalism – albeit not with this "unpleasant" terminology (an echo of Wittgenstein's complaint) – already in his *General Theory of Knowledge* of 1918:[92] "spatio-temporal concepts may be used to describe *any arbitrary* reality, without exception, including the reality of consciousness."[93] Schlick's original argument for this conclusion was more or less repeated in his 1935 article, namely that: "Physics is the system of exact concepts that our knowledge correlates to *all* reality, since according to our hypothesis the entire world is open to designation by that conceptual system."[94] In his 1935 article, Schlick made two points: first, he argued that the application of color concepts in our subjective experience can be correlated with physical determinations concerning the frequency of light and the state of the nervous system at the time; second, that the application of the concept of mourning can be correlated with certain behavioral traits. Schlick had already employed the first type of argument in 1918,[95] clearly anticipating Carnap's argument in 1932 for the intersensuality and intersubjectivity of the physical language.[96] Only the latter of the two points made in 1935, the one that turned on behaviorist application conditions of psychological terms (employed by Carnap in 1932 to establish the universality of the physical language and independently gaining increasing importance in Wittgenstein's thought), was not used by Schlick in 1918 when the universality of the physical language was simply asserted.

His correspondence with Neurath makes clear that Carnap was deeply embarrassed[97] and that he accepted that Schlick first formulated the thesis of physicalism.[98] Though not put in terms of the formal/material mode distinction, Schlick's early formulation turned on different conceptual ways of "designating the real."[99] While in fact violating the strictures on material mode talk, Schlick may be seen to have anticipated the metalogical rendering of the thesis Carnap placed so much weight on (translation relations) by his talk of correlations. Applying the strict criteria for the identity of doctrines used here, however, we may hesitate to follow Carnap's estimation.[100]

Still, the parallels between Carnap's and Schlick's conceptions of physicalism in the 30's run deep indeed. Both stress that physicalism is an empirical thesis and does not express the necessity that the phenomenal and the physicalist languages are intertranslatable[101] – nor, it is implied, that a phenomenal language was impossible. Both stress, moreover, that 'physicalism' does not represent a philosophical movement or a comprehensive viewpoint, but merely the thesis of the 'universal' character of the physical language.[102] It is this latter point in particular which once again brings out the difference between Carnap's (and Schlick's) physicalism and Neurath's conception: for Neurath, radical physicalism included the thesis that it was impossible to leave the realm of language (and confront statements with reality) and issue pronouncements on the nature of Reality.[103] (Similarly, his private language argument sought to establish that a phenomenal language was a human impossibility.) Quite consistently then, Neurath denied Carnap's conclusion that Schlick anticipated (what he understood by) 'physicalism,' though he noted in his report on the Paris Congress later that year that Schlick "developed important starting points for physicalism already in his theory of knowledge."[104]

In their exchange on this matter, we can see the gradual recognition by Neurath and Carnap of the deep differences between their guiding philosophical conceptions. (Had Wittgenstein been on speaking terms with them, it is likely that a similar clarification would have obtained between them.) Not only is it true that each disputant "was convinced his task [initiating, publicizing, detailing the thesis of physicalism] was the really important one"[105] – but the thesis they sought to promote was not one but many, even different between Carnap and Neurath.[106]

IX

If this diagnosis is accepted another conclusion follows immediately. Note that it was the *basis* of empirical knowledge claims – traditionally understood to be the immediate given – that was called into question, more or less radically, in these varying physicalist ways. This fact supports a view of the philosophical developments taking place around the turn of the 30's that contradicts the still common lore of the Vienna Circle. It is a mistake to view the theorists of the Vienna Circle as merely "logistically reconstructed foundationalists." Not only the justification for high-level generalizations, but also the basic theorem of empiricism had come under increasing pressure within the empiricist camp: Neurath's radical position and argument is but the most striking proof of this fact. And as the misguided priority disputes investigated here indicate, however unique the final contribution of Wittgenstein may be thought to be, it must be noted that at least some of his concerns were independently shared by proponents of the "scientism" his philosophy is often thought to tell against.[107]

But this is still not the whole lesson of the priority dispute over physicalism, for we also have to ask the question: what emboldened Neurath to adopt his private language argument? Two observations are relevant here. The first is that, since the early 1910's, Neurath had relied on the conception of language put forth in his mentor Ferdinand Tönnies' "Philosophical Terminology".[108] For Tönnies language was essentially social – a view that in the Vienna Circle Neurath felt called upon to defend once he was confronted with Carnap's individualist conception of linguistic competence in the *Aufbau* and *Unity*. The second point to note is that in two publications of the year 1931 – the year he first employed his private language argument – Neurath quoted the following passage from Marx and Engel's *German Ideology*.

> . . . Only now, after we have considered the primary historical relationships, do we find that man also possesses "consciousness"; but, even so, not inherent, not "pure" consciousness. From the start the 'spirit' is afflicted with the curse of being 'burdened' with matter, which here makes its appearance in the form of agitated layers of air, sounds, in short, of language. Language is as old as consciousness, *language is practical consciousness that exists also for other men and for that reason alone it really exists for me personally as well*: language, like consciousness, only arises from the need, the necessity of intercourse with other men . . .[109]

Of this passage, it has been noted that it "vigorously attacked" the "idea of a private language" and that it may have influenced Wittgenstein's thinking at Trinity College, Cambridge through the intermediary of Piero Sraffa, the Marxist economist (himself one of the very few influences Wittgenstein ever acknowledged).[110] We must leave this issue of influence as obscure as it is, even though it is clear that Marx and Engels only leave us with the common language. Indeed, some years earlier, Robert Cohen had already drawn attention to the parallel between "Marx' aperçu: language is practical consciousness" and Wittgenstein's "to imagine a language is to imagine a way of life".[111]

There can be a little doubt, however, that Neurath did not fail to see the implication of Marx and Engels' passage when he quoted it twice in 1931. One of his quotations occurs in a paper dedicated to convincing his comrades in the Austrian Socialist Party that all the "philosophy" Marxists ever needed was – physicalism! Thus placed in context, there remains little mystery why Neurath should have come up with his private language argument as early as he did.[112] Rather, it further illumines Carnap's remark that Neurath

> believed that our [sic!] form of physicalism was an improved, non-metaphysical and logically unobjectionable version which today should supersede both the mechanistic and the dialectical forms of nineteenth century physicalism.[113]

Neurath's own theory of science may be joined to the debate between Marxism and Logical Empiricism that Professor Cohen himself (re)initiated in his contribution to the Carnap *Festschrift*.[114] Indeed, the editorial suggestion of his that I have followed up here nudges us gently to continue such explorations.[115]

London School of Economics

NOTES

[1] E.g. alphabetically: Alberto Coffa, "From Geometry to Tolerance: Sources of Conventionalism in 19th Century Geometry", in *From Quarks to Quasars*, ed. by R. G. Colodny (Pittsburgh: University of Pittsburgh Press, 1986), pp. 3–70, fn. 56; idem, *The Semantic Tradition from Kant to Carnap*, ed. by L. Wessels (Cambridge, England: Cambridge University Press, 1991), pp. 407–8; Rudolf Haller, "Geschichte und wissenschaftliches System bei Otto Neurath", in *Wittgenstein, der Wiener Kreis und der kritische Rationalismus*, ed. by H. Berghel, A. Hübner, E. Köhler (Vienna, Austria: Hölder-Pichler-Tempsky, 1979), tr. "History and the System of Science in Otto Neurath", in *Rediscovering the Forgotten Vienna Circle. Austrian Studies on Otto Neurath and the Vienna Circle*, ed. by T. E. Uebel (Dordrecht, The Netherlands: Kluwer, 1991), p. 36; idem, "Wittgenstein and Physicalism", *Critica* **21** no. 63 (1989), pp. 17–32; *idem*, "Ansichten und Deutungen des Physikalismus", forthcoming; Michael Heidelberger, "Zerspaltung und Einheit: vom logischen Aufbau der Welt zum Physikalismus", in *Philosophie, Wissenschaft, Aufklärung*, ed. by H.-J. Dahms (Berlin: de Gruyter, 1985), pp. 144–189, p. 177 fn.; Merril B. Hintikka & Jaakko Hintikka, *Investigating Wittgenstein* (Oxford: Blackwell, 1986), pp. 145 ff.; Jaakko Hintikka, "Ludwig's Apple Tree: Evidence Concerning the Philosophical Relations between Wittgenstein and the Vienna Circle", in *Traditionen und Perspektiven der analytischen Philosophie. Festschrift für Rudolf Haller*, eds. W. L. Gombocz *et al.* (Vienna: Hölder-Pichler-Tempsky, 1989), pp. 187–202; Brain McGuinness, "Wittgensteins Beziehungen zum Wiener Kreis", in *Jour Fixe der Vernunft. Der Wiener Kreis und die Folgen*, eds. P. Kruntorad, R. Haller, W. Hochkeppel (Vienna: Hölder-Pichler-Tempsky, 1991), pp. 108–126; Ray Monk, *Ludwig Wittgenstein. The Duty of Genius* (1990, New York: Viking Penguin, 1991), p. 324; Michael Nedo & Michele Ranchetti, *Ludwig Wittgenstein. Sein Leben in Bildern und Texten* (Frankfurt, Germany: Suhrkamp, 1983), pp. 254–5, 381–2; Thomas Oberdan, "The Concept of Truth

in Carnap's *Logical Syntax of Language", Synthese* **93** (1992), 239–260, fn. 12; David Pears, *The False Prison*, vol. 2 (Oxford: Clarendon Press, 1988), pp. 302, 316; Neil Tennant, "The Life and Work of the Early Carnap", in N. Rescher, ed., *Scientific Inquiry in a Philosophical Perspective* (Lanham, MD: University Press of America, 1987), pp. 272–74; Friedrich Stadler, "Wittgenstein und der Wiener Kreis – zwischen Rezeption und Plagiat", in *Philosophie, Psychoanalyse, Emigration. Festschrift für Kurt Fischer zum 70. Geburtstag*, ed. P. Muhr, P. Feyerabend, C. Wegeler (Vienna: Universitätsverlag), pp. 398–414.

[2] See Otto Neurath, "Protokollsätze", *Erkenntnis* **3** (1932/33), tr. "Protocol Statements", in *Philosophical Papers*, eds. Robert S. Cohen and Marie Neurath (Dordrecht: Reidel, 1983) (hereafter cited as "*PP*"), p. 97 line 2 – "Therefore it does not make sense to speak of monologising *[private]* languages, as Carnap does . . ." (my italics) – and Robert S. Cohen, "Dialectical Materialism and Carnap's Logical Empiricism", in P. A. Schilpp, ed., *The Philosophy of Rudolf Carnap* (LaSalle, Ill.: Open Court, 1963), pp. 99–158, at 149–151. For Cohen's important role in the Neurath rediscovery, see my "Otto Neurath and the Neurath Reception: Puzzle and Promise", in *Rediscovering*, pp. 3–22.

[3] The sequence of events (accusations and responses sent via Schlick, etc.) can be reconstructed from the letters by Wittgenstein to Carnap, August 20, 1932, and to Schlick, August 8 and 21, 1932, in Nedo & Ranchetti, pp. 254–5 and 381–2; cf. Hintikka & Hintikka, *op. cit.*, p. 146 and Hintikka, *op. cit.*, pp. 191–2, 202, McGuinness, *op. cit.*, Stadler, *op. cit.*

[4] Rudolf Carnap, "Die physikalische Sprache als Universalsprache der Wissenschaft", *Erkenntnis* **2** (1931/2), tr. of rev. version by M. Black, *The Unity of Science* (London: Kegan Paul, Trench Trubner & Co., 1934, hereafter: *Unity*), p. 74; cf. p. 95.

[5] Tennant, *op. cit.*; Oberdan, *op. cit.*; Coffa, *Tradition*, p. 408.

[6] Rudolf Haller and Heiner Rutte, "Gespräch mit Heinrich Neider", *Conceptus* **28–30** (1977), pp. 29–30; Moritz Schlick, "De la relation entre les notions psychologiques et les notions physiques", *Revue de Synthèse* **10** (1935), transl. P. Heath "On the Relation between Psychological and Physical Concepts", in Schlick, *Philosophical Papers Vol. 2*, eds. H. L. Mulder and B. F. B. van de Velde-Schlick (Dordrecht: Reidel, 1979), p. 426. Neider's claim was noted in passing by Heidelberger, *op. cit.*

[7] E.g., Tennant, *op. cit.*, and Monk, *op. cit.*, refrain from a full assessment though their tone implies they side with Carnap and Wittgenstein, respectively. Still stronger, Pears, *op. cit.*, seems to echo Wittgenstein's complaint by asserting that (not only Schlick's but also) Carnap's ideas about language and the incommunicability of intuitive mental content were "developed out of the ideas put forward by Wittgenstein" (pp. 302, 316). Pears has in mind as sources Wittgenstein's notes posthumously published as *Philosophical Remarks*, ed. R. Rhees (Chicago: University of Chicago Press, 1975), §58, and his remarks in Friedrich Waismann, *Wittgenstein and the Vienna Circle. Conversations Recorded*, tr. J. Schulte and B. McGuinness, ed. B. McGuinness (Oxford: Blackwell, 1967), pp. 49–50.

[8] Coffa, *Tradition*, p. 408, Oberdan, *op. cit.* Coffa also considered Wittgenstein's priority claim there and in "Tolerance", *op. cit.*, fn. 56, and called it "ludicrous" and "absurd," respectively. Coffa's concern there, however, was Carnap's metalogic and conception of the a priori, so it not pertinent to the present topic.

[9] Haller, "History," p. 36, and "Wittgenstein," respectively. For a further differentiation of Neurath's physicalism see his "Ansichten".
[10] Hintikka & Hintikka, *op. cit.*, p. 147, and Hintikka, *op. cit.*, p. 193, respectively.
[11] McGuinness, *op. cit.*, p. 122, Stadler, *op. cit.*, p. 410.
[12] Hintikka & Hintikka, *op. cit.*, p. 146, and Hintikka, *op. cit.*, p. 193. The Hintikkas' thesis of Wittgenstein's physicalism has not remained uncontested. On the one hand, McGuinness, *op. cit.*, p. 121 fn., detects a broad form of physicalism already in the *Tractatus*, while on the other Pears, *op. cit.*, p. 316, fn. 33, sees in Wittgenstein's work at the time only the demand for physicalistic *criteria* for sensation concepts. I will not engage with these exegetical matters, since for present purposes the Hintikka's reading certainly makes the strongest case for Wittgenstein's priority claim.
[13] This point has long been noted by Hector-Neri Castaneda, "Private Language Problem", in *Encyclopedia of Philosophy*, eds. P. Edwards *et al.* (New York: Macmillan and Free Press, 1967), vol. 6, pp. 459–464, but not explored further; see below note 80.
[14] Carnap, *Der Logische Aufbau der Welt* (Berlin, Germany: Bernary, 1928), tr. R. George *The Logical Structure of the World* (Berkeley: University of California Press, 1967), §§54–60, 64 (hereafter cited as "*Aufbau*").
[15] Moritz Schlick, *Allgemeine Erkenntnislehre* (Berlin, Germany: Springer, 1918, rev. ed. 1925), transl. A. E. Blumberg *General Theory of Knowledge* (LaSalle, Ill.: Open Court, 1974), pp. 29, 38 (hereafter referred to as "*GTK*").
[16] Carnap, *Aufbau*, §66.
[17] E.g. Alberto Coffa, "Idealism and the Aufbau", in *The Heritage of Logical Positivism*, ed. N. Rescher (New York: University Press of America, 1985), pp. 133–156; Michael Friedman, "Carnap's *Aufbau* Reconsidered", *Nous* **21** (1987), pp. 521–545, "Epistemology in the *Aufbau*", *Synthese* **93** (1992), pp. 15–57; Joelle Proust, *Questions de Forme: Logique et proposition analytique de Kant a Carnap* (Paris, France: Fayard, 1986), tr. A. A. Brenner *Questions of Form: Logic and the Analytic Proposition from Kant to Carnap* (Minnesota: University of Minnesota Press, 1989); Alan Richardson, "How not to Russell Carnap's Aufbau", in *PSA 1990*, eds. A. Fine, M. Forbes, L. Wessels (East Lansing, Mich.: Philosophy of Science Association, 1990), vol. 1, pp. 3–14; cf. Richard Creath, "Carnap, Quine and the Rejection of Intuition", in *Perspectives on Quine*, R. Barret and R. Gibson, eds. (Oxford, England: Blackwell, 1990), pp. 55–66; and Thomas A. Ryckman, "Coordination and Convention: A Chapter of Early Logical Empiricism", in *PSA 1990, op. cit.*, vol. 2, 1991, pp. 149–158. The failure of Carnap's reduction need not interest us here (see, e.g., Friedman, "*Aufbau* Reconsidered".).
[18] Carnap, *Unity*, p. 42.
[19] *Ibid.*, p. 96.
[20] *Ibid.*, p. 38 fn.
[21] *Ibid.*, pp. 38–39.
[22] *Ibid.*, p. 37.
[23] *Ibid.*, p. 66.
[24] *Ibid.*, pp. 50–52.
[25] *Ibid.*, p. 76.
[26] For another recent treatment see Thomas Oberdan, "Positivism and the Pragmatic Theory of Observation", *PSA 1990, op. cit.*, vol. 1, pp. 25–37. The failure of Carnap's "syntacticism" need not interest us here either (see, e.g. Michael Friedman, "Logical Truth

and Analyticity in Carnap's *Logical Syntax of Language*", in *History and Philosophy of Mathematics*, eds. P. Aspray, P. Kitcher (Minneapolis: University of Minnesota Press, 1988), pp. 82–94.

27 Carnap, *Unity*, pp. 80, 81.

28 *Ibid.*, pp. 87–88; cf. 60–66.

29 *Ibid.*, p. 91.

30 Brian McGuinness, "Editor's Preface", in Waismann, *op. cit.*

31 Waismann, *op. cit.*, p. 45.

32 Hintikka & Hintikka, *op. cit.*, pp. 138–140.

33 Coffa pointed out (*Tradition*, p. 406) that they even contain a *mis*representation of an argument from *Philosophical Remarks*, namely in the very parallel exploited by Pears (cf. note 7 above).

34 Hintikka & Hintikka, *op. cit.*, p. 141 n. 12 notes that in manuscripts from May 1932 Wittgenstein still toyed with the idea of a phenomenalistic system of representation.

35 Hintikka & Hintikka, *op. cit.*, pp. 166, 168.

36 *Ibid.*, p. 241 (cf. also Coffa's treatment of Wittgenstein in transition in *Tradition*, ch. 14).

37 *Ibid.*, p. 243. The Hintikkas refer to passages in the second part of MS 116 which – according to by G. H. V. Wright, "The Wittgenstein Papers", rev. and repr. in *Ludwig Wittgenstein. Critical Assessments, vol. 5*, eds. V. A. and S. G. Shanker (London: Croom Helm), pp. 1, 12 – "contains revisions from the academic year 1937–8". See also Hintikka & Hintikka, *op. cit.*, p. 243, fn. 4 for reference to an earlier (abandoned) attempt to introduce the notion of language game and rule-following simultaneously in 1931.

38 Carnap, *Unity*, p. 44.

39 Carnap to Schlick, July 17 and September 28, 1932 (RC 029-29-09, 029-29-03). Carnap merely sent Wittgenstein a note of acknowledgement and referred him to his simultaneous letter to Schlick (September 28, 1932, RC 102-78-01). (All documents cited with prefix "RC" or "HR" are to be found in the Archive of Scientific Philosophy, Hillman Library, University of Pittsburg. I wish to thank Gerald Heverley, its curator, for his help and the Archive for permission to quote from its holdings.)

40 Here Carnap made good a printer's omission of a late revision to the German version. See note 49.

41 See note 5.

42 Seven letters, postcards and telegrams to Carnap and Phillip Frank between January and March 1932 (RC 029-12-69, 029-12-68, 029-12-67, 029-12-65, 029-12-64, 029-12-62, 029-12-59).

43 Carnap to Neurath, January 16, 1932 (RC 029-12-70).

44 "Soziologie im Physikalismus", *Erkenntnis* **2** (1931/32), tr. "Sociology in the Framework of Physicalism", in *PP*, pp. 58–90.

45 Neurath to Carnap, February 17: "I am not usually a hunter for prorities – but that I should appear as a unimaginative fool ["*matter Tepp*"] who is only able to badly reproduce Carnap's sublimely formulated thesis – that went too far, I felt." (RC 029-12-62, cf. 029-12-59).

46 Neurath to Carnap, January 25 and 28 (RC 029-12-69, 029-12-68).

47 *Unity*, p. 74 (cf. Carnap to Neurath, February 2 and 10, RC 029-12-66, 029-12-63).

[48] Carnap to Neurath, March 2 (RC 029-12-60/61) acknowledged Neurath's influence (amongst others: conversations with Feigl and Hempel and his own work on metalogic, to which the distinction of the formal and material mode is owed) yet only speaks of Neurath's "theses" or "instinctive aperçus" which, he held, were left "unargued for" and were mostly "not understood in the Circle."

[49] For instance, in his response of February 17, Neurath insisted on his "poignant formulation", and again (RC 029-12-62; cf. the earlier RC 029-12-62, 029-12-64 and later 029-12-59) asked for quotation from his "Physikalismus" (*Scientia* **50** (1931), tr. "Physicalism", in *PP*, pp. 52–57) – in addition to Carnap's references in his footnote to Neurath's *Empirische Soziologie* (Vienna: Springer, 1931, part tr. "Empirical Sociology" (hereafter *Empirical Sociology*), in Neurath, *Empiricism and Sociology*, ed. by R. S. Cohen and M. Neurath (Dordrecht: Reidel, 1973; hereafter: *ES*) pp. 319–421) and "Sociology." Carnap did not know this paper (RC 029-12-63) and did not accede to Neurath's request to quote from it, but agreed to add a reference to it in proof which the printers promptly forgot to insert – much to his embarrassment (Carnap to Reichenbach, June 7, 1932, HR 013-41-38). Carnap made good this printer's error in the English translation. Carnap's additional suggestion to note the date of receipt of their two essays at the end of their published version (reported in RC 029-12-60/61) was apparently dropped after discussion with Reichenbach (Carnap to Reichenbach, March 13, 1932, HR 013-41-40). As late as October 1932, Neurath complained that the order of publication was not complimentary to his originality (RC 029-12-24).

[50] See Wittgenstein to Carnap, August 20, 1932, in Nedo & Ranchetti, *op. cit.*

[51] Haller and Rutte, *op. cit.*, pp. 29–30. This was not the first time Neider made this claim, though on the previously documented occasion, the basis for his claim was even less clear; see letter to Maria Reichenbach and Wesley Salmon, March 1, 1971 (HR 032-02-029).

[52] Carnap, *Aufbau*, §59.

[53] The intersubjectivity ascribed to the "world of physics" in §136 was held to be reconstructible in methodologically solipsist terms in §148.

[54] Rudolf Carnap, "Einheitswissenschaft auf physischer Basis", *Erkenntnis* **1** (1930), p. 77.

[55] Rudolf Carnap, "Die alte und die neue Logik", *Erkenntnis* **1** (1930), tr. I Levi, "The Old and the New Logic" in *Logical Positivism*, ed. A.J. Ayer (New York: Free Press), p. 47.

[56] This is claimed, e.g., in Haller, "Wittgenstein", p. 21.

[57] *Aufbau*, p. 107; translation changed. George's translation ("Certain structural properties are analogous for all streams of experience") introduces the notion of 'analogy' not yet employed in §66, but only in §§146–9; since Carnap *may* be suggesting two different conceptions of intersubjectivity in these different sections, this translation *may* be misleading. Note that on my reconstruction, Neider's diagnosis of Carnap's failure goes through in either case.

[58] Carnap, *Aufbau*, p. 229.

[59] For the difference between the 1930 and 1932 positions of Carnap – not relevant here – see my "Rational Reconstruction as Elucidation? Carnap in the Early Protocol Sentence Debate", *Synthese* **93** (1992), 107–140, and *Overcoming Logical Positivism From Within. The Emergence of Neurath's Naturalism in the Vienna Circle's Protocol Sentence Debate* (Amsterdam-Atlanta: Rodopi, 1992), chap. 6.

[60] Otto Neurath, *Lebensgestaltung und Klassenkampf* (Berlin: Laub, 1928), tr. "Personal Life and Class Struggle" in *ES*, p. 287.

[61] Otto Neurath, "Wege der wissenschaftlichen Weltauffassung", *Erkenntnis* **1** (1930), tr. "Ways of the Scientific World-Conception", in *PP*, pp. 47, 44 respectively.

[62] Otto Neurath, "Wissenschaftlichen Weltauffassung", *Arbeiterzeitung* October 13 1929, Wien, repr. in Neurath, *Gesammelte philosophische und methodologische Schriften*, eds. R. Haller and H. Rutte (Wien: Hölder-Pichler-Tempski, 1981) (hereafter *GpmS*), p. 347, my italics.

[63] See, e.g., Neurath, "Weltanschauung und Marxismus", *Der Kampf* **24** (1931), repr. in *GpmS*, p. 411.

[64] Carnap states the date for the completion of the first version in various letters, e.g. to Neurath, January 16, 1932, RC 029-12-70. Arthur J. Benson, "Bibliography of Rudolf Carnap", in *The Philosophy of Rudolf Carnap, op. cit.*, pp. 1023–4, states that it served as the basis of a lecture given on March 1, 1931 in Vienna. Virtually the same argument also underlay his "Psychologie in physikalischer Sprache", *Erkenntnis* **3** (1932/33) tr. G. Schick "Psychology in Physical Language" (hereafter referred to as "Psychology") in *Logical Positivism* (*op. cit.*) which was developed more or less in parallel to *Unity* (but revised and published later).

[65] Carnap, "Einheitswissenschaft", p. 77.

[66] Carnap to Neurath, December 23, 1933 (RC 029-03-06); my italics.

[67] Which is not wholly unlikely. Consider Neider's cryptic testimony as amended by my explication above: "Philosophy can consider [sentences about the auto-psychological] in explaining how [the protocol sentences] come about, and once they are formulated they must be intersubjectively verifiable."

[68] See my "Rational Reconstruction" for examples of an earlier version of Carnap's *Unity* argument without the material/formal mode distinction.

[69] Carnap to Neurath, March 2, 1931 (RC 029-12-60/61).

[70] Wittgenstein to Schlick, August 8, 1932, in Nedo & Ranchetti. In fact, Wittgenstein would have done better to refer to propositions 3.33-3.334, which explicitly propose a "logical syntax" that only describes the expressions but does not mention their meaning; thus Wittgenstein's strictures on metalinguistic speech only concern semantics, not syntactical talk (cf. Hintikka & Hintikka, *op. cit.*, ch. 1 sects. 4 and 5, and ch. 6 sect. 5).

[71] Rudolf Carnap, *Die logische Syntax der Sprache* (Vienna: Springer, 1934), rev. ed. tr. A. Semeaton *The Logical Syntax of Language* (London: Kegan, Paul & Trubner, 1937; repr. Paterson, N.J.: Littlefield, Adams & Co., 1959), pp. 282–284.

[72] E.g., one affirming, according to Waismann's Wittgensteinian Thesis (said to date from "ca. 1930"), that "ostension steps outside of language and connects signs with reality" (Waismann, *op. cit.*, p. 246), and one denying it (in Cambridge lectures, Michaelmas term 1930): "ostensive definition does not get us away from symbolism. All we can do in ostensive definitions is to replace one set of symbols by another" (*Wittgenstein's Lectures 1930–1932*, ed. by D. Lee (Oxford: Blackwell, 1980), p. 23).

[73] Haller, "Wittgenstein", p. 29.

[74] "*Billige Verschrottung*", Wittgenstein to Carnap, August 20, 1932, repr. in Nedo & Ranchetti, *op. cit.*

[75] *Unity*, p. 44/5; cf. "Psychology", *op. cit.*, p. 166.

[76] Carnap, "Psychology", p. 170. (Note also that the English translation, *Unity*, p. 93,

dropped the confession of continued allegiance to methodological solipsism of the corresponding passage in the original (*Erkenntnis* **2** (1931/32), p. 461).)

77 *Ibid.*, p. 191.

78 Rudolf Carnap, "Über Protokollsätze" *Erkenntnis* **3** (1932/33), tr. R. Creath and R. Nollan "On Protocol Statements", *Nous* **21** (1987), pp. 462 ff., see esp. pp. 458, 469.

79 In their order of composition: "Sociology", pp. 62–3, 65; "Physicalism", pp. 54–5; "Das Fremdpsychische in der Soziologie" *Erkenntnis* **3** (1932/33), pp. 105–106; "Protocol Statements", *PP*, p. 96. Later occurrences or references to the argument are "Radikaler Physikalismus und 'wirkliche Welt';", *Erkenntnis* **4** (1934), tr. "Radical Physicalism and 'the Real World", *PP*, p. 110, and "Universal Jargon and Terminology", *Proceedings of the Aristotelian Society* n.s. 41 (1941), *PP*, pp. 228–9. The first appearance of Neurath's argument was in a discussion with Carnap and Hahn on March 4, 1932, a rough transcript of which is preserved as *Besprechung über Physikalismus* (RC 029-17-03); cf. my "Rational Reconstruction . . .", and *Overcoming*, chap. 6.

80 Note that Hector-Neri Castaneda, "Private Language Problem", in *Encyclopedia of Philosophy* eds. P. Edwards *et al.* (New York: Macmillan and Free Press, 1967), vol. 6 pp. 459–464, lists Carnap's "Psychology" and Julius Weinberg, *An Examination of Logical Positivism* (London: Kegan Paul, 1936), as pre-Wittgensteinian discussions of the private language problem, and that Haller, "Wittgenstein", p. 27, calls Carnap's argument "*in nuce* one of the first versions of the argument for the primacy of a public language against a private language." I may repeat that the argument in Carnap's "Psychology" is basically the same as his argument turning on statements about minds in his *Unity* – with the mere addition of the sentence: "The situation is the same with sentences about one's own mind" and the renewed application of the same argument ("Psychology", p. 191).

81 Neurath, "Physicalism", p. 54/5; italics restored, translation amended from "language of physics" to "physicalist language". (A similar, yet fragmented, version appears in "Sociology", pp. 62, 63, 65, which was written before his "Physicalism" yet published later.)

82 See also the gloss of this argument by Michael Heidelberger, "Zerspaltung und Einheit", p. 177, who does not, however, designate it as a private language argument (nor note other occurrences). Remarks by Heiner Rutte, "Über Neuraths Empirismus und seine Kritik am Empirismus", *Grazer Philosophische Studien* 16/17 (1982), tr. "On Neurath's Empiricism and his Critique of Empiricism", in *Rediscovering*, p. 177, about Neurath's physicalism in general, and by Dirk Koppelberg, *Die Aufhebung der analytischen Philosophie* (Frankfurt a.M.: Suhrkamp, 1987), p. 28, about a similar passage in Neurath's "Protocol Sentences" (without noting other occurrences), suggest a vague proximity to Wittgenstein's argument; but Rutte equates Neurath's and Carnap's arguments and Koppelberg clearly shies away from attributing a private language *argument* to Neurath.

83 The former point was stressed explicitly in "Das Fremdpsychische", pp. 105–106; that the latter point lay behind Neurath's private language argument is made clear (though not wholly explicitly) in his "Universal Jargon", pp. 228/9.

84 See my "Neurath vs. Carnap: Naturalism vs. Rational Reconstructionism Before Quine", *History of Philosophy Quarterly* **9** (1992), pp. 445–470, and *Overcoming*, ch. 10.

85 Neurath, "Physicalism", p. 55, "Protocol Statements", pp. 91–92; cf. Haller, "Ansichten."

86 Note, again, the long laudatory footnote in *Unity*, p. 74, claiming, in the original, agreement between himself and Neurath in "all essential respects," and in the 1934 translation even in "all respects."
87 Carnap to Neurath, March 2, 1932 (RC 029-12-60/61).
88 Neurath, "Physicalism", p. 53.
89 *Wittgenstein's Lectures, op. cit.*, p. 82.
90 Haller, "Wittgenstein", p. 24.
91 See *Besprechung, op. cit.*, and Neurath, "Soziologie in physischer Sprache", *Erkenntnis* **2** (1931), p. 312. This lecture, incidentally, belonged to the series "Probleme der Einheitswissenschaft" which also included early versions of Carnap's "Unity" and "Psychology" (*ibid.*, p. 311).
92 Schlick to Carnap, December 3, 1932 (RC 029-29-01); Schlick, "Relation", p. 426; Wittgenstein to Schlick, August 8, 1932, Nedo & Ranchetti, *op. cit.*
93 *GTK*, p. 295.
94 *Ibid.*, p. 296.
95 *Ibid.*, pp. 285–289.
96 Carnap, *Unity*, §4.
97 "Why did we not notice this earlier on? And wasn't Schlick's attitude opposed to this thesis when we two began to discuss physicalism in the Circle?" Carnap to Neurath, March 26, 1935 (RC 029-09-66). Carnap's embarrassment is understandable. He had referred to Schlick's *Erkenntnislehre* as a whole in his *Physikalische Begriffsbildung* (Karlsruhe, 1926), in which he already had claimed that qualitative statements were uniquely correlatable with quantitative statements (p. 60), and to specific doctrines in it in his "Eigentliche und uneigentliche Begriffe" (*Symposion* **1** (1927), pp. 355–374) and in *Aufbau* – none of these references, however, concern the sections on the mind-body problem where Schlick's "physicalism" is first announced.
98 Carnap to Neurath, May 15, 1935 (RC 029-09-54); note also his remarks on *GTK* in his "Intellectual Autobiography", in *The Philosophy of Rudolf Carnap, op. cit.*, pp. 21–22, and Herbert Feigl, "Physicalism, Unity of Science and the Foundations of Psychology", *ibid.*, fns. 32 and 36.
99 *GTK*, pp. 295–296.
100 Given his willingness to accept Schlick's claim in 1935, Carnap could have accepted Neurath's of 1932 also more wholeheartedly than he did. Perhaps in 1935 he may have done so in retrospect, if by then old differences had not become clear and new ones were becoming apparent.
101 Schlick, "Relation", pp. 426–427, Carnap, *Unity*, p. 61.
102 Schlick, "Relation", p. 427; Carnap to Neurath, June 2 and July 4, 1935 (RC 029-09-49; 029-09-37).
103 Neurath to Carnap, May 28 and June 21, 1935 (RC 029-09-50; 029-09-45).
104 Neurath to Carnap, April 19, 1935 (RC 029-09-60) and Neurath, "Erster Internationaler Kongress für Einheit der Wissenschaft", *Erkenntnis* **5** (1935), repr. *GpmS*, p. 659.
105 McGuinnes, "Beziehungen", p. 122.
106 Koppelberg, *Aufhebung*, distinguished two senses of "physicalism" in Neurath, which we may render as a metalinguistic thesis in the sense of Carnap and a distinct meta-epistemological thesis of his own. This, his naturalism, of course alters the theoretical valency of the view shared with Carnap. On Neurath's "radical physicalism", see also

my "Neurath's Programme for Naturalistic Epistemology", *Studies in History and Philosophy of Science* **22** (1991), pp. 623–646, and *Overcoming,* chapts. 10–12.

[107] See also Karl H. Müller, "Neurath's Theory of Pictorial-Statistical Representation", in *Rediscovering, op. cit.*, pp. 223–243.

[108] Ferdinand Tönnies, "Philosophical Terminology", *Mind* **8** (1899), pp. 289–332, 467–491 and **9** (1900), pp. 46–61. This, the Welby Prize essay, was pointed to as of great importance to his development by Neurath in "Universal Jargon", p. 217. For a discussion of Tönnies' influence see my "On Neurath's Boat", in Cartwright *et al., Otto Neurath Philosophy between Science and Politics* (Cambridge: Cambridge University Press), forthcoming.

[109] Neurath, *Empirical Sociology, op. cit.*, p. 51, italics added: "Weltanschauung und Marxismus", p. 411, quotes all but the first of these sentences. Note that the long quote (*"Blütenlese"*) in *Empirical Sociology* represents Neurath's arrangement of phrases obviously taken from the first publication of Part 1 of the *German Ideology*, edited by Riazanov and published in 1926.

[110] Ferrucio Rossi-Landi, *Il linguaggio come lavoro e come mercato* (Milan: Bompiani, 1968), tr. *Language as Work and Trade* (Sth. Headley, Mass.: Bergin & Garvey, 1983), p. 27; this chapter 1, "Towards a Marxian Use of Wittgenstein", also in *Austrian Philosophy Studies and Texts*, ed. by J. C. Nyiri (Munich: Philosophia, 1981), pp. 113–149. For a brief discussion, see Alan Janik, "Wittgenstein, Marx and Sociology", *Essays on Wittgenstein and Weininger* (Amsterdam-Atlanta, GA: Rodopi, 1985), pp. 136–157.

[111] Cohen, "Dialectical Materialism", p. 151.

[112] Of other contemporary arguments or views about language to this effect – John Dewey's *Experience and Nature* (LaSalle, Ill.: Open Court, 1925, 2nd. ed. 1919), once mentioned as such by Quine, for instance – it is less clear whether Neurath was then acquainted with them. For some further remarks, concerning Mauthner and also Bühler – to whose importance for Wittgenstein Stephen Toulmin has drawn attention ("Der Metaphysiker Wittgenstein", *Neues Forum* **17** (1970), pp. 699–703, quoted in Achim Eschbach, "Karl Bühler und Ludwig Wittgenstein", in *Karl Bühler's Theory of Language*, ed. by A. Eschbach (Amsterdam: Benjamins, 1988), pp. 385–406 – see *Overcoming, op. cit.*, chap. 10, fn. 11.

[113] Carnap, "Intellectual Autobiography", in *The Philosophy of Rudolf Carnap, op. cit.*, p. 24.

[114] Cohen, "Dialectical Materialism", *op. cit.* On Neurath's own earlier, ultimately rejected attempt to engage in dialogue with Horkheimer and the Frankfurt School, see Hans-Joachim Dahms, "Die Vorgeschichte des Positivismus-Streits: von der Kooperation zur Konfrontation. Die Beziehungen zwischen Frankfurter Schule und Wiener Kreis 1936–1942", in *Jahrbuch für Soziologiegeschichte 1990*, ed. by H.-J. Dahme *et al.* (Opladen: Laske & Budrich), pp. 9–78. On the double embedding of Neurath's physicalism in the contemporary debates of the Vienna Circle and the Austrian Socialist Movement, see Cartwright and Cat in Cartwright *et al., Empiricism Without Tears.*

[115] I wish to thank Sylvain Bromberger, Richard Creath, Arthur Fine, Michael Friedman, Adolf Grünbaum, Rudolf Haller, David Hull, Thomas Oberdan, Ulrich Röseberg, Thomas A. Ryckman, Friedrich Stadler and Meredith Williams for remarks and advice.

ROSHDI RASHED

CONIC SECTIONS AND BURNING MIRRORS: AN EXAMPLE OF THE APPLICATION OF ANCIENT AND CLASSICAL MATHEMATICS

1. INTRODUCTION: ON THE APPLICATION OF MATHEMATICS.

What does it mean, 'to apply mathematics?' Even if mathematics could 'speak' with phenomena immediately and directly, it would be of no help, for this would mean that everything could be expressed in mathematical terms and that mathematical structures were an exact reflection of shapes carved in space and forever fixed in time. We would grasp the ideal functioning of every phenomenon, and our power over things would be infinite. But it does not take an epistemologist or an historian of science to realize that nothing could be further from the truth. A veil hangs between mathematics and things, in the case of most of the sciences, a thick veil. Over millennia, this veil has become more transparent, first in astronomy and then in optics, and more recently in mechanics and the other disciplines of physics. But despite this veil, there have always been applications of mathematics that force us to learn about phenomena and things in terms of mathematics. Clearly, as the applications grew in number and impact, so our question claimed the attention of philosophers, until – starting from the second half of the eighteenth century – it became the very heart of modern philosophy.

To apply is also to act, that is, to draw from objects only those useful traits that might be expressed in the language of mathematics; other traits must be set aside, or, at the very least, remain in the background. It is in the choice of these traits, in their delineation, that the difficulty, but also the value, of an application lies. To examine the application of mathematics is, first of all, to stress this reductive, simplifying, and schematic task as well as its means and its residue. But no sooner is this question raised than – provided one is an historian of science – one sees that the term 'application' covers a broad spectrum and has several meanings. At one end of the spectrum, application can try to enhance understanding; at the other end, it tries to clarify conduct in order to guide our actions. In this last instance, the goal of mathematics is not only to grasp functioning but also to perform an act effectively.

K. Gavroglu et al. (eds.), Physics, Philosophy and the Scientific Community, 357–376.

The two meanings are not separate, even if philosophical discourse, ever anxious about radicalism, privileges the first.

Philosophers have all recognized that the nub of the problem of applying mathematics lies in the identification of the traits of the phenomena that are receptive to mathematical information. The relevance of the latter depends, in turn, on its power to explain, predict, and finally, act. In a word, rather than apply mathematics to things themselves, we limit ourselves to applying it to the concepts attached to things. Stemming from the very generality of mathematics, this approach is more pronounced when influenced by the language needed to discuss it. But the concepts that are introduced between mathematics and things appear necessary to transmit mathematical information to them. Philosophers have not failed to raise the questions summarized here nor to highlight the difficulty. We can find two examples in Leibniz and Hume. And we have all read at least once Kant's famous formula: "I maintain, however, that in every special doctrine of nature only so much science proper can be found as there is mathematics in it" (Kant 1970, p. 6). Despite its specific historical determinations, Kant's thesis, which has been the subject of gifted work, became the most general and the most commonly shared. But equally as important as Kant's position was the insistence on the link between the application of mathematics and precise prediction. This was enough to dispel any illusion that might crop up and relegate mathematics to the level of simple language. Prediction is nothing of the sort: mathematics delivers more than experience gives to it. All our thoughts on the application of mathematics must take account of this fact.

The extension of the application of mathematics to other, initially resistant fields has raised a second question: just what does 'applying mathematics' mean when, instead of one theory suitable for conveying mathematical information we come across only doctrinal considerations or crude doctrines, when we consider, for example, Tartaglia's dynamics, Descartes's physical optics, and the social sciences today? I will not pursue this line of questioning here, but it does have the merit of bringing a heretofore marginal question to the fore, namely the variety of meanings of "application" and their essential disparities. And, in fact, applications are valid neither in their absolute incontestability nor in their efficacy.

It is clear that in examining the applications of mathematics we are led to analyse scientific knowledge itself, something we are neither pretentious nor immodest enough to try. We propose to delve as deeply as possible into history in order to find the earliest applications of mathematics and to understand their significance. We shall examine what distinguishes them from other forms of application, both contemporary and subsequent. From this historical perspective we shall pay particular attention to the contribution of these applications to the history of mathematics. Our goal is to demonstrate that some applications are as old as mathematics itself and that in other respects, without the history of its applications, the history of mathematics would be deprived of an essential dimension. I shall confine myself to a single example, that of the burning mirrors.

2. POETIC SCIENCES AND THE APPLICATION OF MATHEMATICS.

Beginning with the second half of the eighteenth century, particularly with d'Alembert and his successors, the potential of any one science was often reduced to the potential application of mathematics to the phenomena in question. This attitude was foreign to the scholars of antiquity, and particularly so to the dominant philosophy, that of Aristotle. For him, mathematics and physics are separate: the first deals with knowledge that can be proven, while the second deals with the knowledge of becoming. But this opposition of principle was not the radical schism that some commentators, beginning with Alexander, made it out to be. Aristotle reserved at least three uses for mathematics. Before getting to those, let us remember that for the philosopher, mathematics concerns the physical, but by no means a separate world. Nonetheless, it has its own objects, which are not physical objects. These mathematical objects, which are not detached from physical objects, exist after them while in definition they precede them. If we add to this the celebrated doctrine of potentiality, we can no longer speak of the application of mathematics in this context. This problem is completely alien to Aristotelian philosophy, but it is not opposed to it. Didn't Aristotle resort to mathe-matics in his physics? Didn't he use the theory of proportions in his discussion of motion? In other respects he acknowledged that mathematics had another, instrumental use, namely in the poetic sciences, in the productive knowledge intended for the manufacture of useful or beautiful objects. Here is what he wrote in the famous passage of the *Nicomachean Ethics*: "For a carpenter and a geometer investigate the right angle in different ways; the former does so insofar as the right angle is useful for his work, while the latter inquires what it is or what sort of thing it is; for he is a spectator of the truth" (Book 1, Chap. 7, Sect. 1098a, 11. 28–32). Finally, for Aristotle, there are "the more physical of the branches of mathematics" (*Physics*, Book II, Chap. 2, Sect. 194a) of which the example of optics is of particular interest to us here. Despite its distinctive place as the so-called most physical part of mathematics, optics still keeps its distance; or, as Aristotle writes in the *Physics*: "While geometry investigates physical lines, but not *qua* physical, optics investigates mathematical lines, but *qua* physical, not *qua* mathematical" (Book II, Chap. 2, Sect. 194a, 11. 9–11). If we compare this passage with that of the *Metaphysics* we cannot but be astonished. In effect, neither, harmonics nor optics, writes Aristotle: "considers its objects *qua* sight or *qua* voice, but *qua* lines and numbers" (Book 13, Chap. 3, Sect. 1078a, 11. 14–16). And our surprise is linked to Aristotle's idea of the subordination, of the relation between arithmetic and geometry, on the one hand, and harmonics, optics, astronomy, and mechanics on the other. We are dealing with the relation of superior to inferior described in the *Posterior Analytics* (Book I, Chap. 7, 75b; Chap. 9, 76a; Chap. 13, 78b–79a). In brief, as long as we are dealing with physical rays in optics, we study them geometrically only by considering them as straight lines. Thus geometry serves to describe the figure of a phenomenon, to trace its contour; it is not responsible for either explaining or grasping its essence. This is precisely what we see at work in the geometric study of the rainbow at the end of the *Meteorology*, assuming that this passage on the rainbow is genuine.

The question of the application of mathematics, in the sense that will appear later, does not arise in this context. But mathematics still plays two roles, one instrumental in the poetic sciences and the other in the determination of the contour of a phenomenon. Perhaps it is because of these two roles assigned to mathematics that the first Greek mathematicians concerned with the problems of the application of mathematics did not declare themselves anti-Aristotelians. Let us turn now to the latter, but limit ourselves to that privileged discipline, optics. Two great traditions stand out: that carried on under the banner of Archimedes on burning mirrors and that of Euclid, and later Ptolemy, on a geometry of vision. As for Hero of Alexandria, he concerned himself particularly with the application of catoptrics to useful or amusing problems.

We are considering a single tradition, that of research on burning mirrors, confining ourselves, moreover, to the study of conic sections. We shall see, on the one hand, that the application of mathematics takes on a particular meaning by no means opposed to Aristotle's philosophy, and that, on the other hand, it is mathematically productive. As we said earlier, this application was carried out on *organons*, on instruments, just as it was done elsewhere on machines. It made possible the elaboration of a mechanical model that allowed one to speak mathematically of a localized phenomenon, for example, the propagation of rays parallel to the axis of a paraboloid mirror. In the field of mathematics this application developed an entire branch of research on conics, which can be neither identified nor understood without it. Nevertheless, its apparent physical sterility would mask a potential productivity, which becomes effective when one moves from mirrors to lenses. To demonstrate these assertions, let us return briefly to the origins of the theory of conic sections as they relate to burning mirrors.

3. THE TRADITION OF CATOPTRICS IN THE HISTORY OF THE THEORY OF CONIC SECTIONS.

The origins of the theory of conic sections are obscure, as origins often are in mathematics. Some facts and, above all, some conjectures and commentaries derived from later evidence really do add up and these can be found in some still unsurpassed books: Zeuthen's *Die Lehre von den Kegelschnitten im Altertum* ['The theory of conic sections in antiquity'], published in 1886, and Heath's edition of Apollonius's *Treatise on Conic Sections* (1896). To these we should add two studies of Otto Neugebauer (1932; 1959). In the shorter article, which was published in 1959 and is of greater interest (although it is entirely conjectural since his hypothesis is not supported by any fact), Neugebauer tries to show that the theory of conic sections was invented by Eudoxus in order to conceive and construct a certain type of sundial, thus confirming the astronomical origin of this theory. The other, much longer article deals with the concept of focus, which I will discuss more fully below. My goal here is certainly not to present what would of necessity be a very short summary of what has been well expounded at great length elsewhere. I only want to recall some salient facts that need to be borne in mind in order to understand the application of mathematics to burning mirrors.

3.1. Menaechmus, a student of Eudoxus, is credited with discovering the theory of conic sections in the middle of the fourth century B.C.E. The conics were obtained from the plane sections of a cone with a circular base. According to later evidence, it seems that Menaechmus was dealing with a right-angle cone, which he cut by a plane perpendicular to one of its generators. The three conics are obtained by varying the angle at the apex of the cone, hence the three terms (which were subsequently modified by Apollonius): Menaechmus's right-angle conic section becomes Apollonius's parabola, an obtuse-angle section becomes a hyperbola, and an acute-angle one becomes an ellipse. It is precisely this perpendicularity of the cutting plane to one of the generators that provided Neugebauer with his conjecture about the sundial.

But this discovery attributed to Menaechmus by Geminus (according to statements of Proclus of the fifth century C.E.) was not the only one. According to Eratosthenes (by way of the sixth century mathematician, Eutocius), it was again Menaechmus who, using two conic sections – two parabolas or one parabola and one hyperbola – solved the problem of determining the two means in order to solve the problem of doubling the cube. From the start, the theory of conic sections might have been tied directly to the solution of problems of solid geometry and thus, applied.

3.2. Whatever the historical path that led to about the fourth century B.C.E., it is still Aristaeus the Elder and Euclid who are credited with the first two treatises on *Conics*. Conic sections would have been considered explicitly as means to obtain the plane loci needed to solve problems of solid geometry. But although Aristaeus's book is mentioned in the *Collection* of Pappus of Alexandria (1986, I, Book 7, pp. 114–115; Pappus 1965, pp. 672–673; Pappus 1933, II, p. 503), Archimedes, six centuries earlier in his *Quadrature of the Parabola* (167), mentions a certain 'Elements Relating to Conic Sections' ('εν τοῖς κωνικοῖς) which may well have been Aristaeus's work or that of another mathematician such as Euclid.

3.3. A privileged moment in research on conic sections is to be found at the beginning of the third century B.C.E., with Conon of Samos, Trasydeus, Nicoteles of Curene, and Archimedes. This intense activity is not only attested to by Archimedes's works but it is also evoked by the evidence of Apollonius, particularly in his Preface to Book 4 of the

Conic Sections. Younger than Archimedes by one or two generations, Apollonius pursued this tradition of the Alexandrian mathematician and, in a way, he perfected it. His treatise, *Conic Sections* – eight books devoted entirely to this subject – is the only one to survive antiquity. The first four books have come to us in a Greek as well as an Arabic version; the next three exist only in an Arabic translation, while the eighth book was lost very early on. In this treatise Apollonius expounds most of the elementary properties of conic sections in a purely synthetic manner, without the slightest mention of earlier analysis. It is Apollonius as well who modified the terminology and imposed the use of the terms 'parabola', 'hyperbola', and 'ellipse'. The two terms, 'parabola' and 'ellipse', appear in Archimedes's *Method* (Book III. 87) and in his *On Conoids and Spheroids* (Prop. 9) respectively. But Heiberg is of the opinion, based on a statement of Pappus (Pappus 1986 I, pp. 114–117) that it was Apollonius who had opted for the new terminology, that these terms might have been interpolated. The definition covered is in any case more general. Unlike the situation for Euclid (Book XI, definition 18 of the *Elements*), for Apollonius a conic surface is no longer generated by the rotation of a right triangle, but by the displacement of an infinite straight line passing through a fixed point and meeting a fixed circle, that is, through an apex and a circle directrix.

3.4. In order to characterize the three conic sections, Apollonius, at the beginning of his treatise, defines the diameter and the associated conjugate direction, in his language, the ordinate. He then takes these as the reference point for coordinates in order to obtain the fundamental property (the *symptoma*)[1] of the points of each of the three conics. He is led to differentiate them and in each case to characterize a segment called the *latus rectum* (the double of the parameter) which depends on the cone and the secant plane.

After having recalled some salient facts that mark the high road of research on conic sections starting with the fourth century B.C.E., we can better frame our question. Had there ever been another research tradition on conic sections – connected to that just mentioned – whose roots could be found in application? And, on such a hypothesis, where would we find the contributions of its representatives, and what meaning should be applied to the term *application*? These are the questions that take us to the history of burning mirrors.

From the Latin rhetorician and philosopher, Apuleius (125–180), we know that Archimedes wrote a treatise on catoptrics. Two centuries later, Theon of Alexandria confirmed this attribution. On the other hand, a legend that spread from the sixth century on, revived by Olympiodorus and by a certain Fūthiṭos, echoed Archimedes's interests. According to the well-known legend (Rashed 1992), Archimedes set fire to Marcellus's fleet during the siege of Syracuse by using burning mirrors. But this interest in burning mirrors that history attributes to Archimedes does not distinguish him in particular. We know in fact from another anonymous Greek – the author of the famous Bobbio fragment – that Apollonius himself might also have written on burning mirrors. If we add to this the work of pseudo-Euclid on the same subject, we see that this research was present relatively early. But this evidence and these legends, whose reliability is difficult to evaluate, have nonetheless one common element: research on burning mirrors is essentially the work of the theorists of conic sections. But if the evidence suggests such a conjecture to us, common sense proposes a second: works on burning mirrors had to tackle the problems of the construction of figures insofar as it was necessary to construct these mirrors, or at the very least indicate a method for fashioning their models. Thus if this research existed and if one considered conic mirrors, one could not avoid the problem of application from the start, a problem that lies at the very heart of theoretical research.

Today we are in a position to confirm these conjectures and to establish the existence of this tradition of research on conic sections, starting with several texts, one from Archimedes's successor and a probable contemporary of Apollonius, namely Diocles, and others – difficult to date – from Apollonius's successors, who pursued this research until the end of the tenth century, at least. Let us begin with a somewhat detailed examination of the compilation of Diocles's book that has come down to us, called *On Burning Mirrors* (Περὶ Πυρίων).

To begin with, Diocles's book is presented in a paradoxical manner, namely 16 propositions divided into six relatively independent subgroups. The first subgroup includes three propositions, 1, 4, and 5, which deal with the paraboloidal burning mirror. Following the sequence of propositions and explicit references from one to another and to the preliminaries, these propositions, together with the Preface, reveal a certain unity. It is this group that is of definitive interest to us here. The second group is made up of the second and third propositions, which deal with the spherical mirror and are inserted abruptly and surprisingly between propositions of the first group. These two propositions are nonetheless connected, at least indirectly, with the Preface and, together with the first group, constitute all that the book contains on burning mirrors. Proposition six deals with apparent diameters. The other groups deal with geometrical constructions. Here is the outline of the book:

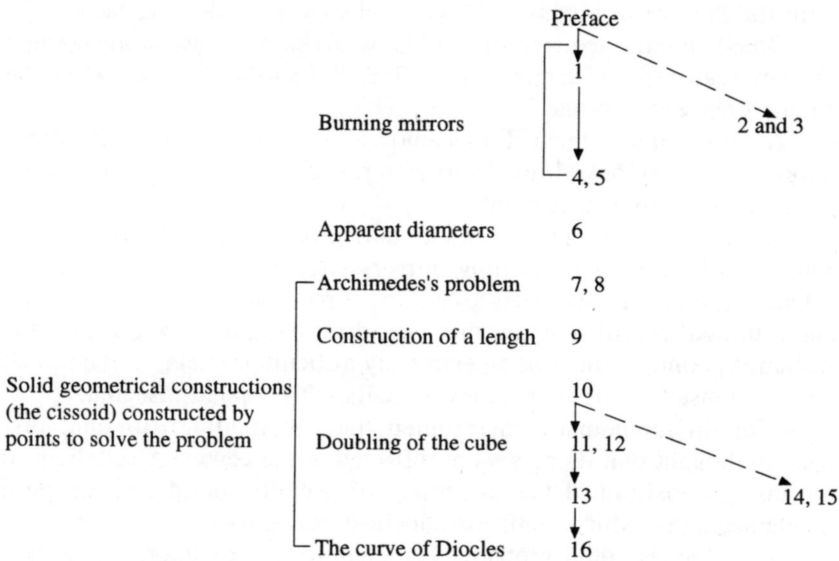

Beneath this diversity of themes presented and their heterogeneity, we see a certain unity of intention and homogeneity of approach. In effect, Diocles, in most of the groups, works by applying geometry, particularly the theory of conic sections, either to catoptric problems or to the problems of geometric constructions. One could say that his approach is a deliberately applied one, and when compared with famous mathematical compositions of the period, his book is profoundly informed by this characteristic. We may even say that, at least in the writings that have come down to us, Diocles was not at all tempted by research into geometric properties for its own sake, but wanted essentially to achieve the requisite constructions. Although it has never been stressed enough, this trait of Diocles's work is nevertheless doubly fruitful for us, for it allows us from now on to grasp certain aspects of this work, in particular, the author's silence concerning the proof of certain results of the theory of conic sections that he brings up – for example, the properties of the sub-tangent and the sub-perpendicular for the parabola. On the other hand, Diocles shows us traces of a research tradition on conic sections directed precisely by the same goal of application. Thus he will be our principal guide along this path.

In the Preface to his book, Diocles tells us the following facts:

– One Pythion, a geometer from Thasos, about whom we know nothing further, searched for a mirror to reflect the sun's rays following the circumference of a circle.

– An astronomer named Hippodamos, about whom we have no further information, also looked for a mirror to reflect the sun's rays at a single point and to burn at that point.

– Dositheus, a disciple of Conon and correspondent of Archimedes, concerned himself with burning mirrors after the death of his master, solved the problem, and tried to build a mirror that would burn without being turned toward the sun, that would be fixed and would tell time without a gnomon. This was an extremely difficult problem, which placed him by himself at the highest level attained by applied research.

– Certain mathematicians studied the spherical mirror and mistakenly thought that its rays were reflected at the center. A scholium to the last proposition of the catoptrics of pseudo-Euclid echoed these mathematicians, which confirms Diocles's statement.

– Archimedes could probably be added to these authors, as he was in contact with some of them.

All these facts converge and are evidence that, beginning around the middle of the third century B.C.E., there was a wave of research devoted to burning mirrors that used conic sections. According to his own words, Diocles was not the first to have undertaken such research. The essential point of his task was obviously to finish the study of the issues raised by his predecessors, either because they were still unresolved or to offer new solutions. But this wave of applied research, in its turn, carved out a new field of research into the theory of conic sections, namely the optical properties of conics. Let us take a closer look at the result for the parabola, the only case studied by Diocles. The ellipse will appear in the writings of Anthemius of Tralles, while in the case of the hyperbola, we must wait for the birth of anaclastics and the tenth-century mathematician, Ibn Sahl (Rashed 1990, 1993).

Diocles states that the problem posed by the astronomer, Hippodamus, namely, to construct a mirror to reflect the suns's rays at a single point, "was solved by Dositheus." Nevertheless, we do not know whether Dositheus built this mirror nor whether he demonstrated the focal property. It is Diocles who leaves us uncertain by claiming for himself "the composition of the proofs" and "the classification of the problem", assertions that are far from clear: how could the student and successor of Conon and the correspondent of Archimedes have left out the proof? Moreover, how could he have built such a mirror – hardly an easy construction – without having first teased out the optical property of the parabola, that is, without having proven the existence of an observable point on the axis at which all rays parallel to the axis meet after reflection? If this were so, it would suggest that research into the optical properties of certain conic sections had already been carried out in the third century, and in the school of Conon. Now we need to know whether this research had seen the light of day by the time of research into the burning mirrors or independently of it. And this leads us to take up the history of the notion of the focus of the parabola.

4. THE PARABOLIC BURNING MIRROR AND THE FOCUS OF THE PARABOLA.

The notion of the focus of the parabola (without being given any particular name; for this we must wait for Kepler) arises twice in Diocles: the first time in the study of the parabolic mirror and the second time in a problem on geometrical construction. In order to answer the question raised above, let us retrace the history of this idea as well as that of the directrix, so that we can appreciate more precisely the contribution of the research on the burning parabolic mirror to the understanding of this notion.

In a text of Pappus – the one cited endlessly for a century – the Alexandrian mathematician tells us that Euclid was already familiar with these notions in the case of the three conic sections. According to Pappus, Euclid would have considered these notions to determine the loci of the points. But is Pappus's evidence – he is the historian on this occasion – the effect of a late reading of Euclid's *On Surface Loci* or is it a definite statement? Whatever it is, this testimony is not confirmed by a mathematician who knew Euclid's books better than anyone and, what's more, lived about six centuries before Pappus – I mean Apollonius himself. In the *Conics* one looks in vain for the notion of the focus of the parabola or that of the directrix, for the three conic sections.

In the third book of the *Conics*, Apollonius (in propositions 45 to 52) determines the focal properties of central conics. These appear in the course of the proof of the eight propositions dealing with the relations between the positions of the tangents and the rays from the foci to the points of contact. The notion of focus for these central conics already appears under the heading of "points stemming from the application" ("τὰ 'εκ τῆς παραβολῆς γινηθέντα σημεῖα"); see Apollonius 1891–1893, I, 424, line 10). But how do we then explain the absence of the focus of the parabola and the directrices of the three conic sections? There are two types of answers to this question. One could invoke Apollonius's own method for determining the foci of central conics, which is understood in the terms given these remarkable points, namely the method of applying areas (Apollonius 1891–1893 I, p. 432; Apollonius 1923, p. 263; Apollonius 1896, pp. 113–114). We are dealing with the application of a rectangle which has as its sides the principal diameter and the associated right side. This method of applying areas undoubtedly clarifies for us the reasoning that led Apollonius to consider the foci of central conics, but it is not enough to explain why the notions raised are absent.

Neugebauer proposes another type of explanation. According to the eminent historian, the issue is the immediate character of this notion. It is as if Apollonius had never judged it necessary to bring up such a notion in the *Conics*. In his study Neugebauer expounds a principle which, according to him, was followed by Apollonius in his book. According to this principle, Apollonius's approach consisted of obtaining the properties of a hyperbola by means of generalizing propositions about the parabola. If one then tries to reduce the propositions on the hyperbola in the *Conics* to those of the parabola, Neugebauer writes that the "existence of a focus of the parabola follows directly and trivially." Neugebauer concludes "that we can practically rule out the fact that Apollonius could miss that implication." This is certainly very interesting, but by Neugebauer's own admission, this thesis cannot explain the absence of the notion of the directrix, which does not follow from generalizing the properties of the parabola to obtain those of the hyperbola. But there is more: this thesis assumes that what Apollonius had in mind, implicitly or not, was a common definition of the parabola and the hyperbola and, consequently, of the three conic sections. Now this does not seem to be the case. In a word, since the appearance of Zeuthen's famous book of 1886, all historians of conic sections share a position that is not remarkable for its clarity: while not ignoring the notion of the focus of the parabola, Apollonius utters not a word about it. And the same goes for the directrix. But this bias has not succeeded in quelling the astonishment of the very people who adopt it.

We freely admit that it is really difficult to answer these negative questions with certainty. Nevertheless, let us note that this notion of focus was never treated in itself nor as a means to characterize the curve, not only where the parabola is concerned but also in the very place where it appears in Apollonius's *Conics*. As Zeuthen himself has already remarked, even if Apollonius teased out some of the most important focal properties of central conics, it is as steps in a proof of the loci of points whose distance to these foci have a given sum or difference. In this context Apollonius is not at all driven to reconsider the parabola as a conic section whose center and second focus go to infinity. In the *Conics*, he does not seem to be looking for a common definition for the three sections, that is, a definition for each as the locus of points where their distances to a fixed point and to a fixed straight line are constant and equal to the eccentricity. Apollonius does not seem to be trying to characterize the optical properties of these curves.

Let us turn now to the tradition of the catoptricians, who concerned themselves with burning mirrors, in order to see these notions emerge in the course of the application of the theory of conic sections to mirrors. We have said that this tradition is attested to by Dositheus, Diocles, and many others, and it is the survival of a compilation of Diocles's writings that allows us to make a first identification of this trend. A good tactic will be to examine quickly the book that has survived.

In the first part (the first five propositions, whose organization we have already outlined), Diocles's goal is clear: to know the optical properties of two kinds of mirrors, the parabolic and the spherical, and to make models of them. It is understood that this new task cannot be limited to a theoretical study, or, as Archimedes wrote first and Diocles later, an apodictic one: one must "master the problem and make it usable," while assuring it a practical dimension. It is thus necessary to adapt the knowledge of conic sections to this goal in the case of the parabolic mirror.

In the first proposition, Diocles moves from the traditional definition of the parabola, by means of the abscissa and the ordinate, in order to study the focal property; for this he has recourse to the properties of the sub-tangent and the sub-normal. Next he shows that a light ray falling parallel to the axis on a parabolic surface is reflected toward the focus. In the second part of the same proposition, Diocles passes to the paraboloid of revolution and shows that, if its inner surface is reflecting and if its axis is turned toward the sun, every solar ray streaming toward the mirror is reflected toward the focus. He shows that the larger the parabolic arc generating the paraboloid, the more numerous the rays reflected toward the focus.

In the third part of the same proposition, Diocles considers a chord that is perpendicular to the axis of the parabola and shows that by rotating around the chord, the corresponding arc generates a circle whose center is the middle of the chord. He then takes as the mirror the inner surface of half of the figure obtained and states that if the mirror is turned toward the sun, the solar rays are reflected toward the points of a semi-circle described by the focus. We can show that this last statement of Diocles is not precise, except during equinoxes, that is, when the declination of the sun is null or close to zero. In this last case, there is a concentration of rays reflected in the proximity of a point that changes its place from hour to hour on the circle described by the focus of the parabola.

As we have said, the next two propositions – the second and third – are devoted to the spherical mirror. In the fourth proposition, Diocles returns to the determination of the parabola, generator of a mirror, which is a paraboloid of revolution, such that the reflected rays are concentrated in a point situated at a given distance from the center of the mirror. This leads to the construction of a parabola if the focus and apex are known. In this proposition, Diocles proceeds by construction, using points of the figure, with the focus and directrix being known. This figure provides a model for the sought-for mirror. In the fifth and last proposition, devoted to burning mirrors, Diocles shows that the figure obtained earlier proves the fundamental property of the parabola, that is, the *symptoma*: the square constructed on the ordinate is equal to the rectangle constructed on the right side, having the abscissa for its width, or, in our language, the equation of the parabola.

Note that in the course of applying conics to the burning instrument, Diocles teases out the optical property from the focus and the focus-directrix property from the parabola. Now this last property, understood in the course of research on the optical properties of the curve, will be used by Diocles in a purely geometric application, namely the problem of the doubling of the cube. In this second context, Diocles's approach is identical to the first.

Let us summarize the steps of this approach. We have just seen that Diocles begins by using points to draw a curve defined in terms of its apex and its focus. He next justifies his outline by constructing a straight line – the directrix – and shows that each point obtained is equidistant from this straight line and the focus. But in the course of this construction Diocles never has to *characterize* the parabola by means of this focus-directrix property. He undoubtedly knows that the parabola constitutes the locus of points that have this property and yet, for him, as for his successors down through the centuries, the true characterization of the curve must come about through the *symptoma*, as it does for all mathe-maticians, even before Apollonius. The *symptoma* allows one to locate the points of the curve in a system of axes constituted by the diameter and the tangent at the apex of the parabola. And because, according to him, the focus-directrix property does not allow him to characterize the curve, Diocles then had to show that the curve outlined, whose construction is justified afterwards, is indeed a parabola. In short, he had to deduce the *symptoma* from the focus-directrix property, which he does in the fifth proposition. This sheds light on the justification and placement of this proposition. It is in fact this proposition that finally allows him to refer back to the first proposition, since he has demonstrated that the curve is really a parabola.

Diocles takes exactly the same approach in another group of propositions concerning the doubling of the cube. Thus in the tenth proposition, where this property appears again, Diocles goes through the same procedure, step-by-step: using points he constructs two curves with the same apex and different foci. He then justifies this construction by showing that every point of each of the two curves has the corresponding focus-directrix property. He deduces the *symptoma* from each of the curves and he writes: ". . . it is equally clear that the two lines . . . are two sections of a right-angled cone," in Apollonius's terminology, two parabolas.

The whole question of the application of conic sections to mirrors boils down to this: did Diocles – or one of his predecessors – discover the optical properties of conics as the result of this application or did he have them beforehand as a geometrical result, only exploiting them in his research on optics? In other words, are these optical properties of conic sections the results of the application of conics to optics, or are they the means that made this same application possible?

No response is satisfactory *a priori*. Let us return to history and to the text of Pappus mentioned above (*Collection*, Book VII, Prop. 238). According to Pappus, Euclid was aware of this property. Pappus gives as a lemma in Euclid's *On Surface Loci* a proposition that can be written as follows: the locus of points whose distance to a given point is related to the distance to a given straight line is a conic section. It is an ellipse, a parabola, or a hyperbola, depending on whether the relation is less than, equal to, or greater than one. Unless one can prove him wrong, Pappus's assertion suggests that Euclid used this proposition. Nevertheless the question remains: why would Euclid not have proven it? Every time historians respond, they only put forward more or less arbitrary conjectures. Zeuthen states that "Euclid has used a previously known proposition." Heath pushes the same conjecture further, starting with the fact that the proposition cannot be found in the four books of the *Conics* attributed to Euclid, since Apollonius's *Conics*, which recapitulate Euclid's, do not contain precisely this property. Heath concludes with these words: "If Euclid did not take it from his own *Conics*, what is more likely is that it was contained in Aristaeus's *Solid Loci*." If we wanted to avoid such conjectures, we would have to take Pappus at his word and compare what he reports from Euclid with Diocles's text.[2]

The analysis of the text lets us show that, after having assumed the existence of the points that prove the focus-directrix property (still without constructing them), Pappus shows that these points have the fundamental property of the parabola. This step corresponds to the one where Diocles constructs points that have the focus-directrix property and shows that they equally obey the *symptoma*. A considerable difference is already apparent between the two mathematicians: Diocles uses the property mentioned above to construct effectively the points of the curve, while Pappus assumes their existence and shows that if they exist they belong to one parabola. This is understandable since Pappus is led to demonstrate a converse: every point of this parabola demonstrates the focus-directrix property. This split between Diocles and Pappus is not pure and simple: it is not just a matter of the process of the proof, but shows the difference in the goals they strove for. Diocles's problem is not, as it is for Pappus, to determine the locus of the points but to construct the points independently of the conic sections in order to show subsequently that these points belong to a parabola traditionally characterized as such by mathematicians of conics, that is, defined in terms of the *symptoma*. This proof does not introduce any concept that was unknown to Diocles's predecessors, but the fact remains that Diocles, as far as we know, was the first to have formulated it.

But clearly Diocles's problem is by no means the general problem of reducing the construction of a curve capable of reflecting rays parallel to a given direction toward a given fixed point to that of a curve defined in terms of a focus-directrix property. Such a question, which is not the one raised by Diocles, would lead in effect to the solution of a differential equation, which would give a family of curves. Only Anthemius of Tralles would treat a particular instance of this general problem eight centuries later. He assumes as known two points A and B of the curve and fixed point C on the mediatrix of AB, the direction of the rays being that of the mediatrix. In this case, a single parabola solves the problem.

5. THE FRUITFULNESS OF THE APPLICATION OF CONIC SECTIONS TO BURNING INSTRUMENTS.

As we have just seen, during Archimedes's time and later, the application of conic sections to the study of burning mirrors was fruitful not only for the knowledge of the anaclastic properties of mirrors, but also for the theory of conics itself. The catoptricians had a whole tradition of research on conics at the same time that the high road led from Aristaeus the Elder to Euclid, Conon, Archimedes, and Apollonius, to cite only the most salient stages. The descendants of the catoptricians include the Alexandrians in the school of Conon, undoubtedly Archimedes himself, Dositheus, all those who cite Diocles, and finally, Diocles himself. This application of conics permitted raising some previously barely understood notions to the level of a real mathematical tool, thus providing them with an operating dimension, that is, the introduction of the process of construction by points (or later, by continuous plotting) of conic curves among other mathematical results.

But this tradition lasted long after Diocles. It can be found with the Byzantines, in Arabic, in the seventeenth century. And better yet: without any fundamental changes in character, this research passed to a higher level when scientists raised the still unasked question about burning by refraction. And from mirrors they passed to lenses, particularly to the plane-convex and biconvex lens. This is where the discovery of Snell's law comes in: to make a lens, they undertook completely new research on the hyperboloid, its plane, tangents, etc. And it is at this juncture that scientists tackled different mechanical processes for the continuous plotting of the hyperbola, as well as for the other conic sections. As we have shown, this important discovery took place during the last decade of the tenth century and was made by Ibn Sahl. These questions will be raised again in the seventeenth century by Kepler, Snell, and many other scientists in Mersenne's circle.

In this tradition, which dealt with mirrors or lenses by applying mathematics, it was less a matter of quantifying a physics of light in order to establish a science of light phenomena, than of conceiving an organon, which is nothing but mathematical theory incarnate. Mathematics embraces fewer physical meanings than the very figure of this instrument and the mode of its functioning. But in order for this application to be possible, the mathematician finds himself forced to postulate a certain number of physical hypotheses, not at all on the nature of light nor its movement, but on propagation, reflection, and heat, in the case of the mirrors, and on the medium, refraction, and constancy of the refractive index for the lenses. But whatever the physical poverty of such an application, it nevertheless expresses a new epistemic conception: it becomes admissible that the goals of mathematical knowledge can lie outside of itself, in the realization of a technical object. This conception does not appear to contradict the Aristotelian conception formulated for the poetic sciences, but it undoubtedly has not had the philosophical grounding that it deserves. Perhaps it is for this reason that it has remained veiled before the eyes of the majority of the historians of science.

The situation is completely different when, instead of organons, there are the ideas of phenomena that are treated mathematically. Thus one draws another conception of the application as a 'composition' of mathematical and physical notions. This conception and the project corresponding to it are explicitly formulated as such by Ibn al-Haytham at the beginning of the eleventh century. But that is another story (Rashed 1990, 1993).

CNRS – Paris

[Translated from the French by Carolyn R. Fawcett.]

NOTES

[1] A *symptoma* is the equation of a conic curve.

[2] Pappus investigates the points on a curve Γ whose distance to a fixed point and a fixed straight line is given; let it be k. The case $k = 1$ is dealt with in Proposition 238 of his *Collection* (Pappus 1986 **I**, pp. 368–371), where it is reduced to the problem treated in Proposition 236 (*ibid.*, pp. 362–367), which is as follows:

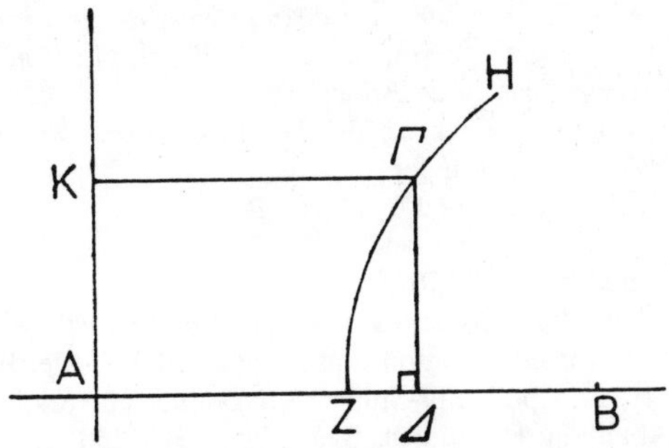

Given a segment AB with midpoint Z, we try to calculate points Γ projecting to Δ on AB and satisfying

$$(1) \qquad \Gamma\Delta^2 + \Delta B^2 = \Delta A^2.$$

From (1) we conclude $\Gamma\Delta^2 = \Delta A^2 - \Delta B^2$

$$(2) \qquad \Gamma\Delta^2 = 2AB.\ \Delta Z;$$

this equality (2) shows that Γ belongs to the parabola with apex at Z, axis AB and straight side 2 AB, parabola ZH.

Pappus then writes the converse: (2) ⇒ (1); each point that satisfies (2) also satisfies (1), and the parabola ZH is the geometric locus of the points that satisfy (1).

Note that (1) is written $\Gamma B^2 = \Delta A^2$, or $\Gamma B = \Delta A$ and if K is the orthogonal projection of Γ on the perpendicular to AB at A, we get $\Gamma B = \Gamma K$, that is

$$(3) \qquad \Gamma B/\Gamma K = 1$$

Now it is precisely in Proposition 238 (*ibid.*, pp. 368–371) that Pappus poses the following problem: given a straight line and a point B that is projected from A on that straight line, we look for the points Γ that satisfy (3). He then shows that Γ satisfies (1) and thus belongs to parabola ZH.

Pappus then shows the converse: each point of the parabola that satisfies (2) also satisfies (3). Finally, we get (1) ⇔ (2) ⇔ (3).

REFERENCES

1. Apollonius of Perga. 1891–1893. *Apollonii Pergaei quae graece exstant cum commentariis antiquis*, edidit et latine interpretatus est I. L. Heiberg. Leipzig: Teubner. 2 v.
2. Apollonius of Perga. 1896. *Treatise on conic sections*, edited in modern notation, with introductions including an essay on the earlier history of the subject, by T. S. Heath. Cambridge: University Press.
3. Apollonius of Perga. 1923. *Les coniques d'Apollonius de Perge*, oeuvres traduites pour la première fois du grec en français avec une introduction et des notes par Paul ver Eecke. Bruges: Desclée de Brouwer.

[The following three items are taken from *The Basic Works of Aristotle* edited by Richard McKeon. New York: Random House.]

4. *The Nicomachean Ethics*, translated by W. D. Ross.
5. *Physics*, translated by R. P. Hardie and R. K. Gaye.
6. *Metaphysics*, translated by W. D. Ross.
7. Kant, Immanuel. 1970. *Metaphysical Foundations of Natural Science*, with introduction and essay, by James Ellington. Indianapolis: Bobbs-Merrill.
8. Neugebauer, Otto. 1932. "Apollonius-Studien," in *Quellen und Studien zur Geschichte der Mathematik*, Abteilung B: Studien, Band 2, pp. 215–253.
9. Neugebauer, Otto. 1959. "Eccentric and Epicyclic Motion according to Apollonius", in *Scripta mathematica* **24**, pp. 5–21.
10. Pappus of Alexandria. 1933. *La collection mathématique*, oeuvre traduite pour la première fois du grec en français par Paul ver Eecke. Paris: Desclée de Brouwer. 2 v. [This is a French edition of Pappus of Alexandria 1965.]
11. Pappus of Alexandria. 1965. *Pappi Alexandrini Collectiones que supersunt*, e libris manuscriptis edidit latina interpretatione et commentariis instruxit Fridericus Hultsch. Amsterdam: Hakkert. 4 v. in 3. (Reprint of 1875–1878 edition.)
12. Pappus of Alexandria. 1986. *Book 7 of the Collection*, edited with translation and commentary by Alexander Jones. New York: Springer-Verlag. 2 v.
13. Rashed, Roshdi. 1990. "A Pioneer in Anaclastics. Ibn Sahl on Burning Mirrors and Lenses", *Isis* **81**, pp. 464–491.
14. Rashed Roshdi. 1992. "Fūthiṭos (?) et al-Kindi sur 'l'illusion lunaire'", in ΣοφιΗε ΜΑΙΗΤΟΡΕΣ, *"Chercheurs de sagesse": Hommage à Jean Pépin*, edited by Marie-Odile Goulet-Caze, Goulven Madec, and Denis O'Brien (Paris: Institut d'Études Augustiniennes), pp. 533–559.
15. Rashed Roshdi. 1993. *Géométrie et dioptique au Xe siècle: Ibn Sahl, al-Qūhī et Ibn al-Haythan*. Paris: Les Belles Lettres.
16. Zeuthen, Hieronymus Georg. 1886. *Die Lehre von den Kegelschnitten im Altertum*. Copenhagen: Host & Sohn.

DIRK J. STRUIK

SOME SOCIOLOGICAL PROBLEMS IN THE HISTORY OF MATHEMATICS

1

The relations between economic production and scientific research have become increasingly close in the present period, and with them the relations between government, industry and the social structure in general on the one hand, and mathematics on the other hand. There exists plenty of activity in what we may call the sociological aspects of contemporary mathematics, both in research and education.

Compared to this we must state that the sociology of mathematics created or prevalent in previous periods has received far less attention. True, it has been long recognized that slavery in ancient Greece enabled a section of the ruling elite to indulge in abstract social, political and ethical discussions inclusive those leading to the abstract type of mathematics we know from Euclid. It has also been well understood that merchants and bankers of Renaissance days in Europe promoted an arithmetic that with its Hindu-Arabic position system led to new discoveries, also in algebra.

However, it was not until the Russian physicist Boris Hessen, in 1931 at the London conference on the history of science, placed Newton into the framework of a budding mercantilist society, and with him the whole Scientific Revolution in the 17th century Europe with its novel mathematics into this framework, that we can speak of a serious beginning of the sociology of mathematics of the past.[1]

There have been several studies, but most of them do not go as deep as I believe the subject deserves. There have been exceptions, such as the relation of science and society, inclusive of mathematics, in England from Bacon to Newton, and the mathematics of ancient China, as well as of Mesopotamia. But there are still great gaps, and fruitful controversies may well be coming.

The fact that only a beginning has been made is not so much due to lack of interest in a field where many of its cultivators still believe in mathematics as a free domain of the mind unsullied by bread-and-butter interests. The point is rather that sociology of mathematics of the past

K. Gavroglu et al. (eds.), Physics, Philosophy and the Scientific Community, 377–383.

as a serious doctrine requires knowledge and understanding not only of that type of mathematics, but also of political and economic history as well as of astronomy and physics, not to speak of archaeology or occasionally also of religion, art and anthropology. And who, in this time of specialization, can indulge in such varied studies? Academic training, as a rule, does not encourage this type of approach.

Some of the most important, yes, exciting, problems in the sociology of the mathematics in past periods, have therefore been pretty much, though not entirely, neglected. They are worth being discussed – stone age, bronze age empires, Greece, Islam, Renaissance, Scientific Revolution, 19th century. I will here give two examples of what I have in mind.

I have already mentioned the existence of slavery as a supposedly primary factor in the emergence of the Greek "Golden Age" of mathematics. As an explanation it is unsatisfactory. Slavery in the latifundia of the Roman Empire brought no new mathematics; slavery, prevalent in the Ante-Bellum society of the Southern United States, brought interesting political thought-think of Jefferson and Madison, but little mathematics.

Speaking of Golden Ages, let us compare the Greek Golden Age with the Spanish one of the 16th and 17th century and the Dutch one of the 17th century. The Greek Golden Age brought great art and great mathematics. The Dutch one did the same, it was the age of Rembrandt and Huygens. The Spanish one brought great art – Cervantes, Murillo – but very little mathematics. Here you have a nice sociological problem in the history of mathematics, the more so since the new Greek and the new Dutch mathematics had an entirely different character.

Let us put this in another way. Europe has three Southern peninsulas, the Balkan, the Italian and the Iberian one, all in the age-old Mediterranean cultural region.

The Balkan peninsula brought the "Greek miracle" and after that hardly any more really new mathematics. The Italian one, from the time of Fibonacci or even earlier (and let us not forget Pythagoras and Archytas), was always in the forefront or next to it, although with ups and downs. And the Iberian peninsula, after a productive Moorish period, also in mathematics, produced (apart from Alfonso el Sabio perhaps) only one figure who has entered our history books, but Pedro Nuñes was hardly an innovator. Why were the Alps a transmission belt, and the

Pyrenees a wall? Again, a pretty sociological problem for the historian of mathematics – and of science in general.

2

Back to the Greeks. Understanding of what happened to thought in general and mathematics in particular during these crucial years between 600 and 300 B.C. looks more hopeful, I believe, if we concentrate on that unique phenomenon, the city state, the *polis*. Here we meet for that time an entirely new social structure, developing in sharp class struggles and debates its own cultural and philosophical ideas, leading to a new kind of mathematics. True, many of the mathematical concepts originated in Egypt, Mesopotamia, perhaps also India. But the Greeks, in their own new independent setting did not just borrow, they transformed. "Whatever the Greeks take from Foreigners", we read in a contemporary text, "they change by making it something finer." So also their mathematics.

Merchants, throughout the ages, have a practical eye for numbers and ciphering, hence for computational arithmetic. In the *polis* the philosophizing sophists passed from this to the discussion of number in the abstract, as they also did with concepts like virtue and the good. In the same way they also transformed the geometry of the Egyptians (which still carries in its name the memory of practical surveying) into the abstract, "Greek" type of an axiomatically based science.

At this point, slavery may have played a role, not only by freeing an elite from hard labor and enabling it to discussing *politics* – mark the name – including art and mathematics, but also by making manual work despicable, *banausic*. It is all in Aristotle. Thus the existence of slavery inside the setting of the *polis* may help to understand how Greek mathematics was founded. There may have been other factors, such as geographical ones.

When the *polis* declines, creative mathematics changes character in Alexandria and eventually peters out when the Romans change Egypt into their colony. Mathematics never flourished under colonial conditions, even if it produced at one time a Ramanujan.

It is along these lines that, I think, a sociology of Greek mathematics may be successful. Helpful studies exist, such as those by George Thompson from a literary, Benjamin Farrington from a general scientific, Winspear from a philosophical, or Van Der Waerden from a mathematical point of view, especially in his studies on the Pythagoreans.[2]

The Ptolemies headed an ancient empire with an irrigation economy. Here its own traditional mathematics, dating from the Bronze Age, blended with the new *polis* mathematics. This brings us to another problem: that of the mathematics of the ancient empires of Mesopotamia and Egypt and perhaps related to their bureaucratic and engineering problems. We think here of the controversial work of Wittfogel.[3] But this is a subject I can not tackle here. I know that Jens Høyrup has something to say on this subject.[3]

3

The second case I would like to draw attention to is that of the relation between the Scientific Revolution of the 17th century and its new mathematics – calculus, coordinate geometry, theory of probability – and the transformation of society in parts of Europe from late feudalism to early capitalism. Some will deny that such a relation exists. However, both shattering changes occurred in the same period at the same time and at the same rather small spot on the earth – incidentally together with a third, equally shattering, phenomenon, the Reformation. The way I see it is that we must analyze the relation between the three explosions and their influence on mathematics. At the bottom of this whole series of revolutions lies, I think, the transition from feudalism to capitalism.

Curiously enough, in the preamble to the Scientific Revolution as well as in the first period of capitalism, we meet once again the *polis*, the city state of the Renaissance in Italy. Here again bitter internal struggles and an emerging mercantile class with new ideas on art, science and mathematics. Again, what is borrowed, now from ancient Greece and Islam, is *transformed* – compare the David of Michel Angelo with the Apollo of Praxiteles, or the Algebra of Cardan with the Algebra of Al-Khwarizmi. But here the analogy ends.

We now have a society without slavery and with a spate of new inventions, notably printing. The mercantile class, already engaged in extensive industrial enterprise, develops a new type of economy, that of early capitalism. This creates an atmosphere of search for profit combined with a spirit of investigation. When the city states of Italy decline under foreign rule the new spirit continues in force across the Alps, going to the cities around the North Sea. Here, where under a declining feudal order an aggressive merchant class, again with increasing industrial interest, had long been growing. And so we see all over that section of Europe what may be called the capitalist spirit: it combines a search for profit through domination of nature through science and technology with an encouragement of the quest for knowledge for its own sake. But behind this there is always an eye either for possible profit or for the betterment of mankind, though mankind usually only understood as the more or less well-to-do in one's own country.

Such a spirit can not be satisfied with Aristotelian principles, nor can it be satisfied with the essentially static mathematics of the Greeks, though it can build on it. A dynamic society here demands a mathematics of motion. The new philosophy, from Bacon and Descartes to Newton and Leibniz, is closely connected with the new mathematics. Bacon misses this particular point, but his call for torturing nature like you torture witches, to get at the truth, expresses the spirit of the age.

Thus, by combining the algebra of Islam with the mathematics of the Greeks, and especially with its most advanced type, that of Archimedes, the mathematics of change is born, the calculus. It is a necessary factor in the whole Scientific Revolution of that age, which, as I already said, is developed just in these countries – France, England, the Dutch Republic, where a militant early bourgeois class is creating the manufacturing form of capitalism.

It seems to me that it is the task of a sociology of mathematics to analyze as carefully as possible the often very subtle, often very obvious, way by which the social process has been stimulating the growth of this new mathematics. Understand me right: I do not claim that economic factors directly influenced the mathematics of this period. Against a Huygens, whose search for better clocks as aid to oceanic navigation stimulated some of his best mathematics, we find a Fermat, seemingly entirely wrapped up in the search for truth for truth's sake, yet a truth fitting into the thought of the age. Next to a Descartes, whose mathematics was closely related to his philosophy of nature mastery, we find a Pascal, whose mathematical thought was an aspect of his religious search for the understanding of the infinite. All they have in common is that they worked and thought in a region of Europe where a merchant-industrial class was flourishing.

Incidentally, they also were all male, white – and almost all – bachelors. Also they were Christians, and in their philosophy of life and work influenced by their religion at a time that this religion also was influenced by deep shocks. This brings us to another problem in which we can also see strong sociological elements, namely that of the influence of Protestantism, especially of its Calvinistic variety, on the course of the Scientific Revolution – a highly controversial subject. But I cannot touch upon this aspect of our subject.

There already exists considerable material helpful in the study of this relation between the new mathematics, the Scientific Revolution and the transition from feudalism to capitalism in Western Europe. Far more material than in the case of the Greeks. After all, we are living closer to the one than to the other, and while the cultural climate of ancient Greece is remote to us, what I have called "the capitalist spirit" is still very much with us and easily understandable, at least in principle. I have mentioned Hessen, who has been followed by others studying the century of the British Revolution, authors such as S.F. Mason, Christopher Hill and Robert Merton. Sombart already discussed the *Rechenhaftigkeit* of the early bourgeoisie. Work has also been done on the Dutch Republic, Germany and Italy and probably on other countries. We should not forget the studies of E. Zilsel, the Vienna historian.[4]

A last remark – Another illustration for the existence of a relation between a powerful mercantile class on the way to capitalism in the 17th century and the Scientific revolution is the case of Spain and Portugal, where the merchant class, after the rebellions and expulsions in the 16th century, was entirely subdued by bureaucratic feudalism and was unable to inspire new scientific ideas.

NOTES

[1] This article is based on a lecture given at the XVIIIth International Congress of History of Science held in Munich, August 1989, see further; D. J. Struik, "Sociology of Mathematics Revisited", *Science and Society* **50** (1986) 288–299. On Hessen and his influence see P. G. Werskey, *The Visible College*, London, 1970.

[2] G. Thomson, *Studies in Ancient Greek Society II. The First Philosophers*, London, 1955; B. Farrington, *Greek Science*, 2 vols, Penguin 1944, 1949; A. Winspear, *The Genesis of Plato's Thought*, New York, 1940.

[3] K. A. Wittfogel, *Oriental Despotism*, Yale Univ. Press, 1957, see J. Needham's critique in *Science and Society* **23** (1959), 58–65; J. Høyrup, *Mathematics and Early State Formation*, Roskilde University, 1989.

[4] R. K. Merton, "Science, Technology and Society in Seventeenth Century England", *Osiris* **4** (1938), 360–632, reprinted as book, New York, 1970; S. F. Mason, "Some Historical Roots of the Scientific Revolution", *Science and Society*" **14** (1950), 237–264; E. Zilsel, "The Sociological Roots of Science", *American Journal of Sociology* **47** (1942), 544–562.

Boston Studies in the Philosophy of Science

105. F. Burwick (ed.): *Approaches to Organic Form.* Permutations in Science and Culture. 1987 ISBN 90-277-2541-1
106. M. Almási: *The Philosophy of Appearances.* Translated from Hungarian. 1989 ISBN 90-277-2150-5
107. S. Hook, W.L. O'Neill and R. O'Toole (eds.): *Philosophy, History and Social Action.* Essays in Honor of Lewis Feuer. With an Autobiographical Essay by L. Feuer. 1988 ISBN 90-277-2644-2
108. I. Hronszky, M. Fehér and B. Dajka: *Scientific Knowledge Socialized.* Selected Proceedings of the 5th Joint International Conference on the History and Philosophy of Science organized by the IUHPS (Veszprém, Hungary, 1984). 1988 ISBN 90-277-2284-6
109. P. Tillers and E.D. Green (eds.): *Probability and Inference in the Law of Evidence.* The Uses and Limits of Bayesianism. 1988 ISBN 90-277-2689-2
110. E. Ullmann-Margalit (ed.): *Science in Reflection.* The Israel Colloquium: Studies in History, Philosophy, and Sociology of Science, Vol. 3. 1988 ISBN 90-277-2712-0; Pb 90-277-2713-9
111. K. Gavroglu, Y. Goudaroulis and P. Nicolacopoulos (eds.): *Imre Lakatos and Theories of Scientific Change.* 1989 ISBN 90-277-2766-X
112. B. Glassner and J.D. Moreno (eds.): *The Qualitative-Quantitative Distinction in the Social Sciences.* 1989 ISBN 90-277-2829-1
113. K. Arens: *Structures of Knowing.* Psychologies of the 19th Century. 1989 ISBN 0-7923-0009-2
114. A. Janik: *Style, Politics and the Future of Philosophy.* 1989 ISBN 0-7923-0056-4
115. F. Amrine (ed.): *Literature and Science as Modes of Expression.* With an Introduction by S. Weininger. 1989 ISBN 0-7923-0133-1
116. J.R. Brown and J. Mittelstrass (eds.): *An Intimate Relation.* Studies in the History and Philosophy of Science. Presented to Robert E. Butts on His 60th Birthday. 1989 ISBN 0-7923-0169-2
117. F. D'Agostino and I.C. Jarvie (eds.): *Freedom and Rationality.* Essays in Honor of John Watkins. 1989 ISBN 0-7923-0264-8
118. D. Zolo: *Reflexive Epistemology.* The Philosophical Legacy of Otto Neurath. 1989 ISBN 0-7923-0320-2
119. M. Kearn, B.S. Philips and R.S. Cohen (eds.): *Georg Simmel and Contemporary Sociology.* 1989 ISBN 0-7923-0407-1
120. T.H. Levere and W.R. Shea (eds.): *Nature, Experiment and the Science.* Essays on Galileo and the Nature of Science. In Honour of Stillman Drake. 1989 ISBN 0-7923-0420-9
121. P. Nicolacopoulos (ed.): *Greek Studies in the Philosophy and History of Science.* 1990 ISBN 0-7923-0717-8
122. R. Cooke and D. Costantini (eds.): *Statistics in Science.* The Foundations of Statistical Methods in Biology, Physics and Economics. 1990 ISBN 0-7923-0797-6

Boston Studies in the Philosophy of Science

123. P. Duhem: *The Origins of Statics.* Translated from French by G.F. Leneaux, V.N. Vagliente and G.H. Wagner. With an Introduction by S.L. Jaki. 1991 ISBN 0-7923-0898-0
124. H. Kamerlingh Onnes: *Through Measurement to Knowledge.* The Selected Papers, 1853-1926. Edited and with an Introduction by K. Gavroglu and Y. Goudaroulis. 1991 ISBN 0-7923-0825-5
125. M. Čapek: *The New Aspects of Time: Its Continuity and Novelties.* Selected Papers in the Philosophy of Science. 1991 ISBN 0-7923-0911-1
126. S. Unguru (ed.): *Physics, Cosmology and Astronomy, 1300-1700.* Tension and Accommodation. 1991 ISBN 0-7923-1022-5
127. Z. Bechler: *Newton's Physics on the Conceptual Structure of the Scientific Revolution.* 1991 ISBN 0-7923-1054-3
128. É. Meyerson: *Explanation in the Sciences.* Translated from French by M-A. Siple and D.A. Siple. 1991 ISBN 0-7923-1129-9
129. A.I. Tauber (ed.): *Organism and the Origins of Self.* 1991 ISBN 0-7923-1185-X
130. F.J. Varela and J-P. Dupuy (eds.): *Understanding Origins.* Contemporary Views on the Origin of Life, Mind and Society. 1992 ISBN 0-7923-1251-1
131. G.L. Pandit: *Methodological Variance.* Essays in Epistemological Ontology and the Methodology of Science. 1991 ISBN 0-7923-1263-5
132. G. Munévar (ed.): *Beyond Reason.* Essays on the Philosophy of Paul Feyerabend. 1991 ISBN 0-7923-1272-4
133. T.E. Uebel (ed.): *Rediscovering the Forgotten Vienna Circle.* Austrian Studies on Otto Neurath and the Vienna Circle. Partly translated from German. 1991 ISBN 0-7923-1276-7
134. W.R. Woodward and R.S. Cohen (eds.): *World Views and Scientific Discipline Formation.* Science Studies in the [former] German Democratic Republic. Partly translated from German by W.R. Woodward. 1991 ISBN 0-7923-1286-4
135. P. Zambelli: *The Speculum Astronomiae and Its Enigma.* Astrology, Theology and Science in Albertus Magnus and His Contemporaries. 1992 ISBN 0-7923-1380-1
136. P. Petitjean, C. Jami and A.M. Moulin (eds.): *Science and Empires.* Historical Studies about Scientific Development and European Expansion. ISBN 0-7923-1518-9
137. W.A. Wallace: *Galileo's Logic of Discovery and Proof.* The Background, Content, and Use of His Appropriated Treatises on Aristotle's *Posterior Analytics.* 1992 ISBN 0-7923-1577-4
138. W.A. Wallace: *Galileo's Logical Treatises.* A Translation, with Notes and Commentary, of His Appropriated Latin Questions on Aristotle's *Posterior Analytics.* 1992 ISBN 0-7923-1578-2
Set (137 + 138) ISBN 0-7923-1579-0

Boston Studies in the Philosophy of Science

139. M.J. Nye, J.L. Richards and R.H. Stuewer (eds.): *The Invention of Physical Science.* Intersections of Mathematics, Theology and Natural Philosophy since the Seventeenth Century. Essays in Honor of Erwin N. Hiebert. 1992 ISBN 0-7923-1753-X
140. G. Corsi, M.L. dalla Chiara and G.C. Ghirardi (eds.): *Bridging the Gap: Philosophy, Mathematics and Physics.* Lectures on the Foundations of Science. 1992 ISBN 0-7923-1761-0
141. C.-H. Lin and D. Fu (eds.): *Philosophy and Conceptual History of Science in Taiwan.* 1992 ISBN 0-7923-1766-1
142. S. Sarkar (ed.): *The Founders of Evolutionary Genetics.* A Centenary Reappraisal. 1992 ISBN 0-7923-1777-7
143. J. Blackmore (ed.): *Ernst Mach – A Deeper Look.* Documents and New Perspectives. 1992 ISBN 0-7923-1853-6
144. P. Kroes and M. Bakker (eds.): *Technological Development and Science in the Industrial Age.* New Perspectives on the Science–Technology Relationship. 1992 ISBN 0-7923-1898-6
145. S. Amsterdamski: *Between History and Method.* Disputes about the Rationality of Science. 1992 ISBN 0-7923-1941-9
146. E. Ullmann-Margalit (ed.): *The Scientific Enterprise.* The Bar-Hillel Colloquium: Studies in History, Philosophy, and Sociology of Science, Volume 4. 1992 ISBN 0-7923-1992-3
147. L. Embree (ed.): *Metaarchaeology.* Reflections by Archaeologists and Philosophers. 1992 ISBN 0-7923-2023-9
148. S. French and H. Kamminga (eds.): *Correspondence, Invariance and Heuristics.* Essays in Honour of Heinz Post. 1993 ISBN 0-7923-2085-9
149. M. Bunzl: *The Context of Explanation.* 1993 ISBN 0-7923-2153-7
150. I.B. Cohen (ed.): *The Natural Sciences and the Social Sciences.* Some Critical and Historical Perspectives. 1994 ISBN 0-7923-2223-1
151. K. Gavroglu, Y. Christianidis and E. Nicolaidis (eds.): *Trends in the Historiography of Science.* 1994 ISBN 0-7923-2255-X
152. S. Poggi and M. Bossi (eds.): *Romanticism in Science.* Science in Europe, 1790–1840. 1994 ISBN 0-7923-2336-X
153. J. Faye and H.J. Folse (eds.): *Niels Bohr and Contemporary Philosophy.* 1994 ISBN 0-7923-2378-5
154. C.C. Gould and R.S. Cohen (eds.): *Artifacts, Representations, and Social Practice.* Essays for Marx W. Wartofsky. 1994 ISBN 0-7923-2481-1
155. R.E. Butts: *Historical Pragmatics.* Philosophical Essays. 1993 ISBN 0-7923-2498-6
156. R. Rashed: *The Development of Arabic Mathematics: Between Arithmetic and Algebra.* Translated from French by A.F.W. Armstrong. 1994 ISBN 0-7923-2565-6

Boston Studies in the Philosophy of Science

157. I. Szumilewicz-Lachman (ed.): *Zygmunt Zawirski: His Life and Work.* With Selected Writings on Time, Logic and the Methodology of Science. Translations by Feliks Lachman. Ed. by R.S. Cohen, with the assistance of B. Bergo. 1994 ISBN 0-7923-2566-4
158. S.N. Haq: *Names, Natures and Things.* The Alchemist Jābir ibn Ḥayyān and His *Kitāb al-Aḥjār* (Book of Stones). 1994 ISBN 0-7923-2587-7
159. P. Plaass: *Kant's Theory of Natural Science.* Translation, Analytic Introduction and Commentary by Alfred E. and Maria G. Miller. 1994 ISBN 0-7923-2750-0
160. J. Misiek (ed.): *The Problem of Rationality in Science and its Philosophy.* On Popper vs. Polanyi. The Polish Conferences 1988–89. 1995 ISBN 0-7923-2925-2
161. I.C. Jarvie and N. Laor (eds.): *Critical Rationalism, Metaphysics and Science.* Essays for Joseph Agassi, Volume I. 1995 ISBN 0-7923-2960-0
162. I.C. Jarvie and N. Laor (eds.): *Critical Rationalism, the Social Sciences and the Humanities.* Essays for Joseph Agassi, Volume II. 1995 ISBN 0-7923-2961-9
Set (161–162) ISBN 0-7923-2962-7
163. K. Gavroglu, J. Stachel and M.W. Wartofsky (eds.): *Physics, Philosophy, and the Scientific Community.* Essays in the Philosophy and History of the Natural Sciences and Mathematics. In Honor of Robert S. Cohen. 1995 ISBN 0-7923-2988-0
164. K. Gavroglu, J. Stachel and M.W. Wartofsky (eds.): *Science, Politics and Social Practice.* Essays on Marxism and Science, Philosophy of Culture and the Social Sciences. In Honor of Robert S. Cohen. 1995 ISBN 0-7923-2989-9
165. K. Gavroglu, J. Stachel and M.W. Wartofsky (eds.): *Science, Mind and Art.* Essays on Science and the Humanistic Understanding in Art, Epistemology, Religion and Ethics. Essays in Honor of Robert S. Cohen. 1995 ISBN 0-7923-2990-2
Set (163–165) ISBN 0-7923-2991-0

Also of interest:

R.S. Cohen and M.W. Wartofsky (eds.): *A Portrait of Twenty-Five Years Boston Colloquia for the Philosophy of Science, 1960-1985.* 1985 ISBN Pb 90-277-1971-3

Previous volumes are still available.

KLUWER ACADEMIC PUBLISHERS – DORDRECHT / BOSTON / LONDON